Translating Early Modern Science

Intersections

INTERDISCIPLINARY STUDIES IN EARLY MODERN CULTURE

VOLUME 51 – 2017

The titles published in this series are listed at *brill.com/inte*

Translating Early Modern Science

Edited by

Sietske Fransen
Niall Hodson
Karl A.E. Enenkel

BRILL

LEIDEN | BOSTON

Cover illustration: Adriaen Lommelin, frontispiece of Noël de Berlaimont, *Dictionariolum et colloquia octo linguarum, Latinae, Gallicae, Belgicae, Teutonicae, Hispanicae, Italicae, Anglicae et Portugallicae* (Antwerp, Hendrick Aertsen: 1662). Engraving. The Hague, Koninklijke Bibliotheek. Image © Koninklijke Bibliotheek.

Library of Congress Cataloging-in-Publication Data

Names: Fransen, Sietske, editor. | Hodson, Niall, editor. | Enenkel, K. A. E., editor.
Title: Translating early modern science / edited by Sietske Fransen, Niall Hodson, Karl A.E. Enenkel.
Description: Leiden ; Boston : Brill, 2017. | Series: Intersections : interdisciplinary studies in early modern culture, ISSN 1568–1181 ; volume 51 | Includes bibliographical references and index.
Identifiers: LCCN 2017026216 (print) | LCCN 2017029152 (ebook) | ISBN 9789004349261 (E-book) | ISBN 9789004349254 (hardback : acid-free paper)
Subjects: LCSH: Science—Europe—History—16th century. | Science—Europe—History—17th century. | Science—Europe—History—18th century. | Scientific literature—Translating—Europe—History. | Translators—Europe—History. | Scholars—Europe—History. | Translating and interpreting—Europe—History. | Knowledge, Sociology of—History. | Europe—Intellectual life.
Classification: LCC Q127.E8 (ebook) | LCC Q127.E8 T725 2017 (print) | DDC 418/.035—dc23
LC record available at https://lccn.loc.gov/2017026216

Typeface for the Latin, Greek, and Cyrillic scripts: "Brill". See and download: brill.com/brill-typeface.

ISSN 1568-1181
ISBN 978-90-04-34925-4 (hardback)
ISBN 978-90-04-34926-1 (e-book)

This book is printed on acid-free paper and produced in a sustainable manner.

Contents

Acknowledgements

This volume is the result of an initial conference, and several conference panels. On 28 June 2013, we organized a one-day conference 'Translation and the Circulation of Knowledge in Early Modern Science' at the Warburg Institute in London; we subsequently participated in two panels 'Translating How-To' at the European Society of the History of Science in Lisbon from 4–6 September 2014; and we organized two more panels on our initial conference theme at the Meeting of the Renaissance Society of America in Berlin, 26–28 March 2015.

Our first thanks must go to our authors, for their engaging presentations, their excellent contributions, and their patience in the publication process. The conference at the Warburg Institute would not have been possible without the support of its then director, Peter Mack, nor the support and help of Joanna Woodall in helping us organize, and the support (and key-note address) from Sven Dupré, as well as Guido Giglioni and Eric Jorink. We would like to thank Sven Dupré and Elaine Leong for inviting us to speak in their panels in Lisbon, and Monica Azzolini (RSA representative for the History of Science) and Dario Tessicini (University of Durham) for supporting our panels at the RSA meeting in Berlin.

To organize our conference in London we received generous funding from the Warburg Institute, the Society for Renaissance Studies, the Institute of Medieval and Early Modern Studies and Department of English at the University of Durham, and the Royal Historical Society. We thank all institutions and societies for their support.

During the editing of the volume we have had all kinds of help from Florence Hsia, Elaine Leong, and Sachiko Kusukawa—for which our grateful thanks.

We have both been lucky with the institutional support to finish this volume. Niall would like to thank Durham University, while Sietske would like to thank the Warburg Institute, the Max Planck Institute for the History of Science in Berlin, and the Centre for Research in the Arts, Social Science and Humanities (CRASSH) at the University of Cambridge, especially the AHRC-funded project 'Making Visible: the visual and graphic practices of the early Royal Society' (AHRC Grant Reference: AH/M001928/1).

Sietske Fransen (*Cambridge*) *and Niall Hodson* (*Durham*)
February 2017

Notes on the Editors

Sietske Fransen
is a postdoctoral research associate in the AHRC-funded project 'Making Visible: the visual and graphic practices of the early Royal Society', at the Centre for Research in the Arts, Social Sciences and Humanities (CRASSH), University of Cambridge. She received her Ph.D. from the Warburg Institute in London with a thesis entitled *Exchange of Knowledge Through Translation: Jan Baptista van Helmont and his Editors and Translators in the Seventeenth Century*. She was a postdoctoral fellow at the Max Planck Institute for the History of Science in Berlin (2014–2015) where she started her current research project on visualization as a form of translation in early modern scientific practices. She has published articles on science and translation in *Morgen-Glantz, Literature Compass*, and *ISIS*.

Niall Hodson
is a cultural historian whose current research focusses on translation at the early modern Royal Society and the role of its Secretary, Henry Oldenburg, as a translator and intermediary in the Republic of Letters. He received his M.A. from the Warburg Institute, and has since undertaken research at Durham University and held fellowships at the Edward Worth Library and Utrecht University.

Karl A.E. Enenkel
is Professor of Medieval Latin and Neo-Latin at the University of Münster (Germany). Previously he was Professor of Neo-Latin at Leiden University (Netherlands). He is a member of the *Royal Netherlands Academy of Arts and Sciences*. He has published widely on international Humanism, early modern culture, paratexts, literary genres 1300–1600, Neo-Latin emblems, word and image relationships, and the history of scholarship and science. Among his major book publications are *Die Erfindung des Menschen. Die Autobiographik des frühneuzeitlichen Humanismus von Petrarca bis Lipsius* (2008), and *Die Stiftung von Autorschaft in der neulateinischen Literatur (ca. 1350–ca. 1650)* (2015). He has (co)edited and co-authored some 30 volumes on a great variety of topics, among others, *Modelling the Individual. Biography and Portrait in the Renaissance* (1998), *Recreating Ancient History* (2001), *Mundus Emblematicus* (2003), *Cognition and the Book* (2004), *Petrarch and his Readers* (2006), *Early Modern Zoology* (2007), *The Sense of Suffering. Constructions of Physical Pain in Early Modern Culture* (2009), *The Neo-Latin Epigram* (2009), *Portuguese*

Humanism (2011), *The Authority of the Word* (2011), *Discourses of Power. Ideology and Politics in Neo-Latin Literature* (2012), *The Reception of Erasmus* (2013), *Transformation of the Classics* (2013), *Die* Vita *als Vermittlerin von Wissenschaft und Werk* (2013), *Neo-Latin Commentaries and the Management of Knowledge* (2013), *Zoology in Early Modern Culture* (2014), *Iohannes de Certaldo. Beiträge zu Boccaccios lateinischen Werken und ihrer Wirkung* (2015), *Discourses of Anger in the Early Modern Period* (2015), and *Jesuit Image Theory* (2016). He has founded the international series *Intersections* (Brill); *Proteus. Studies in Early Modern Identity Formation*; *Speculum Sanitatis: Studies in Medieval and Early Modern Medical Culture (500–1800)* (both Brepols), and *Scientia universalis. Studien und Texteditionen zur Wissensgeschichte der Vormoderne* (LIT-Verlag).

Notes on the Contributors

Michael Bycroft
is a Leverhulme Early Career Research Fellow at the University of Warwick. He received his Ph.D. in History and Philosophy of Science from the University of Cambridge in 2013. He then spent one year as a research fellow at the Max Planck Institute for the History of Science. The theme of his research is role of material goods in the development of the natural sciences in early modern Europe. His main current projects are a monograph on the origins of gemmology and an edited collection on gems in the early modern world. He has published papers on a range of topics, including Enlightenment physics, twentieth-century social science, and the symmetry principle in science studies.

Meghan C. Doherty
is the Director and Curator of the Doris Ulmann Galleries and Assistant Professor of Art History at Berea College in Berea, KY. Her research focuses on the connections between art and science as seen in the visual culture of the early Royal Society of London, and her current book project, *Carving Knowledge*, features studies of primary visual and written materials related to Robert Hooke's *Micrographia*, Francis Willughby's *Ornithology* and the *Philosophical Transactions of the Royal Society*.

Rodolfo Garau
Ph.D. in Philosophy (2015), is presently guest lecturer at Bard College, Berlin. His research focuses on the history of early modern philosophy and science. He is particularly interested in the cross-use of mechanical concepts in early modern biology and philosophy, and in the history of early modern materialism. Along with Justin E.H. Smith, he is currently translating from Latin to English Gassendi's *Syntagma Philosophicum*.

Felicity Henderson
is Lecturer in Archives and Material Culture at the University of Exeter. She is currently editing Robert Hooke's diary for publication with Oxford University Press, and she is particularly interested in the links between science and literature in the early-modern period.

Charles van den Heuvel
is Head Research of History of Science and Scholarship at the Huygens Institute for the History of the Netherlands. Furthermore, he holds the chair *Digital*

Methods and Historical Disciplines at the University of Amsterdam. His most relevant publication is *'De Huysbou', A reconstruction of an unfinished treatise on architecture, town planning and civil engineering by Simon Stevin, History of science and scholarship in the Netherlands* (2005).

Jan van de Kamp
Ph.D. (2011) in theology, is a post-doctoral researcher at the Institute for Religious Studies at the University of Bremen (Germany). He has published on the exchange of devotional literature in early modern Europe and on the Reformation in North-West Europe. The title of his doctoral dissertation was *„auff bitte und einrahten etzlicher frommen Menschen ins hochteutsche übersetzet". Deutsche Übersetzungen englischer und niederländischer reformierter Erbauungsbücher 1667–1697 und die Rolle von Netzwerken* ['Rendered into High German at the request and recommendation of certain pious persons'. German translations of English and Dutch Reformed devotional literature from 1667 to 1697 and the role of networks].

B. Harun Küçük
is Assistant Professor of History and Sociology of Science at the University of Pennsylvania. He has published on science and religion and, on the history of science, medicine and technology in the early modern Ottoman Empire. He is currently completing a book on the history of science in the Ottoman Empire between 1650 and 1750. He has received his Ph.D. from UC San Diego's History Department and Science Studies Program and was formerly a pre- and post-doctoral fellow at the Max Planck Institute for the History of Science.

Joyce van Leeuwen
is a Postdoctoral Research Scholar at the Max Planck Institute for the History of Science in Berlin. She obtained her doctorate from the Humboldt University of Berlin in 2012. Her research interests lie in Greek paleography, diagrammatic reasoning, history of mechanics, and early modern science. In 2016 she published a monograph, titled *The Aristotelian Mechanics: Text and Diagrams*, in Springer's *Boston Studies in the Philosophy and History of Science*.

Thomas Morel
Ph.D. (2013) in History of Science, is *Maître de conférences* at the Université Lille Nord-de-France (France) after a post-doc at the Technische Universität and Max Planck Institute for the History of Science in Berlin. His research themes include social history of mathematics and history of mathematical practices in Europe during the early modern period.

Richard J. Oosterhoff
Ph.D. (Notre Dame, IN, 2013) in history and philosophy of Science, is a research fellow at St Edmund's College, Cambridge, and a fellow on *Genius Before Romanticism: Ingenuity in Early Modern Art and Science*, an ERC project at CRASSH, University of Cambridge. He is finishing a book on the circle of Jacques Lefèvre d'Étaples, and is beginning another on the growing interest in naïve knowers and the "untutored mind" in early modern Europe.

Iolanda Plescia
(Ph.D. in English Studies) is Lecturer in English Linguistics and Translation at Sapienza University of Rome, where she teaches history of the English language and translation studies. She has most recently published essays on Shakespeare's language and produced the first Italian edition of Henry VIII's *Letters to Anne Boleyn* (Rome 2013) as well as an edited Italian translation of Shakespeare's *Troilus and Cressida* (Milan 2015).

Fabien Simon
is Associate Professor in Early Modern History at the University Paris-Diderot-Paris 7. He has published various papers concerning a social and cultural history of universal languages and their creators in early modern Europe and his researches deal more generally with the history of knowledge, concerning languages especially (for example a chapter on early modern collections of languages from all over the world in the collective volume *Textology via Scientific Writings*). He recently edited, with Liliane Pérez and Marie Thébaud-Sorger *L'Europe des sciences et des techniques (XV*e*–XVIII*e*): Un dialogue des savoirs* (2016).

List of Illustrations

Introduction: Translators and Translations of Early Modern Science*

Sietske Fransen

In the introduction to the German translation of Jan Baptista van Helmont's (1579–1644) originally Latin *Ortus medicinae* [The Rise of Medicine], the translator Christian Knorr von Rosenroth (1636–1689) explains that his chief aim was to make the text understandable to the reader. To accomplish that level of comprehensibility he had to make some choices:

> In der Ubersetzung nun hab ich mich bey einem so schweren Werck nach äusserister Müglichkeit befliessen alles verständlich zu machen [...] auch etwan neue und ungewöhliche Worte brauchen müssen, weil es die neue und ungewöhliche Lehr-Arten nicht anderst wollen zu lassen. Und um dieser Ursach willen bin ich auch mit Neben-Setzung des Niederländischen bißweilen etwas sorgfältiger gewesen, als mancher erachten dörffte nothwendig zu seyn: damit ja bey so schweren Dingen auch schier nicht ein einziges Wort hinschleichen dörffte, woraus einige Deutlichkeit zu schöpffen.[1]

> In the translation of such a difficult work I have strived with utmost ability to make everything understandable [...] also I had to use some new and uncommon words, because the new and uncommon teachings did not allow me otherwise. And for this reason, I have also been more precise with the comparison to the Dutch than many may deem necessary, so that with the complicated things, not one word would escape from which clarity could be created.[2]

---

* I would like to thank Felicity Henderson, Niall Hodson, Sachiko Kusukawa, Andrew McKenzie-McHarg, and Katherine M. Reinhart for their useful comments on earlier versions of this introduction.

1 Knorr von Rosenroth Christian, "Andere Vorrede", in Jan Baptista van Helmont, *Aufgang der Artzney-Kunst*, trans. Christian Knorr von Rosenroth (Sulzbach: Johann Andreae Endters Sel. Söhne, 1683; reprint, with contributions by W. Pagel – F. Kemp, 2 vols. (Munich: 1971)), fol.)()()(2ᵛ.

2 All translations are my own unless otherwise stated.

 | DOI 10.1163/9789004349261_002

The 'new and uncommon' medical theories of Van Helmont were hard to understand for his contemporaries, as we can gather from this comment.[3] Fortunately for German readers, Knorr von Rosenroth did not only use the Latin text written by Van Helmont (published posthumously in Amsterdam in 1648), but compared it with the much shorter Dutch version *Dageraed* [Daybreak] (Amsterdam: 1659).[4] The publication of the German translation of Van Helmont's works can therefore be seen as a comparative translation of the Latin and Dutch source texts—both prepared by Van Helmont himself, and both published posthumously—giving the German reader all the tools to make sense of Van Helmont.

Knorr von Rosenroth's introduction to the reader gives us a rare insight into the reasoning of a translator at work.[5] His apology for new words is not uncommon, but his solutions—often giving a variety of suggested translations in German—are all the more idiosyncratic when compared with other translators of Van Helmont's texts.[6] After his studies in philosophy, philology, theology, and law at the universities of Leipzig, Wittenberg and Leiden, Knorr von Rosenroth was the privy counsellor of Christian August, Count Palatine of Sulzbach (1622–1708) from 1668 until his death. In Sulzbach Knorr assembled many (natural) philosophers and theologians around him, and is remembered especially for the publication of many translations and collections of cabbalistic and natural philosophical texts. He had met Van Helmont's son, Franciscus Mercurius (1614–1698) in the Netherlands in the 1660s and would maintain a life-long friendship with him. During one of the latter's many visits to Sulzbach, Knorr von Rosenroth and Franciscus Mercurius collaborated on the full translation of all the works of Van Helmont the Elder. Knorr von Rosenroth translated works from Latin, Dutch and English into German, but he also translated Hebrew into Latin, and could be described as a professional translator.[7]

3 See on Van Helmont and his new medical theories Hedesan G.D., *An Alchemical Quest for Universal Knowledge: The 'Christian Philosophy' of Jan Baptist Van Helmont (1579–1644)* (Abingdon – New York: 2016).

4 See for a comparison of the Dutch and Latin text, Van Helmont's use of language and the translators' dealings with his language, Fransen S., *Exchange of Knowledge Through Translation: Jan Baptista van Helmont and his Editors and Translators in the Seventeenth Century*, Ph.D. dissertation (University of London, The Warburg Institute: 2014).

5 On Knorr von Rosenroth, see Kemp F., "Christian Knorr von Rosenroth: Sein Leben, Seine Schriften, Briefe und Übersetzungen", in Van Helmont *Aufgang der Artzney-Kunst* vol. 2, xxi–xxxviii; and Battafarano I.M. (ed.), "Special Issue: Christian Knorr von Rosenroth: Werk und Wirking", *Morgen-Glantz* 2 (1992).

6 See Fransen, *Exchange of Knowledge Through Translation* 174–192.

7 For a full list of his translations and publications, see Kemp, "Christian Knorr von Rosenroth" xxxii–xxxvii.

Due to his court position, and sometimes with the financial help of Franciscus Mercurius van Helmont, Knorr von Rosenroth was in the position to publish his translations, often by the Sulzbach printer Abraham Lichtenthaler. In contrast to some of the translators we will encounter in this volume, Knorr von Rosenroth seems to have acted on his own behalf—he was not requested to produce translations, nor did he need them to gain access to a patron, as he already had one in Count Christian August. His practices as a translator arose from his interests in certain topics and his eagerness to compare (as in Van Helmont's Dutch and Latin), compile (as in the case of his impressive *Kabbala denudata*), and comment (in all cases) on the texts of interest.[8] In making them available to new groups of readers, whether that was Dutch and Latin texts for German readers, or Hebrew texts for Latin readers, he took on an active role in the circulation of knowledge to new audiences.[9]

Both Jan Baptista and Franciscus Mercurius van Helmont, and Knorr von Rosenroth can be seen as products of their time, who were tackling common issues that arose from science and language in the seventeenth century. In the early modern period many new scientific and medical theories were developed, many things were observed and discussed for the first time, and at the same time more and more people were writing about and publishing their ideas and observations in both Latin and the European vernaculars. New ideas and discoveries called for new words and new ways to describe them across the breadth of European vernaculars. As we will see in due course, the role of the many translators in the early modern period was instrumental to the way in which theories, ideas, and discoveries would travel and spread from one language and one audience to another.

Translation and Science

The history of early modern science is strongly connected to translation. As we shall see in this volume, translation was at the core of scientific exchange in this period. However, early modern science in Europe could not have existed without several translation movements during the middle ages. Many Greek scientific and philosophical texts were translated into Arabic during the eighth and ninth centuries, which were then translated into Latin in the eleventh to

8 Schmidt-Biggemann W., "Christliche Kabbala oder Philosophia Hebraeorum: Die Debatte zwischen Knorr von Rosenroth und Henry More um die rechte Deutung der Kabbala", *Morgen-Glantz* 16 (2006) 285–322.

9 On Knorr von Rosenroth and *Kabbala denudata*, see Schmidt-Biggemann W., *Geschichte Der Christlichen Kabbala*, 4 vols. (Stuttgart: 2012–2015) vol. 3, Ch. 3.

the thirteenth centuries.[10] Many of the texts translated into Latin immediately became part of the university curriculum, such as Euclid's *Elements*, Ptolemy's *Almagest*, and several Aristotelian texts. Greek, Arabic, and Latin were all, at different times, hegemonic as languages of science, although they were never the only languages read, written and spoken by those people engaging with science at any one time.[11] The translation movements from one dominant language to another were, however, of utmost importance for the delivery of science to the different linguistic audiences. The translators, such as Constantine the African, Gerard of Cremona, and Michael Scotus who were major actors in the process of language change from Arabic to Latin, had their own programme or rationale for choosing texts for translation. And in this way it became the translator's decisions that determined the core texts of Western science.[12]

The increased interest in classical texts amongst men of learning in the late fourteenth and fifteenth centuries meant that many originally Greek writings were rediscovered and translated into Latin. A text would often circulate in many different translations. In the time of the humanists there was simultaneously an increased awareness of European vernacular languages as potential languages for poetry, prose, and science.[13] The sixteenth and seventeenth centuries saw a dramatic rise in the production of printed books in general but also specifically in scientific texts in both Latin and vernacular languages. Educational systems—Latin schools and universities—had thus far been conducted mainly in Latin, meaning that every educated man was literate in Latin before becoming knowledgeable in any particular subject. However,

10 For Greek to Arabic translation, see Daiber H., "Die griechisch-arabische Wissenschaftsüberlieferung in der arabisch-islamischen Kultur in Übersetzungen des 8.–10. Jahrhunderts", in Kittel H. – House J. – Schultze B. (eds.), *Übersetzung: Ein Internationales Handbuch zur Übersetzungsforschung = Translation: An International Encyclopedia of Translation Studies = Traduction: Encyclopédie internationale de la recherche sur la traduction*. 3 vols. (Berlin: 2004) vol. 2, 1206–1217.

For translations from Arabic into Latin see Burnett C.S.F., "The Coherence of the Arabic-Latin Translation Program in Toledo in the Twelfth Century", *Science in Context* 14 (2001) 249–288. And in general: Montgomery S.L., *Science in Translation: Movements of Knowledge through Cultures and Time* (Chicago – London: 2000).

11 On that topic see Gordin M.D. (ed.), "FOCUS: Hegemonic Languages in Sciences", *ISIS* 108, 3 (forthcoming: 2017).

12 Goyens, Michéle – De Leemans P. – Smets A. (eds.), *Science Translated: Latin and Vernacular Translations of Scientific Treatises in Medieval Europe*, Mediaevalia Lovaniensia ser. 1, 40 (Leuven: 2008).

13 Copenhaver B.P., "Translation, Terminology and Style in Philosophical Discourse", in Schmitt C.B. – Skinner Q. – Kessler E. – Kraye J. (eds.), *The Cambridge History of Renaissance Philosophy* (Cambridge: 1988) 77–110.

post-Reformation Europe saw a growth in vernacular education, and in literacy more generally.[14] With literacy, vernacular languages became a more important vehicle of transferring knowledge than before, and the standardization and defence of the use of vernacular languages for science, was therefore especially strong in the sixteenth century.[15]

The greater production of books written in Latin as well as vernacular languages also increased the need for translators. As with the growth of literacy in the vernacular, there was also a diversification of language skills. To put it differently, the acceptance of science written in vernacular languages had two sides: on the one hand the information was now more readily available for those who were literate but did not read Latin. On the other hand the diversification of languages meant that Latin, which had been the dominant language of science for more than a thousand years, lost ground. One suddenly needed to know many different languages to keep up with all the scientific literature that was published. The increasing influence of these European vernaculars alongside Latin thus gave rise to the importance of translators for immediate and direct communication between the different linguistic regions of Europe. Inside these linguistic borders the vernacular languages started to become more important as utilitarian languages for science. In institutions such as the Royal Society in London and the Académie des sciences in Paris, English and French respectively were the main languages of communication and administration. Whether voluntarily or involuntarily, national language were promoted as appropriate for conducting and communicating scientific research. However, there was still a choice of which language to use, especially in the circles of well-educated men and women.[16]

This volume thus concentrates on the sixteenth, seventeenth, and eighteenth centuries to investigate the role of translators in early modern science. Translation has become the focus of historical studies from many different disciplinary backgrounds in recent years. The relatively young field of Translation Studies has fruitfully drawn on the field of linguistics and literary studies, providing terminology and a theoretical framework necessary for the comparison of textual practices. The main focus of scholars of Translation Studies has been literary translation and translation theory, and the study of scientific

14 Waquet F., *Latin, or the Empire of a Sign: from the sixteenth to the twentieth centuries*, trans. J. Howe (London – New York: 2002).

15 Burke P., *Languages and Communities in Early Modern Europe* (Cambridge: 2004) esp. chapters 2 and 3.

16 On the the choice between Latin and vernaculars amongst seventeenth-century practitioners of science Fransen S., "Latin in a Time of Change: The Choice of Language as a Signifier of New Science?", *ISIS* 108, 3 (forthcoming: 2017).

texts is relatively understudied.[17] The interest in the historical branch of this field has increased over the last twenty years, clearly visible in an impressive three-volume encyclopaedia of translation studies, covering translation theory, cultures of translation in all historical periods and geographical regions, as well as many case studies of biblical and literary texts.[18] For the early modern period, the important work of Theo Hermans on Dutch sixteenth- and seventeenth-century texts helpfully reflects on translation, as well as on the role of translators and translation practices more generally.[19]

Peter Burke, a cultural historian, has been of great importance in leading historians in the direction of translation.[20] With his book on European language societies and his various edited volumes and articles on translation he has started a continuing discussion. In Burke's 2007 edited volume with Ronny Po-chia Hsia we see a shift occurring: away from the comparing of texts, originals and copies, and whether or not the translator has done a 'good' or 'faithful' job. Instead, Burke examined 'cultural' translation which entails questions such as who made the translation, for whom, and what was its impact.[21] These questions resulted in a shifting focus of research from a purely textual analysis to the role and function of translations, and explicitly also to the role of the translator. In addition to the growing literature on the topic, there have been several large research projects within the past ten years, focussing on inventorizing, cataloguing, and analysing translation processes. These projects include the Renaissance Cultural Crossroads Project that produced a catalogue of all 'translations out of and into all languages printed in England, Scotland, and Ireland before 1641'.[22] Also worth mentioning is the project on Renaissance Aristotelianism in Renaissance Italy that catalogued all Italian translations

17 Bermann S. – Porter C. (eds.), *A Companion to Translation Studies* (Chichester: 2014) 2–6.

18 Kittel – House – Schultze, *Übersetzung = Translation = Traduction.*

19 Hermans T., *Door eenen engen hals: Nederlandse beschouwingen over vertalen, 1550–1670* (The Hague: 1996); Hermans T., "The Task of the Translator in the European Renaissance: Explorations in a Discursive Field", in Bassnett S., *Translating Literature* (Cambridge: 1997) 14–40. See also his articles in Kittel – House – Schultze, *Übersetzung = Translation = Traduction.*

20 Burke, *Languages and Communities.*

21 Burke P., "Cultures of Translation in Early Modern Europe", in Burke P. – Po-chia Hsia R. (eds.), *Cultural Translation in Early Modern Europe* (Cambridge: 2007) 7–38.

22 https://www.hrionline.ac.uk/rcc/, accessed on 15 February 2017; and see also Barker S. – Hosington B.M. (eds.), *Renaissance Cultural Crossroads: Translation, Print and Culture in Britain, 1473–1640*, Library of the Written Word 21 (Leiden-Boston: 2013). See also Coldiron A.E.B., *Printers without Borders: Translation and Textuality in the Renaissance* (Cambridge: 2014). Bistué B., *Collaborative Translation and Multi-Version Texts in Early Modern Europe* (Farnham: 2013).

of Aristotelian works between 1400 and 1650, which transformed the hitherto Latin-dominated study of early modern Aristotelian reception into a more complicated and linguistically diverse story of reception.[23] The edited volume on translation and the book trade by José María Pérez Fernández and Edward Wilson-Lee has added a layer of social-economical history to the role of translation.[24] Karen Newman and Jane Tylus recently asked whether there would have been a Renaissance without translation, which they answered with 'a resounding no' from the perspective of literary texts and their translations.[25]

For the field of history of science, James Secord's article "Knowledge in Transit" started a more intensive discussion about the role of translation in the transmission, circulation, and making of knowledge.[26] Burke and Po-chia Hsia paved the way for more discussions about science and translation, also paying attention to the hitherto understudied translations from vernacular languages into Latin, emphasizing that the move away from Latin as the language of science was not a simple, one-directional movement.[27] A volume on the translation of knowledge in the early modern Low Countries edited by Harold Cook and Sven Dupré comprises studies not only of textual translation, but broadens the scope of research into the translation of images, objects, and ideas.[28]

Aims of the Volume

Against this backdrop of the recent work on translation, the current volume seeks to provide a point of entry into this varied and inter-disciplinary subject as it is emerging: exploring the role of translation in early modern science, and analysing the nature of translations and practices of translators. It also seeks to explore the part played by translators as mediators, agents, and interpreters in the intellectual history of the period.

---

23 http://vari.warwick.ac.uk, accessed on 15 February 2017: this website contains a searchable catalogue of 'Aristotelian works written or published in Italian between 1400 and 1650.'

24 Pérez Fernández J.M. – Wilson-Lee E. (eds.), *Translation and the Book Trade in Early Modern Europe* (Cambridge: 2014).

25 Newman K. – Tylus J., "Introduction", in Newman K. – Tylus J. (eds.), *Early Modern Cultures of Translation* (Philadelphia: 2015) 3.

26 Secord J.A., "Knowledge in Transit", *Isis* 95, 4 (2004) 654–672.

27 Burke – Po-chia Hsia, *Cultural Translation*; Pantin I., "The Role of Translations in European Scientific Exchanges in the Sixteenth and Seventeenth Centuries", in Burke – Po-chia Hsia, *Cultural Translation* 163–79.

28 Cook H.J. – Dupré S. (eds.), *Translating Knowledge in the Early Modern Low Countries*, Low Countries Studies on the Circulation of Natural Knowledge 3 (Zurich – Münster: 2013).

By focussing on particular translators, networks of translators, translated texts, and the reception of sources by translators, this volume tackles a number of important questions that shape our understanding of the circulation of knowledge in the early modern world. Starting with Peter Burke's questions about who translated scientific texts—how, why, and for whom—we take the investigation further by looking at how translations were distributed, read, used, and understood. Exploring these questions across the translation of texts, images, and ideas offers fresh insights into how the process of translation worked, and how essential it was to the daily life of physicians, natural philosophers, and other practitioners of science. It also shows that translators often had reasons to translate beyond a simple desire to bring the content of a certain text to a new linguistic audience. Rather, social and economical profit, as well as a veiled means of bringing the translator's own thoughts to an audience, were often the motivating factors behind publishing translations of scientific texts.

By investigating translators and translations of scientific texts in the early modern period, the editors and contributors of this volume aimed to further our understanding of the circulation of knowledge, and the way such knowledge is transformed by the actions—intentionally or not—of the persons involved in the process. In this volume early modern translators of scientific texts have been examined by scholars with a wide variety of methods and theoretical background. As historians of science, literature and linguistics, theology, philosophy, art, and language, the authors show that translation is a powerful tool for historical research. Across all fields of historical research, whether in literature, science, or theology, translation played an important role in the distribution of knowledge, and the wide range of disciplinary backgrounds of the authors proves that the study of translation is a useful way to transcend differences in approach and disciplinary boundaries to interpret and understand the making of knowledge and science. Although all articles are published in English, the authors discuss translations from, in, and between Dutch, English, French, German, Greek, Italian, Latin, Turkish, and universal languages. In addition to the historical translations that are studied by the authors, the variety of early modern languages in the source texts has naturally brought up the additional issue of providing modern translations of these texts to make the articles and comparisons therein understandable for the readers of this volume in English. We hope that the inclusion of all text fragments in the original languages will help those readers who might want to have a look at the primary source material themselves. Since the current academic world has adopted English as its hegemonic language, this seemed a workable compromise.[29]

29 See on the history of English as a language of science Gordin M.D., *Scientific Babel: The Language of Science* (Chicago: 2015).

This volume is divided into three parts. Contributions in the first part discuss the translation practices of individuals in the context of scholarly societies and networks. Felicity Henderson opens with a chapter on Robert Hooke and the translations he and some of his colleagues at the Royal Society made in order that topics could be discussed in English in the meetings of the Royal Society. Henderson concentrates on the translations that were never printed and have therefore received less attention by historians. Many of these translations were produced in collaboration and the dominant theme of the translated texts is travel. The chapter shows how translations were not only provided for publication, but often on a more informal basis for immediate use. The following chapter by Jan van de Kamp discusses another Fellow of the Royal Society, Theodore Haak, and his translations of theological texts and how these practices were connected, both interfering with and influencing his interests in natural philosophy. Haak's large network of German, Dutch, Danish, and French men of learning often led to him translating texts on request. In Haak's case it seems to have been more important to be part of these networks and circles, than what the exact content of the text for translation was. Michael Bycroft discusses the case of Charles Dufay, a well-known member of the Académie des sciences in Paris. Bycroft investigates the almost entirely unknown translation project that secured Dufay a place in the Académie in the first place. His French translation of the Italian naturalist Filippo Buonanni, *Trattato sopra la vernice* is not only an interesting case of the translation of practical and artisanal vocabulary, it also shows how the production and publication of a translation can give the translator access to certain authorities or networks. In the final article of this section, Meghan Doherty discusses the translation of images between the Royal Society and the Académie des sciences and their respective journals. Both journals consisted mainly of articles in English and in French respectively, and thus, if reproduced in either journal, needed translating. How did the editors of both journals deal with the images? Could they be understood without translation? These questions lead to a wonderful discussion about images as a potential universal language, and the necessity for visual education.

Contributions in the second part of the volume discuss the translation of practical knowledge. Charles van den Heuvel discusses the unpublished writings and drawings of the Dutch mathematician and advocate of the Dutch language, Simon Stevin. Van den Heuvel's chapter looks at the reception of Stevin's texts in various translations. And taking Stevin's perspective, it seems that many of his translators did not always understand his mathematics. Thomas Morel in the next chapter discusses the influx of learned classical mathematics into vernacular manuals of subterranean geometry—how did Euclid end up in mining manuals written in the language used by German miners? Using newly found manuscripts and printed texts, Morel shows how studying translations can tell us when and how classical mathematical texts were added to these manuals, most likely with the aim of explaining the theoretical background of more complex problems. Joyce van Leeuwen's chapter is about the translational practices of Niccolò Leonico Tomeo, an Italian professor in philosophy at the university of Padua. He translated Aristotle's Greek *Mechanics* into Latin and added his own commentaries. Particularly interesting, from the perspective of translation processes, is that we have not only a printed translation produced by Tomeo, but also two manuscript versions of his translation, and a manuscript version of his attempt to reconstruct the original Greek text. Within the text Tomeo included explanatory diagrams, which form, together with the text, the focus of this chapter. In the final chapter of this section, Richard Oosterhoff investigates the potential readership of the French translations of Latin mathematical texts produced by Charles de Bovelles. Oosterhoff discusses how the translation occurs on two levels: between languages, and between spaces of expertise, from the Latin study to the French workshop.

The third part consists of four articles dealing with the translation of more theoretical and philosophical texts. Rodolfo Garau examines Pierre Gassendi's Latin translation of the classical author Epicurus. The whole translation process is, as in several cases discussed in earlier chapters, as much a commentary as a translation, according to Garau, in order to accommodate the readers' understanding and acceptance of the text. Another contemporary English translation (by Walter Charleton) provided the opportunity to examine a double translation and double attempt to adapt the text to a new audience. This chapter is followed by Harun Küçük's discussion of the translation of Copernican science into Turkish by the central-European Ibrahim Müteferrika. This Islamic convert used his translations to implant radical early Enlightenment thought from Western Europe into Ottoman science in the early eighteenth century. His translation practices combined his religious and scientific networks in a similar way as the case of Theodore Haak, even though the content was very different. We return to English soil for the last two chapters. Iolanda Plescia provides a detailed study of the translational practices of the Englishman Thomas Salusbury and his translations of Galileo. Plescia's close analysis of the Italian and English texts demonstrates that the choices of the translator can determine the outcome and understanding of the text. Salusbury used his authority as a translator to decide on the English interpretation of Galileo's work, potentially with the final aim of being accepted as a Fellow of the Royal Society. The final article of the volume by Fabien Simon, deals with the lively seventeenth-century discussion about constructing a universal language. As touched upon in Doherty's chapter, some seventeenth-century authors thought that images could represent a sort of universal scientific language, but many other people at the time thought that the diversification of languages (Latin plus all the vernaculars) called for a new, artificial, elite language of science. Simon argues that the universal languages discussed in England and on the continent were as much a code or a distinction of social status as Latin had been in the past. This also meant that in order to use any of the suggested universal language, one had to learn that language, and translators were required to translate into and from a universal language.

Several themes recur across the contributions, proving that however diverse the scientific texts under investigation may be, and however varied the theoretical backgrounds of the authors are, studying translations and their translators brings forth important points about the circulation and transfer of knowledge in early modern Europe. Several articles argue that the goal of the translator in producing (and publishing) a translation was to gain entry into a network. By translating Filippo Buonanni's *Tratatto sopra la vernice* from Italian into French, Charles Dufay not only demonstrate his linguistic qualities, but especially his understanding of the chemical experiments necessary for the making of Buonanni's varnish. It was these skills, as Michael Bycroft explains in his chapter, that interested the members of the Académy des sciences in Paris: Dufay's translation, more than his previous publications, showed that he was capable of experimenting and would be worthy of a place in the Académie. Thomas Salusbury similarly tried to gain access to the Royal Society by translating Galileo into English; he would, however, never become a Fellow. In Harun Küçük's chapter, we encounter a translator, Ibrahim Müteferrika, who used his position as a printer for the Ottoman Sultan to bring Copernican and Cartesian ideas into discussions on science and religion in the Ottoman Empire.

A second important theme that arises from several articles relates to questions about images: how images were translated, whether they needed to be translated, and how they could be used to overcome linguistic barriers. Research into the use, making, and role of images in early modern science is currently the subject of many projects and publications.[30] The focus of this volume brings an important aspect to the fore: namely the translatability of images, as well as the similarities and dissimilarities in the treatment

30 See for example Daston L., "Epistemic Images", in Payne A. (ed.), *Vision and Its Instruments: Art, Science, and Technology in Early Modern Europe* (University Park, PA: 2015) 13–35; Kusukawa S., *Picturing the Book of Nature: Image, Text, and Argument in Sixteenth-Century Human Anatomy and Medical Botany* (Chicago: 2012); Hunter M., "Introduction", *Huntington Library Quarterly* 78, 2 [Special Issue Henderson F. – Kusukawa S. – Marr A. (eds.), "Curiously Drawn: Early Modern Science as a Visual Pursuit") (2015) 141–155; Lüthy C. – Smets A., "Words, lines, diagrams, images: towards a history of scientific imagery", *Early Science and Medicine* 14 (2009) 398–439; Marr A., "Knowing Images", *Renaissance Quarterly* 69, 3 (2016) 1000–1013.

of translating text and images. Doherty shows how the translations between English and French of articles published in the *Philosophical Transactions* and *Journal des Sçavans* came with a discussion about the images. Are these images universally intelligible, or do they need translating as well? In the chapter on the posthumous publications of some works by Simon Stevin, Charles van den Heuvel argues that Stevin's son adapted the images in such a way that they could stand on their own, whereas the original manuscript had assumed a side-by-side reading of text and image. This shows how images could become themselves independent items of scientific evidence and argument. Still further, Joyce van Leeuwen, in her chapter on Niccolò Leonico Tomeo analyses how Tomeo inserted images into his translations, sometimes as a 'proof' of a mathematical calculation, sometimes as an illustration of the text. As a whole, images could be seen as a visual commentary on the text itself, where Tomeo inserts himself as non-verbal commentator.

This brings us to theme of commentaries and pseudo-translations. In the chapter by Van Leeuwen we encounter visual commentary, and in the chapters by Thomas Morel and Rodolfo Garau we find explicit discussions of a translation as a form of commentary on a source text. Morel does so by labelling the translation of Euclid into German mining texts as 'pseudo-translation', whereas Garau discusses translations as commentaries.[31] In all three chapters translations are problematized to show that this practice allows the translator to explicitly comment on a work of his specific interest.

The problem of language diversity and the lack of a universal language for science was a problem that occupied many early modern authors of scientific texts. Fabien Simon devotes his entire chapter to the quest for a universal language in learned circles, and how the search for language was as much a social endeavour as well as a matter of mutual understanding. Doherty also discusses the idea of a universal language in relation to images, and Bycroft explains how a universal system of measurements can be seen as a universal language of science.

Finally, we see the authority and power of the translator in adapting, changing, and transforming text and image recurring in several articles. The different editions of Charles de Bovelles's mathematical texts show how his reshaping of texts reached multiple audiences. Likewise, the many unpublished translations in the circle of Robert Hooke show that it was through translations that many formal and informal discussions could take place. The careful search for French vocabulary for practical experiments meant that Charles Dufay had an influence on the way these topics were discussed in the Académie de sciences in Paris.

31 Enenkel K.A.E. – Nellen H. (eds.), *Neo-Latin Commentaries and the Management of Knowledge in the Late Middle Ages and the Early Modern Period (1400–1700)* (Leuven: 2013).

Similarly, Müteferrika's translations made it possible to discuss radical religious and political thought in the Ottoman Empire, through discussing Copernican and Cartesian worldviews. In all these examples—and the many more that fill all twelve chapters—the translator has an (often unacknowledged) authority in transferring the particular knowledge from the source text to the translation.

Despite the variety of topics and the many different languages and directions of translation discussed in this volume, all these cases show that translators of scientific texts were dealing with similar problems. By bringing together so many different topics in the realm of early modern history of science, written by scholars with different backgrounds, this volume shows how the study of translation thrives on interdisciplinarity and can bring new insights into the history of science. It is the image on the book cover that embodies these sentiments: whatever linguistic, cultural, or disciplinary background we may come from, let us 'samen spraeken in acht taelen' [speak together in eight languages], or however many more languages we bring together.

Selected Bibliography

Barker S. – Hosington B.M. (eds.), *Renaissance Cultural Crossroads: Translation, Print and Culture in Britain, 1473–1640*, Library of the Written Word 21 (Leiden – Boston: 2013).

Bermann S. – Porter C. (eds.), *A Companion to Translation Studies* (Chichester: 2014).

Bistué B., *Collaborative Translation and Multi-Version Texts in Early Modern Europe* (Farnham: 2013).

Burke P., *Languages and Communities in Early Modern Europe* (Cambridge: 2004).

Burke P., "Cultures of Translation in Early Modern Europe", in Burke P. – Po-chia Hsia R. (eds.), *Cultural Translation in Early Modern Europe* (Cambridge: 2007) 7–38.

Burke P. – Po-chia Hsia R. (eds.), *Cultural Translation in Early Modern Europe* (Cambridge: 2007).

Burnett C.S.F., "The Coherence of the Arabic-Latin Translation Program in Toledo in the Twelfth Century", *Science in Context* 14 (2001) 249–288.

Coldiron A.E.B., *Printers without Borders: Translation and Textuality in the Renaissance* (Cambridge: 2014).

Cook H.J. – Dupré S. (eds.), *Translating Knowledge in the Early Modern Low Countries*, Low Countries Studies on the Circulation of Natural Knowledge 3 (Zurich – Münster: 2013).

Copenhaver B.P., "Translation, Terminology and Style in Philosophical Discourse", in Schmitt C.B. – Skinner Q. – Kessler E. – Kraye J. (eds.), *The Cambridge History of Renaissance Philosophy* (Cambridge: 1988) 77–110.

Daiber H., "Die griechisch-arabische Wissenschaftsüberlieferung in der arabisch-islamischen Kultur in Übersetzungen des 8.–10. Jahrhunderts", in Kittel H.

– House J. – Schultze B. (eds.), *Übersetzung: Ein Internationales Handbuch zur Übersetzungsforschung = Translation: An International Encyclopedia of Translation Studies = Traduction: Encyclopédie internationale de la recherche sur la traduction*, 3 vols. (Berlin: 2004) vol. 2, 1206–1217.

Fransen S., *Exchange of Knowledge Through Translation: Jan Baptista van Helmont and his Editors and Translators in the Seventeenth Century*, Ph.D. dissertation (University of London, The Warburg Institute: 2014).

Fransen S., "Latin in a Time of Change: The Choice of Language as a Signifier of New Science?", *ISIS* 108, 3 (forthcoming: 2017).

Gordin M.D., *Scientific Babel: The Language of Science* (Chicago: 2015).

Gordin M.D. (ed.), "FOCUS: Hegemonic Languages in Sciences", *ISIS* 108, 3 (forthcoming: 2017).

Goyens, Michéle – De Leemans P. – Smets A. (eds.), *Science Translated: Latin and Vernacular Translations of Scientific Treatises in Medieval Europe*, Mediaevalia Lovaniensia, series 1, 40 (Leuven: 2008).

Hedesan G.D., *An Alchemical Quest for Universal Knowledge: The 'Christian Philosophy' of Jan Baptist Van Helmont (1579–1644)* (Abingdon – New York: 2016).

Hermans T., *Door eenen engen hals: Nederlandse beschouwingen over vertalen, 1550–1670* (The Hague: 1996).

Hermans T., "The Task of the Translator in the European Renaissance: Explorations in a Discursive Field", in Bassnett S., *Translating Literature* (Cambridge: 1997) 14–40.

Kittel H. – House J. – Schultze B. (eds.), *Übersetzung: Ein Internationales Handbuch zur Übersetzungsforschung = Translation: An International Encyclopedia of Translation Studies = Traduction: Encyclopédie internationale de la recherche sur la traduction*, 3 vols. (Berlin: 2004).

Lüthy C. – Smets A., "Words, lines, diagrams, images: towards a history of scientific imagery", *Early Science and Medicine* 14 (2009) 398–439.

Montgomery S.L., *Science in Translation: Movements of Knowledge through Cultures and Time* (Chicago – London: 2000).

Newman K. – Tylus J. (eds.), *Early Modern Cultures of Translation* (Philadelphia: 2015).

Pantin I., "The Role of Translations in European Scientific Exchanges in the Sixteenth and Seventeenth Centuries", in Burke – Po-chia Hsia (eds.), *Cultural Translation* 163–179.

Pérez Fernández J.M. – Wilson-Lee E. (eds.), *Translation and the Book Trade in Early Modern Europe* (Cambridge: 2014).

Secord J.A., "Knowledge in Transit", *Isis* 95, 4 (2004) 654–672.

Waquet F., *Latin, or the Empire of a Sign: from the sixteenth to the twentieth centuries*, trans. J. Howe (London – New York: 2002).

PART 1

Translating Networks of Knowledge

CHAPTER 1

Translation in the Circle of Robert Hooke*

Felicity Henderson

At a meeting of the Royal Society on 18 July 1678, after an experiment involving a live eel, the conversation turned to the question of what fish eat. Physician Dr William Croone remarked that 'fishmongers never find any thing in the maws of salmon [...] but that the contrary was observed in most other fish'. Sir John Hoskins reported that there was a species of whale said to feed on flies, that 'vast quantities' of flies had been found in these whales' stomachs, and that they seemed to use their fins to strain the flies from the water. At this point, the Society's curator of experiments and long-time London resident Robert Hooke was able to describe the structure of the 'mouth and fins' of a whale stranded at Greenwich about twenty years earlier. He added that he had recently seen, 'printed in Low Dutch', a voyage to 'Spitzbergen or Greenland' which included a description and illustration of this kind of whale.[1]

The conversation moved on, but Hooke's interest in the whales of Greenland had been piqued, and the following month the Society's minutes record the following:

> Mr Hooke produced and read the preface of a book which he had procured to be translated out of High Dutch, containing a description and natural history of Spitsbergen or Greenland, written by one * * * of Hamburgh, who had been there himself, and, upon occcasion of queries sent out of England, had made it his business to inform himself more particularly concerning all matters therein desired; and by the help of Dr Fogelius of Hamburgh, who had translated and delivered those queries to him, had compiled and methodised the same; and for the better illustration of all particulars, had added a great many copper cuts, containing the pictures of the most remarkable particulars, viz. of the whales and other fishes,

* As well as the volume's editors, I would like to thank Sachiko Kusukawa, Giles Mandelbrote, Noah Moxham, William Poole and Vike Plock for discussing various aspects of this chapter with me.

1 Birch Thomas, *The History of the Royal Society of London*, 4 vols. (London, A. Millar: 1756–1757) vol. 3, 425. The Society had also discussed fish food the previous week (see Birch, *History* vol. 3, 422).

 | DOI 10.1163/9789004349261_003

> together with those of the animals, birds, plants, &c. [Fig. 1.1] Mr Hooke added, that he had delivered the said book to a German, in order to have it translated into English.[2]

This book, probably the same one mentioned previously by Hooke despite his incorrect statement that it was printed in Low Dutch, was Friedrich Martens's *Spitzbergische oder Groenlandische Reise beschreibung gethan im Jahr 1671*, which had been published at Hamburg in 1675.[3] Sir John Hoskins had indeed

FIGURE 1.1 *Unknown artist, illustrations from Friedrich Martens,* Spitzbergische oder Groenlandische Reise beschreibung gethan im Jahr 1671 *(Hamburg, Gottfried Schultze: 1675). Engraving, 325 × 365mm. London, Royal Society Library Tracts* CVII/3.

2 Birch, *History* vol. 3, 431, meeting of 29 August 1678 (the asterisks have been used by Birch to indicate an omission in the original minutes of the Society's meeting). See also John Evelyn's account of the discussion (*The diary of John Evelyn*, ed. E.S. de Beer, 6 vols. (Oxford: 1955) vol. 4, 146–147).

3 The book was also translated into Italian (*Viaggio di Spizberga o Gronlanda*, trans. Jacob Rautenfels (Bologna, Giacomo Monti: 1680)), Dutch (in *De Noordsche Weereld*, trans. Simon de Vries (Amsterdam, Aert Dirkz. Ooszaen: 1685)), and French (*Journal d'un voyage*

drawn up some 'enquiries for such as go to Groenland' for the Royal Society in December 1662. Shortly afterwards a response had come in from Stephen Gray of the Greenland Company which included a detailed description of whale-hunting.[4] However several years later, in November 1667, further enquiries for Greenland were printed in the *Philosophical Transactions*, this time prompted by editor Henry Oldenburg's notice of Justus Klobius's *Ambrae historiam* (Wittenberg: 1666) printed in the previous issue of the journal.[5] It was this second call for information that 'Dr Fogelius', the Hamburg physician Martin Vogel, had pointed out to Martens. Vogel was a regular Royal Society correspondent in the late 1660s and early 1670s and a number of his letters to Oldenburg are extant in the Royal Society's archives.[6]

Despite Hooke's public promise to have the remainder of the book translated into English, the minutes of the Society's meetings do not contain any further mention of Martens's work. Hooke's private diary, though, reveals a little more about the transactions that had taken place in order to get this far, and shows that he did in fact pursue the project further. On 29 July, eleven days after his initial mention of the book at the Society's meeting, he had visited the bookseller Moses Pitt and 'Borrowd [the] Description of Spitsberg'.[7] Pitt at this point was a close associate of Hooke's, and his willingness to lend volumes to Hooke certainly contributed to the spread of information in the period. In this case Hooke seems to have passed the volume on, because three weeks later, on 19 August, he noted 'Received translate of Spitsbergen from Mr Haak'.[8] The

---

au Spitzbergen et au Groenlandt in Bernard Jean-Frédéric (ed.), *Recueil de voiages au Nord*, 8 vols. (Amsterdam, Jean-Frédéric Bernard: 1715–27) vol. 2 (1715), 1–205). For the distinction between high and low Dutch in the period see F. Henderson, "Making 'The Good Old Man' speak English: the reception of Antoni van Leeuwenhoek's letters at the Royal Society, 1673–1723", in Cook H.J. – Dupré S. (eds.), *Translating Knowledge in the Early Modern Low Countries* (Zurich – Berlin: 2012) 243–268, esp. 254 n. 38.

4 Birch, *History* vol. 1, 155–156, 199–202, 324–327 and Royal Society Archives MS/215/12 (1662 enquiries for Greenland), MS/215/47 and RBO/2i/34 (Gray's response).

5 "Enquiries for Greenland", *Philosophical Transactions* 2 (1667) 554–555; "Historia Ambræ [i.e. History of Ambergris], Authore Justo Klobio", *Philosophical Transactions* 2 (1667) 538.

6 For Vogel's identification, and his correspondence with the Society, see *The Correspondence of Henry Oldenburg*, ed. and trans. A.R. Hall – M. Boas Hall, 13 vols. (Madison, WI – London: 1965–86), esp. vol. 4, 303; see also Marten M. – Piepenbring-Thomas C., *Fogels ordnungen: aus der werkstatt des Hamburger mediziners Martin Fogel (1634–1675)* (Frankfurt am Main: 2015). Martens Friedrich, "An den Leser", *Spitzbergische oder Groenlandische Reise*, fol. ()[4]r.

7 Robinson H.W. – Adams W. (eds.), *The Diary of Robert Hooke* (London: 1935) 369 (29 July 1678).

8 Ibidem, 372 (19 August 1678).

German natural philosopher Theodore Haak, who was himself a Fellow of the Royal Society and a professional translator, had clearly been instrumental in acquiring the translation for Hooke, and this must have been the source of the translation read to the Royal Society on 29 August.[9] Finally, two months later, Hooke wrote 'Cramer here. I gave him 6 sheets of Spitzberg'.[10] This, presumably, was the translation of Martens's account, and the involvement of the otherwise unidentified 'Cramer' hints at the context. He was employed on Moses Pitt's *English Atlas* project, a major publication that Hooke and other Fellows of the Royal Society promoted and directed in its early stages, though the project as originally conceived never came to fruition.[11] Nevertheless, the first volume of the *Atlas* does indeed include an account of Greenland drawn from Martens's work, although the anonymous compiler explains that he has omitted such things as he judges 'not so useful', and abridged the rest 'for fear of cloying the Reader'.[12] A full English translation of Martens's account was not published until 1694, as part of a collection entitled *An account of several late voyages and discoveries to the south and north* (London, Samuel Smith and Benjamin Walford: 1694), probably edited by Royal Society Fellow Tancred Robinson.[13] Included in that collection is the map of the North Pole first published in Pitt's *Atlas*, as well as Hooke's translation of Abel Tasman's voyages. Thus it seems likely that Hooke had some involvement in the edition, and although the nature of this is unclear it is possible that he also supplied the text

9 For Haak's translating activities see Poole W., "A fragment of the library of Theodore Haak (1605–1690)", *Electronic British Library Journal* 6 (2007), http://www.bl.uk/eblj/2007articles/pdf/ebljarticle62007.pdf, accessed 19 August 2015; Henderson, "Making 'The Good Old Man" speak English'" 247–248; see also the next chapter by Jan van de Kamp in this volume. We should perhaps be wary of attributing this translation directly to Haak, though, since Hooke has not clearly stated that this was the case either in his diary or at the Royal Society meeting, despite the fact that Haak would have been well known to the assembled Fellows and may even have been present at the meeting.

10 Robinson – Adams, *The Diary of Robert Hooke* 381 (26 October 1678).

11 See in particular Hildyard D., "John Pell's mathematical papers and the Royal Society's English Atlas, 1678–82", *BSHM Bulletin: Journal of the British Society for the History of Mathematics* 29, 1 (2014) 18–31; and Taylor E.G.R., "'The English Atlas' of Moses Pitt, 1680–83", *The Geographical Journal* 95, 4 (1940) 292–299.

12 *The English Atlas. Volume I. Containing a description of the places next the North-Pole* (Oxford, Moses Pitt: 1680) 3–4.

13 Boulger G.S., "Robinson, Sir Tancred (1657/8–1748)", revised K. Bagshaw, *Oxford Dictionary of National Biography* (Oxford: 2004; online ed., May 2010), http://www.oxforddnb.com/view/article/23873, accessed 19 Aug 2015 (hereafter ODNB).

of Martens's Greenland adventures, sixteen years after his initial efforts to have them translated.[14]

Despite, or perhaps because of, its various unknowns and its doubtful conclusion, this story epitomizes some aspects of translation and the circulation of knowledge in early Royal Society circles, particularly those associated with Robert Hooke. A request for information printed in the Royal Society's journal reached Martin Vogel in Hamburg and prompted him to encourage Friedrich Martens to publish his German description of Greenland; Hooke noticed the volume in Pitt's bookshop; a conversation about whales led Hooke to have the text translated; and this translation may later have been printed in one or two Royal Society-associated publications.[15] One of the key features of the Martens translation project is the involvement of a number of participants from various different spheres of activity: Hooke himself; the bookseller and publisher Moses Pitt; the Royal Society Fellows assembled at their weekly meeting; Theodore Haak, translator or intermediary; Cramer, clerk or possibly engraver; the editor Tancred Robinson; and looking further back to the production of the initial publication, Martens himself, and Martin Vogel, intermediary and Royal Society correspondent. Only some of these figures were closely involved with the Royal Society and its work but all were vital to the production of the translation. Another striking aspect is the way in which the material in Martens's account was repackaged several times in different formats for different audiences: initially reported orally to the Fellows in response to an ongoing conversation; then partially translated and presented more formally at a second meeting, this time with further information about provenance; later epitomized and formatted for inclusion in the *English Atlas*; and finally, several years afterwards, published in full with newly-cut copies of the original plates.[16] Different aspects of the work were brought to the fore in each of these repackagings, shaped by the immediate circumstances and the perceived

14 *An account of several late voyages* 143 (map); and 131 ff. (Tasman's journal). The map plate has been cleverly re-purposed by removing the Earl of Plymouth's coat of arms from the lower left corner, and title cartouche and illustration from the upper section of the map, necessitating a very minor amount of re-lettering in the upper right corner of the new version, and the addition of a new title in the lower left corner.

15 The Royal Society as an institution had varying levels of involvement in different publications during the period, from general supportiveness right up to financial backing of publications; 'Royal Society associated' here indicates a middle ground of close involvement by one or more Fellows but limited official corporate backing.

16 Interestingly, a set of the original printed illustrations from Martens's book [Fig. 1.1] is preserved in the library of the Royal Society, bound in with two other German works printed in the seventeenth century. These sheets are accompanied by a list of the contents in a

requirements of different audiences: in its separate incarnations the work is primarily a book about whales, a tribute to the Royal Society, a chorography, and a natural history. The episode also demonstrates the importance of not merely circulating translated information, but the significance of the 'meta-data' that accompanies such information. Here, for example, the collaboration of Martens and Vogel in producing the initial text was highlighted to the Fellows at their meeting, along with the fact that the account had been prompted by enquiries sent from the Society. This latter aspect was emphasized in the *Atlas*, as part of the Society's ongoing mission to promote itself and its activities as economically useful.[17] As I have discussed elsewhere, information about both authors and translators was crucial when it came to making a judgement about the worth of the facts in a translated philosophical work. And it was not just the Society that was keen to point out the genesis of the work: the source of this information is Martens's original preface, which arguably links the book's credibility with the status of England's Royal Society. Both parties gain credit from such connections.[18]

Robert Hooke's circle of scientific colleagues and acquaintances provides an interesting focus for the study of translation in the early Royal Society. Hooke was uniquely placed in that he was both a central participant in the Society's endeavours while at the same time holding a paid position as Surveyor for the City of London. He interacted on a daily basis with people from very different social spheres, and he acted as a conduit for certain types of information, particularly relating to some of his key interests: materials and trades, mechanical knowledge, printed books, and geography (in its broadest sense, including travel accounts, descriptions and specimens of foreign flora and fauna, and languages). We can reconstruct much of Hooke's scientific working life from records in the archives of the Royal Society, but for information about his networks of associates, we need to turn to his personal diaries. These show a constant ebb and flow of information between Hooke and others—and, interestingly, frequent references to translation. Several of Hooke's close friends among the Royal Society fellowship, such as Francis Lodwick, Theodore Haak, Richard Waller and Alexander Pitfeild, engaged in major translation projects. Others undertook shorter projects, or assisted with such endeavours. In surveying the translations of the 'Hooke circle' I will explore the circumstances

---

contemporary hand, and may have been used in the preparation of new plates, although they show no physical signs of the copying process (Royal Society Library, Tracts CVII/3).

17 *The English Atlas. Vol. 1*, 3.

18 Henderson F., "Faithful interpreters? Translation theory and practice at the early Royal Society", *Notes and Records of the Royal Society* 67 (2013) 101–122.

of translations as they were produced as well as examining the final product and its reception, since at this stage much of the evidence of rationale and collaborative activity is no longer evident.[19] Other studies of translation of philosophical texts in the early Royal Society milieu have concentrated on printed materials: one of the purposes of this study is to begin to explore the archival evidence.[20] It shows that translation activities that did not result in print publications were more widespread, and perhaps more significant for the exchange of information among a close-knit circle of philosophers, than those that did lead to a printed work. The emphasis in oral and manuscript translation is on speed and utility rather than the production of a polished whole. By concentrating on this rich but rather neglected sphere of philosophical activity I hope to add to our understanding of scientific communication in the early modern period.

Probably the most active translator among Hooke's close associates was the linguist and cloth merchant Francis Lodwick.[21] Lodwick shared with Hooke an abiding fascination with languages and language planning, but he also seems to have been willing to assist the philosophical world by taking on major translation projects in other areas. He was initially useful to Hooke in translating Dutch letters from the microscopist Antoni van Leeuwenhoek to the Royal Society. Hooke noted in his diary that Lodwick had 'read' Leeuwenhoek's letters, usually either at Hooke's rooms, or not far away in the cosy atmosphere of Jonathan's coffee-house.[22] This suggests that Lodwick was giving Hooke (and possibly others around the table at Jonathan's) an oral account of the letters' contents rather than a full translation—useful for reporting

19 In speaking of a 'Hooke circle' I am creating something that did not exist formally at the time. However Hooke's diaries show that he socialized and philosophized with a fairly static friendship group that only partly overlapped with the fellowship of the Royal Society. Over the years core members included Christopher Wren, Theodore Haak, Sir John Hoskins, Abraham Hill, Francis Lodwick, Sir John Lawrence, John Aubrey, Edmond Halley, John Wilkins, Seth Ward, Richard Waller, and Alexander Pitfeild.

20 Translation in the Royal Society context is relatively under-researched, but as well as essays in this volume see Boschiero L., "Translation, experimentation and the spring of the air: Richard Waller's *Essayes of Natural Experiments*", *Notes and Records of the Royal Society* 64 (2010) 67–83; Henderson, "Faithful Interpreters?"; Johns A., *The Nature of the Book: Print and Knowledge in the Making* (Chicago: 1998) esp. 515–521.

21 For a full account of Lodwick's biography and writings see Lodwick Francis, *On Language, Theology, and Utopia*, ed. F. Henderson – W. Poole (Oxford: 2011).

22 See, for example, Robinson – Adams, *The Diary of Robert Hooke* 349.

the main topics of the letters to the Fellows at a Royal Society meeting.[23] It was probably this initial introduction to microscopy that prompted Lodwick to embark on a much more sustained piece of natural history translation, an English version of Jan Swammerdam's Dutch *Ephemeri Vita* (Amsterdam, Abraham Wolfgang: 1675). Lodwick's translation was published in 1681 under the title *Ephemeri Vita, or, the natural history and anatomy of the ephemeron, a fly that lives but five hours* (London, printed for Henry Faithorne and John Kersey), and it was shortly after this translation was presented to the Society that Lodwick was elected a Fellow. Given that he was neither entomologist nor anatomist, it seems likely that Lodwick had been encouraged to take on this project by Hooke or his Royal Society colleagues.[24] Hooke was certainly an interested spectator, if not instigator. He had bought a Dutch copy of the *Ephemeri Vita* in August 1678, and at a meeting of the Royal Society in February 1680 he 'produced a large discourse about insects, being a translation of the principal things contained in Dr Swammerdam's book'.[25] This was certainly a reference to the *Ephemeri Vita*, rather than Swammerdam's earlier *Historia insectorum generalis* (Utrecht, Meinardus van Dreunen: 1669), and presumably the translation was Lodwick's.[26] The work must have been completed by 4 April 1680, when Hooke noted that he had seen the 'Copy [of] Lodowicks Swammerdam', presumably a fair copy ready for the press. Hooke was with Lodwick again later in the month negotiating publication of the volume. The rather truncated diary entry 'With Lodowick at Childs with Took proferd 50sh for Swammerdam' must mean that the prominent London bookseller Benjamin Tooke had offered to take on Lodwick's Swammerdam translation.[27]

---

23 Lodwick did go on to provide full written translations of other letters some years later: see Henderson, "Making 'The Good Old Man' speak English" 248–250.

24 Lodwick's translations of Leeuwenhoek's letters in the 1680s bear evidence of his continuing struggle to interpret various Dutch entomological words (see Lodwick, *On Language* 13, and Henderson, "Making 'The Good Old Man' Speak English" 250).

25 Robinson – Adams, *The Diary of Robert Hooke* 370 (5 Aug 1678); Birch, *History* vol. 4, 19 (meeting of 26 February 1680).

26 A reference in the Society's minutes to 'the translation of Dr. Swammerdam's description of the insect hemerobius' (which Swammerdam lists as an alternative name for his ephemeron) makes this identification certain (Birch, *History* vol. 4, 29, meeting of 18 March 1680).

27 Robinson – Adams, *The Diary of Robert Hooke* 442 (4 April 1680) and 444 (24 April 1680); for Tooke see Plomer H.R., *A Dictionary of the Printers and Booksellers who were at Work in England, Scotland and Ireland from 1668 to 1725* (Oxford: 1922) 293. There is little evidence of fees paid to authors in the period, but available records suggest that 50 shillings was a reasonable offer for a short text in a rather niche area. See Lindenbaum P., "Authors

It was entirely typical for Hooke (and Lodwick) that this transaction took place at Child's coffee-house in St. Paul's Churchyard, but Tooke was not generally a publisher of philosophical works, and perhaps for this reason the book was eventually published by Henry Faithorne and John Kersey.[28] Hooke was not the only Fellow involved in its preparation: the physician Edward Tyson contributed a preface 'To the reader', in which he cast himself as the producer of the volume, although he clearly stated that the actual translation had been made by 'a person of [his] acquaintance'.[29] Given his interest in comparative anatomy Tyson may well have assisted with the preparation of the text. More difficult to explain is the contribution of Bath physician and antiquary (and occasional poet) Thomas Guidott, in the form of a prefatory verse 'On the history of the *Ephemeron*'. Presumably the connection was Tyson, although Guidott had published on the properties of mineral springs and may have hoped to ingratiate himself into Royal Society circles.[30] The editorial decisions involved in this project are perhaps its most interesting aspect. Swammerdam's original 1675 volume extended to 422 octavo pages: Tyson commented in his preface that 'what made it so large, was [the author's] frequent, Pious Meditations, and

---

and publishers in the late seventeenth century: new evidence on their relations", *Library* s6–17, 3 (1995) 250–269 [doi:10.1093/library/s6–17.3.250], and "Authors and publishers in the late seventeenth century, II: Brabazon Aylmer and the mysteries of the trade", *Library* 3, 1 (2002) 32–57 [doi: 10.1093/library/3.1.32].

28 In the same year Tooke did publish Edward Tyson's *Phocaena, or the anatomy of a porpess* (London: 1680) and also reissued Robert Boyle's *The Sceptical Chymist* (London: 1661; repr. Oxford: 1680), so it is possible he was toying with the idea of building up a scientific stock. He went on to publish, under the Royal Society's imprimatur, William Molyneux's *Dioptrica nova, A treatise of dioptricks in two parts* (London: 1692). Faithorne was later appointed as an official printer to the Royal Society in order to print John Ray's *Historia Plantarum* (London: 1686) (see Rivington C.A., "Early Printers to the Royal Society 1663–1708", *Notes and Records of the Royal Society of London* 39 (1984) 1–27). Hooke at this point obviously fancied himself as something of an agent in the publishing world, as later in the same year he 'treated with Broom and Car for Pappins book but they agreed not' (Robinson – Adams, *The Diary of Robert Hooke* 457, entry of 11 November 1680). These were the publishers Henry Brome and Samuel Carr; the book was Denis Papin's *A new digester or engine for softning bones* (London, printed by J.M. for Henry Bonwicke: 1681).

29 *Ephemeri Vita*, trans. Lodwick, fol. A3r. Based on his preface, Tyson has erroneously been identified as the translator of the volume in the *English Short Title Catalogue* (www.estc.bl.uk) and elsewhere.

30 For Tyson and Guidott see Guerrini A., "Tyson, Edward (1651–1708)", ODNB, http://www.oxforddnb.com/view/article/27961, accessed 19 Aug 2015; and Jenner M.S.R., "Guidott, Thomas (1638?-1706)", ODNB, http://www.oxforddnb.com/view/article/11719, accessed 19 Aug 2015.

Poetry upon the various accidents of the Life, and extraordinary Mechanism of this Creature'. While applauding this natural turn from philosophy to divinity, Tyson blandly stated that 'the Contemplations for some reasons are omitted in this Translation', resulting in a slim quarto of just 44 pages of text that concentrated firmly on the natural history.[31]

The archives of Hooke and Lodwick provide evidence for a number of other translation projects, many clearly collaborative, although none that resulted in a similarly sustained book-length publication. Lodwick embarked on, but seems not to have completed, a narrative of the exploits of the famous English privateer Sir Henry Morgan, translated from Alexandre Exquemelin's account published in Dutch as *De Americaensche zee-rovers* (Amsterdam, Jan ten Hoorn: 1678). The incomplete manuscript of this translation is now among Hooke's papers.[32] Over forty-five folios in length, this is a significant piece of work and signals the strength of Lodwick's commitment to the project. Valued more (by the philosophical community) for its descriptions of South American natural history than the activities of the pirates, Exquemelin's account had been noticed by John Beaumont in his short-lived journal *Weekly memorials for the ingenious*, in which he published reviews of natural philosophical works and translations of articles from the French *Journal des Sçavans*.[33] Though only later elected a Fellow of the Royal Society, Beaumont was a coffee-house associate of Hooke's at this time and it is possible that his review was connected with Lodwick's translation, either inspired by, or prompting Lodwick's work.[34] Further translation work by Lodwick is now lost, but it included the 'Beginning of a translation of Bald[aeus's] Historie of the indies', and some or all of Cornelis Bontekoe's *Tractaat van het excellenste Kruyd Thee* (The Hague,

31 *Ephemeri Vita*, trans. Lodwick A2r-v.

32 London Metropolitan Archives CLC/495/MS01757, item 2; see Lodwick, *On Language* 13–14, 413.

33 Two English versions of Exquemelin's work appeared in 1684: *Bucaniers of America* (London: printed for William Crooke, 1684); and *The history of the bucaniers* (London, printed for Thomas Malthus: 1684). For Beaumont's review see Frohock R., *Buccaneers and Privateers: the Story of the English Sea Rover, 1675–1725* (Newark: 2012) 34 and 47, n. 27; and *Weekly memorials for the ingenious* 15 (19 June 1682) 111–112; see for translations between the *Philosophical Transactions* and the *Journal des Sçavans* also the chapter by Meghan C. Doherty in this volume.

34 For Beaumont see Mandelbrote S., "Beaumont, John (c. 1640–1731)", *ODNB*, http://www.oxforddnb.com/view/article/1876, accessed 19 Aug 2015. Frohock suggests that Beaumont's review prompted bookseller William Crooke to issue his translation (*Buccaneers and Privateers* 52).

Pieter Hagen: 1678).[35] Again, the translation of Bontekoe's work seems to have been prompted (or, indeed, requested) by Hooke, who noted in June 1693 that he had 'deliuerd to Lodwick Bontecoes Book of tea to translate'. Lodwick was an accommodating friend, and the following week he delivered '2 sheets of Bonteco' to Hooke during one of their regular visits to Jonathan's coffee-house.[36] Though nothing more seems to have come of the Bontekoe project, natural historical facts garnered during these acts of translation did occasionally filter back to the Royal Society: for example, at a meeting in October 1690 'Mr Lodwick said, that Mr Baldaeus in his Description of Ceylon, had given an account of Camphire, & particularly that they draw it out of the root of the Cinnamon Tree'.[37] Lodwick seems also to have supplied Royal Society Fellow Martin Lister with translated extracts from the Dutch collections of Nicolaas Jansz. van Wassenaar about an early seventeenth-century expedition to the Amazon.[38] In all this, his extensive private library was perhaps even more significant an asset than his language skills: many of his Dutch books would have been (and still are) extremely rare in England.[39]

If there is one overwhelming theme to the collaborative translations undertaken by Hooke and Lodwick, it is travel.[40] Swammerdam and Leeuwenhoek aside, most of the narratives translated by the two friends are drawn from accounts of voyages. This is presumably a product of Hooke's research interests at the time, which, as well as preparations connected with the *English Atlas* discussed above, included speculations about geological history and also the history of languages (an interest shared with Lodwick). Although the general trend is clear, it is sometimes difficult to pinpoint the rationale behind choices of specific texts to translate. The material might have been new for the Royal

35 Baldaeus Philippus, *Naauwkeurige beschryvinge van Malabar en Choromandel* (Amsterdam, Johannes Janssonius van Waasberge and Johannes van Someren: 1672); Lodwick, *On Language* 414.

36 Gunther R.T., *Early Science in Oxford, vol. X: The Life and Work of Robert Hooke* (Oxford: 1935) 254 (28 June 1693), 256 (3 July 1693, 'Bonteco' here incorrectly printed as 'Bontico').

37 Royal Society Archives JBC/8/10 (meeting of 22 October 1690).

38 Lodwick, *On Language* 12, 413; Bodleian MS Lister 5, fols. 211r–213v.

39 See Henderson F. – Poole W., "The Library Lists of Francis Lodwick FRS (1619–1694): An Introduction to Sloane MSS. 855 and 859, and a Searchable Transcript", *Electronic British Library Journal* (2009), art. 1, http://www.bl.uk/eblj/2009articles/pdf/ebljarticle12009.pdf, accessed 1 December 2016.

40 There has been much recent scholarly interest in the relationship between travel and science in the early-modern period; for a useful list of sources see Carey D., "Inquiries, heads, and directions: orienting early modern travel", in Hayden J.A. (ed.), *Travel Narratives, the New Science, and Literary Discourse, 1569–1750* (Farnham: 2012) 25–51, esp. n. 1.

Society, but the source texts were not necessarily fresh off the press. For example, at a meeting of the Royal Society in March 1682 Hooke introduced a paper containing 'his translation of an account of the discovery to the southward of Nova Hollandia in the East Indies in the year 1643, extracted out of the journal of captain Abel Jansen Tasman, and published in Low Dutch by Dirk Rembrantse'—the original account had come out almost ten years earlier, in 1674. This paper was too long for the time remaining, and according to the minutes of the meeting, it was deferred. If it was read at a later date, its presence was not noted.[41] Given their working relationship at the time, it seems likely that Lodwick assisted with this translation too; he certainly acted as something of a conduit for information emanating from the Dutch settlements in the East Indies.[42] This translation was probably undertaken with the *English Atlas* in mind, but a mixture of geographical and more commercial motives may have lain behind Hooke's interest in Tartary in 1689. A flurry of activity in December of that year began with Hooke borrowing a 'map & Relation of tartary' from the Scottish bibliophile and librarian James Fraser.[43] On the very same day he invited Lodwick to tea, and they 'translated Mullers account of Tartary'.[44] Lodwick's assistance was required again the following day for more translation, this time of Ḟedor Īsakovīch Baĭkov's journey from Moscow to China.[45] Having spent two days on Baĭkov, Hooke moved on to translate a

41 Tasman Abel, *Ontdekking van't onbekende Suit-lant*, printed in Rembrantz van Nierop Dirck, *Tweede deel van enige oefeningen, 'twelk is in geographia ofte aertkloots-beskrijvinge* (Amsterdam, A.S. van der Storck: 1674); Birch, *History* vol. 4, 139 (meeting of 29 March 1682). Presumably this was the text printed in *An account of several late voyages*: see n. 14, above.

42 For example, Lodwick's manuscript extract of a letter between Dutch East India Company personnel based in the East Indies concerning an earthquake in 1673 is now in the Royal Society archives (Cl.P/9/33); he also passed on information gleaned from the Dutch newspapers. Stephen Inwood has suggested that Hooke and Lodwick collaborated on the Tasman translation (*The Man who Knew too Much: the Strange and Inventive Life of Robert Hooke 1635–1703* (London: 2002) 326–327).

43 For Fraser see Evelyn, *Diary of John Evelyn*, ed. de Beer, vol. 4, 330 (14 July 1683); Moffat B., "Fraser, James (1645–1731)", ODNB, http://www.oxforddnb.com/view/article/73233, accessed 21 Aug 2015.

44 The German scholar Andreas Müller had published a number of works on the Chinese language, geography, chronology and culture; it is unclear to which volume Hooke is referring here. On Müller see Lach D.F., "The Chinese studies of Andreas Müller", *Journal of the American Oriental Society* 60 (1940) 564–575, DOI: 10.2307/594086, accessed 23 November 2015.

45 Baĭkov's account was published in *Anhang zwoer reisen: die erste eines Moscowitischen Besandten nach China: die andere herrn Zachariae Wagners* [...] *durch ein gross theil der*

'voyage of cathay 1620', which was the volume borrowed from Fraser.[46] Finally, a couple of days later, he read at the Royal Society his 'Lecture of the Description of Tartary', and 'Shewd Witsens Map'.[47] The minutes of the meeting record this as 'a Description of two or three Voyages through Siberia from Russia to China, wherein [Hooke] shewed, that of old time it hath been performed as well as lately, and that severall of the places found in the Map of Tartary lately published by the Heer Witsen are there mentioned'.[48] This map was Nicolaes Witsen's *Nieuwe Lantkaarte van het Noorder en Ooster deel van Asia en Europa* (Amsterdam: 1687), which represented a key contribution to European mapping of Russia in the seventeenth century, significant for the ongoing quest for new trade routes in the period.[49]

A partial draft of Hooke's lecture on this date, surviving among his papers, sheds further light on the rationale behind the translations of Tartarian travels.

> I indeuoured in my last lecture to shew how vsefully our Inquisitiuenesse might be employed in Collecting the historys of the Inuentions arts and manifactures found out and practised in forrein Parts, whether Diuers of our Merchants yearly trade. Namely by Enquiring after the materialls made vse of and after the ways methods & practises of employing them for perfecting the sayd works [...].[50]

---

welt und unter andern auch nach China (Berlin, Christoff Runge: 1668) and first printed in English under the title "An account of two voyages: The first of Feodor Iskowitz Backhoff, The Muscovite Envoy [...] Translated from the High-Dutch original printed at Berlin", in Churchill Awnsham – Churchill John, *A collection of voyages and travels*, 4 vols. (London: 1704) vol. 2, 545–551.

46 This was a narrative of a journey from Siberia to Peking made by Ivan Petlin in 1618–1619; Hooke notes on his translation that his source was Bergeron Pierre, *Relation des voyages en Tartarie* [...] *plus un traicté des Tartares* (Paris, J. de Heuqueville and L. de Heuqueville: 1634) (see London Metropolitan Archives CLC/495/MS01757, 97v). Despite the fact that Bergeron mentions in a marginal note the English version printed in Purchas (Purchas Samuel, *Purchas his pilgrimes*, 4 vols. (London, William Stansby for Henrie Fetherstone: 1625) vol. 3, book 4, chapter 11), Hooke follows the French account closely.

47 Gunther, *Early Science vol. X* 170–171 (entries for 5–11 December 1689).

48 Royal Society archives, JBO/8/283–4 (meeting of 11 December 1689).

49 Witsen had been elected a Fellow of the Royal Society the previous month, in November 1689; he sent copies of his map to Fellows ("An Account of a large and curious Map of the Great Tartary, lately Publish'd in Holland, by Mr. Nicholas Witsen", *Philosophical Transactions* 16 (1686–1692) 492–494). See Schilder G., "Development and Achievements of Dutch Northern and Arctic Cartography in the Sixteenth and Seventeenth Centuries", *Arctic* 37 (1984) 493–514, esp. 501–502.

50 London Metropolitan Archives CLC/495/MS01757, fol. 98r.

Hooke and his associates formed an active core of Royal Society Fellows who were alert to the economic potential of natural philosophy in terms of improvements to design and manufacturing processes in England; the natural and man-made productions of foreign parts represented an obvious opportunity here, and the ongoing interest in Greenland's whales discussed at the beginning of this chapter was not solely philosophical.[51] But in this lecture Hooke moves on from the discussion of foreign inventions, arts and manufactures to express a desire for more, and more accurate, 'Geographicall & Naturall historys of countryes themselues'. Despite an increased supply of natural histories recently,

> Enough remains vndone to Employ the Industry of Such as will assisting [sic] towards the acquiring soe vsefull a part of knowledge. we are therefore obleiged to all such trauellers who giue vs an account of those things that they haue remarkd in their travells, though possibly they are but very few and those too not very pertinent to the Subjects one would Inquire after, yet since they may giue some vsefull informations of another kind they deserue to be collected and adjoyned to other informations concerning those places.[52]

By translating and collating reports from foreign parts, a more accurate picture will be assembled: matters of fact neglected by one traveller might be mentioned by another; a travel-writer's incidental details might prove to be a philosopher's 'vsefull informations'. On this occasion Hooke's particular interest, signalled by the use of Witsen's map, is the geography of Tartary rather than the natural history or customs of the inhabitants. In his lecture he compared the routes taken by the various travellers, attempting to match them with the newly-available map but generally finding the accounts rather defective, particularly in terms of measuring distance by 'Days Journeys'—'for that sometimes a months time is Spent in passing a Distance that may be travelled in a few days'.[53] Hooke argues that it would have been relatively straightforward for the parties to have observed and recorded longitude, latitude and distance

---

51 The Royal Society minutes are full of discussions that reinforce this point: see, for example, the interest in a 'horn-lantern' presented on 15 and 22 February 1693 (Royal Society Archives JBO/9/111–12). For the connections between travel and economics see Cook H.J., "Moving about and Finding Things out: Economies and Sciences in the Period of the Scientific Revolution", *Osiris* 27 (2012) 101–132.

52 London Metropolitan Archives CLC/495/MS01757, fol. 98r.

53 London Metropolitan Archives CLC/495/MS01757, fol. 98v.

travelled: 'And thereby it would haue been easy to haue placed all the way or Rode with the townes Residences Hills Riuers &c in the Due Situation, but this It seems was not to be expected from a Russe Embassador'.[54]

Examining and comparing printed sources was one of the stated methodologies of the early Royal Society. Contrary to expectation, though, Hooke does not seem to have been collating the travel accounts in order to ascertain their truthfulness as individual accounts. Instead, he is hoping to marry textual and cartographic descriptions in order to build up a composite picture of the Tartarian terrain—indeed, a week later he drew his own 'China Map'.[55] This was the short-term project, but behind it seems to have been the aim of encouraging the Royal Society to issue a 'Due Method' by which land travellers should measure their roads, in the same way that they had issued instructions for sea-men.[56] The emphasis in these nautical instructions was on measurement rather than description, as it tended to be in other sets of instructions issued for travellers, as for example Robert Boyle's printed 'General Heads for a Natural History of a Countrey, Great or small'.[57] Hooke's concern with roads here rather than natural history more generally is consistent with the mercantile outlook noted above.

This was not the first time that Hooke had translated accounts of Asian travels, nor would it be the last. A few years earlier, in 1686, he had printed translations of two letters from the Jesuit missionary Ferdinand Verbiest describing journeys made by the Chinese Emperor into Tartary in the early 1680s.[58] One

54 London Metropolitan Archives CLC/495/MS01757, fol. 98v. Hooke also expresses surprise that none of the 'mathematicall Jesuites' resident in China had ever given such an exact account.

55 20 Dec 1693; Gunther, *Early Science vol. x*, 173. The map seems no longer to be extant.

56 For example, "Directions for Sea-men, bound for far Voyages", *Philosophical Transactions* 1 (1666) 140–143, and Hooke's devices for measuring sea-depth and taking samples of sea-water ("An appendix to the directions for seamen, bound for far voyages", *Philosophical Transactions* 1 (1666) 147–149.

57 Boyle Robert, "General Heads", *Philosophical Transactions* 1 (1666) 186–189. On the Society's instructions for travellers see Carey D., "Inquiries, Heads, and Directions", and Pearl J.H., "Geography and Authority in the Royal Society's Instructions for Travelers", in Hayden (ed.), *Travel Narratives* 71–83.

58 The letters had been printed together in *Voyages de l'Empereur de la Chine dans la Tartarie* (Paris, Estienne Michallet: 1685) with "Eclaircissement necessaire pour justifier la Geographie qui est supposée dans ces Lettres" 75–78; translated as "A voyage of the Emperor of China into the Eastern Tartary, Anno. 1682", "A voyage of the Emperor of China, into the Western Tartary, in the Year, 1683", and "An explanation, necessary to justify the geography supposed in these letters", *Philosophical Transactions* 16 (1686) 39–51, 52–62 and 62–63; accompanied by "The Preface" 36–37, and "Some observations, and

of Hooke's stated aims in publishing these short accounts was to encourage others to investigate Chinese literature, the gateway to 'an *Empire* of Learning, hitherto only fabulously described', and perhaps even more enticingly, to 'a new *Indian* Mine and Treasure [...] which may not be unworthy the consideration of our *Honourable East-India-Company*'.[59] Apart from their intrinsic interest, Verbiest's letters also demonstrated 'the great use of *Mathematical* Knowledge', a branch of European learning which had so captivated the Emperor that he had admitted Verbiest into his presence.[60] This was a rhetorical flourish designed to bolster the Society's claims for the utility of natural philosophy (albeit a rather unexpected one, in that most of his audience would probably have admitted that the veneration of foreign potentates was not high on their list of reasons to study mathematics). Hooke's personal interest was in the Chinese language and chronology.[61] This fascination persisted, as another scrap of translation in Hooke's hand attests. It is a very brief summary of yet another journey from Moscow to China, this one made by the Danish diplomat Eberhard Isbrand Ides in the 1690s. The narration, written in German, was appended to Christian Mentzel's *Kurtze Chinesische chronologia* (Berlin, J.M. Rüdiger: 1696), as Hooke notes in his translation along with the author's claim that he intended to publish a fuller account.[62]

Hooke's extensive archival remains facilitate a more detailed investigation of his translating activities than is possible for some of his close associates, but it is to these that I will now turn because on the whole they engage in a different type of translation project. The brothers-in-law Richard Waller and Alexander Pitfeild were Hooke's constant coffee-house companions from the late 1680s onwards, and they collaborated to produce the most lavish printed translation sponsored by the Royal Society in the period. The French Académie des sciences had published Claude Perrault's *Mémoires pour servir à l'histoire*

---

conjectures concerning the Chinese characters. Made by R. H. R. S. S." 63–78. Despite his initials only being attached to the latter, we can safely attribute the preface and translations to Hooke. A draft of part of the preface and related material in Hooke's hand can be found among his papers in the London Metropolitan Archives (London Metropolitan Archives CLC/495/MS01757, fols. 82r-86r).

59 "The Preface", *Philosophical Transactions* 16 (1686) 37 (italics reversed).

60 Ibidem 36 (italics reversed).

61 See Poole W., "Heterodoxy and Sinology: Isaac Vossius, Robert Hooke and the early Royal Society's Use of Sinology", in Robertson J. – Mortimer S. (eds.), *The Intellectual Consequences of Religious Heterodoxy 1600–1750* (Leiden: 2012) 135–153.

62 Mentzel, *Kurtze Chinesische chronologia* 141–145. The Royal Society's presentation copy with authorial inscription and gold-edged leaves is still in the Society's collections, with a date of 20 October 1697.

naturelle des animaux (Paris: 1671–1676) in a magnificent style but in such limited numbers that 'they became Presents only from the King, or Academy, to Persons of the greatest Quality', as Pitfeild noted in his translator's preface. Thus an English translation was a worthy undertaking, making the book available not only in London but also to the less distinguished citizens of Paris (as Pitfeild also pointed out, in a thinly-veiled dig at the rival academy's publishing practices).[63] Accompanying Pitfeild's edition of the *Memoirs* was Richard Waller's translation of Jean Picard's *Mesure de la terre* (Paris: 1671), another splendid volume produced by the Académie. A letter from Waller to Hooke shows that Hooke was actively involved in sourcing a copy of the *Mémoires*, and that Waller was initially intending to undertake the translation rather than Pitfeild.[64] Hooke's presentation copy of Pitfeild's and Waller's work is extant in the Cambridge University Library.[65]

These were impressive volumes, designed to convince readers that the new philosophy was producing valuable results. Leaving the significance of their subject matter aside, by translating them the English Society emphasized the fact that they were part of a much wider European natural philosophical movement. Hooke also made use of Waller's linguistic expertise for his own projects. In the early 1690s he delivered a series of lectures outlining his geological theories, arguing that the surface of the earth had undergone major change throughout history and that this was largely due to the action of earthquakes. Any news of recent earthquakes was therefore of great interest, and Hooke was no doubt the main beneficiary when Waller translated an Italian account of a terrible earthquake in Sicily in 1693, which had been sent to the Society

63 *Memoirs for a natural history of animals: containing the anatomical descriptions of several creatures dissected by the Royal Academy of Sciences at Paris; Englished by A. Pitfeild. To which is added an Account of the measure of a degree of a great circle of the earth; published by the same Academy, and Englished by R. Waller* (London, Joseph Streater: 1688). For Perrault's *Mémoires* see Guerrini A., "Perrault, Buffon and the Natural History of Animals", *Notes and Records of the Royal Society* 66 (2012) 393–409, and for Waller and Pitfeild see Ezell M.J.M., "Richard Waller, S.R.S.: 'In the Pursuit of Nature'", *Notes and Records of the Royal Society* 38 (1984) 215–233, and Kusukawa Sachiko, "Picturing Knowledge in the Early Royal Society: the Examples of Richard Waller and Henry Hunt", *Notes and Records of the Royal Society* 65 (2011) 273–294.

64 British Library Sloane MS 4067, 197; quoted in Ezell, "Richard Waller" 220.

65 CUL Keynes.Q.6.4. In December 1689 Hooke noted that he gave 'Dr Pragesty' two copies of Waller's *Of the measure of the earth* in return for another book (23 Dec 1693; Gunther, *Early Science vol. X* 174), which at least suggests he had access to a number of copies; unfortunately there are no diary entries extant for the period up to and including publication of Pitfeild's and Waller's translations.

by a correspondent based in Naples. Waller read the account at the Society's meeting on 20 December 1693, and the minutes record that 'it containing several very curious particulars, he was desired to print it in a Transaction'. The translation does appear in the *Philosophical Transactions* but as is often the case the identity of the translator is not stated.[66] Waller had also translated *Saggi di naturali esperienze fatte nell'Accademia del Cimento* (Florence, Giuseppe Cocchini: 1667) from the Italian, a volume describing early experiments conducted by the Italian Accademia del Cimento, published as *Essayes of Natural Experiments made in the Academie del Cimento* (London, Benjamin Alsop: 1684). As Waller pointed out in his preface, the original volume had been presented to the Society by members of the Accademia in March 1668 and apart from some initial interest had lain in the Society's library ever since. Though there is no textual evidence of this, Luciano Boschiero has plausibly suggested that Waller's translation was prompted by Hooke's research interests at the time, which aligned closely with the pneumatic experiments described by the Accademia. However I think we can also view the publication in the same light as the translations of Perrault and Picard discussed above, as part of the Society's drive towards self-promotion.[67]

Given the rather disparate nature of the material Hooke, Lodwick, and their associates translated it is difficult to come to any over-arching conclusions about their methods and rationale, but each project is suggestive in different ways. We get a very rich picture of literary collaboration in the period, going beyond the usual personnel of author, typesetter, engraver, printer and bookseller. But how did this collaboration influence the texts? Translations often required further intermediaries who facilitated the physical circulation of texts, and assisted with their interpretation. The Hooke circle also provided

66 Royal Society Archives JBO/9/147; "A letter from Mr Martin Hartop at Naples, to the Publisher. Together with an account of the late earthquake in Sicily", *Philosophical Transactions* 17 (1693) 827–829; "An extract of the account mentioned in the foregoing letter, taken out of an Italian paper. Written by P. Alessandro Burgos. Printed first at Palermo, and afterwards at Naples. 1693", *Philosophical Transactions* 17 (1693) 830–838. This was Burgos Alessandro, *Distinta relatione delle spaventoso eccidio cagionato da'terremoti ultimamente con replicate scosse, accaduto a' 9 & 11 Gennaro 1693 nel regno di Sicilia* [...] (Palermo: 1693). Waller was at this point editor of the *Philosophical Transactions*.

67 Boschiero, 'Richard Waller's *Essayes*'. Boschiero's argument is convincing although it seems to me that he understates the Society's initial interest in the work, and that although the relevance of the Accademia's experiments to Hooke's and Denis Papin's work may explain the fact of a 1684 translation, other motivations must have been behind the printing of such a high-quality book. Waller attributed the translation to Royal Society president Sir John Hoskins's "Commands" (*Essayes*, fol. [π1]r).

a ready-made audience for translations, and the interest of key members such as Hooke and Lodwick in languages and linguistic theory meant that they were in touch with a wider group of foreign-language speakers and booksellers in London. It is clear that on most occasions, Hooke (in particular) and his colleagues translated texts for which there was a perceived immediate need, either because of their relevance to some ongoing project (such as the *English Atlas*), or to the weekly conversations of the Royal Society Fellows. The exceptions are possibly the book-length publications of Pitfeild, Waller and Lodwick, although these translations are also presented as useful because the originals are unavailable to English readers. Lodwick's Swammerdam translation is notable in that it leaves out a large proportion of the book's contents, the prose and verse material of a theological nature. By ignoring this and translating only the scientific portion of the book, Lodwick and his colleagues were making a clear statement about what they considered to be useful (although time constraints and budget may have been a factor). Were they also saying something about the preferred character of English philosophical treatises? It is interesting to consider the inclusion of Guidott's prefatory poem in the context of the exclusion of Swammerdam's own poetry. Perhaps polite literature is permissible in a scientific context, but only when separated from the main body of the work. Pitfeild's and Waller's printed translations claim simply to be bringing previously unavailable texts to an English audience. However the books' formats suggest they are doing more than this, with the frontispiece of Waller's *Essayes* in particular more reminiscent of classical literature than contemporary science [Fig. 1.2].[68] The publication strategies here may be part of a drive to make natural philosophy more gentlemanly as well as more cosmopolitan.

The bulk of the Hooke circle's translation activity, though, produced shorter pieces of text in the service of ongoing projects, not intended for stand-alone publication. Translators of philosophical texts felt authorized to choose sections of their originals for translation, rather than viewing the work as a unified whole that must be kept intact.[69] As we have seen, this suited the oral proceedings of the Society's meetings, where nuggets of information were traded freely between Fellows but longer accounts were often put aside due to time constraints. More fundamentally, Hooke's philosophy of scientific methodology may also have influenced the choice of material to translate. Put simply, Hooke believed that

68 For a detailed discussion of the frontispiece see Kusukawa, "Picturing Knowledge" 3. Waller also translated and illustrated Maffeo Vegio's addition to Virgil's *Aeneid* (British Library Add. MS 27347).

69 Texts intended for print publication were also subject to translators' manipulation: see Henderson, "Faithful Interpreters?" 113–117.

FIGURE 1.2 *Richard Waller, frontispiece to* Essayes of Natural Experiments made in the Academie del Cimento *(London, Benjamin Alsop: 1684). Engraving, 230 × 173mm. London, Royal Society Library.*

amassing and comparing data was crucial to the project of constructing hypotheses. Scattered throughout his writings are repeated assertions that given enough data, it would be possible to be as certain of facts in the spheres of natural history, medicine and so on, as it was geometrical truths.[70] In part this expectation lay behind his desire to have the Royal Society collect, for example, as many accounts of foreign journeys as possible—a request that the Fellows seem to have accepted and even, at one point, been willing to fund out of their meagre resources.[71] Methodologically, Hooke's use of travel accounts aligns with his (and others') formatting of such data as weather observations so that they are all visible in a single document—except that in the case of the travel accounts, this was a map rather than a table.[72] And because Hooke translated documents that were sometimes decades old, he could trace his Tartarian travels through time as well as space, cross-referencing past and present data.

Over the years, Hooke and his friends embarked on a large number of translation projects in a variety of subject areas. The fact that many of these projects seemingly resulted in partial translations that were neither published nor formally presented at Royal Society meetings means that they have been largely overlooked by modern scholars, but they were clearly valued and discussed at the time and they were an important channel for the circulation of ideas in early-modern Europe. They are an example of the kind of hidden work that went on alongside, and supported, the experiments and observations that underpinned the new philosophy, and they provided an opportunity for a wide range of participants to contribute to the collaborative endeavour.

Selective Bibliography

British Library

Sloane MS 4067

London Metropolitan Archives

CLC/495/MS01757

70 See, for example, Hooke Robert, *Micrographia* (London, J. Martyn and J. Allestree: 1665), fol. d[1]r.

71 Birch, *History* vol. 4, 139 (meeting of 29 March 1682).

72 See Daston L., "Super-Vision: Weather Watching and Table Reading in the Early Modern Royal Society and Académie Royale des Sciences", *Huntington Library Quarterly* 78 (2015) 187–215. Daniel Carey notes that material elicited by Boyle's "General Heads" could easily be presented in tabular form ("Inquiries, Heads and Directions" 50).

Royal Society Archives

MS/215/12

MS/215/47

RBO/2i/34

JBC/8/10

JBO/8/283–4

JBO/9/111–12

JBO/9/147

"Enquiries for Greenland", *Philosophical Transactions* 2 (1667) 554–555.

"Historia Ambræ, Authore Justo Klobio", *Philosophical Transactions* 2 (1667) 538.

"The Preface", *Philosophical Transactions* 16 (1686) 36–37.

An account of several late voyages and discoveries to the south and north (London, Samuel Smith and Benjamin Walford: 1694).

The English Atlas. Volume I. Containing a description of the places next the North-Pole (Oxford, Moses Pitt: 1680).

Birch Thomas, *The History of the Royal Society of London*, 4 vols. (London, A. Millar: 1756–1757).

Boschiero L., "Translation, Experimentation and the Spring of the Air: Richard Waller's *Essayes of Natural Experiments*", *Notes and Records of the Royal Society* 64 (2010) 67–83.

Boulger G.S., "Robinson, Sir Tancred (1657/8–1748)", revised K. Bagshaw, *Oxford Dictionary of National Biography* (Oxford: 2004), http://www.oxforddnb.com/view/article/23873, accessed 19 August 2015.

Carey D., "Inquiries, Heads, and Directions: Orienting Early Modern Travel", in Hayden J.A. (ed.), *Travel Narratives, the New Science, and Literary Discourse, 1569–1750* (Farnham: 2012).

Cook H.J., "Moving about and Finding Things out: Economies and Sciences in the Period of the Scientific Revolution", *Osiris* 27 (2012) 101–132.

Daston L., "Super-Vision: Weather Watching and Table Reading in the Early Modern Royal Society and Académie Royale des Sciences", *Huntington Library Quarterly* 78 (2015) 187–215.

Evelyn John, *The Diary of John Evelyn*, ed. E.S. de Beer, 6 vols. (Oxford: 1955).

Ezell M.J.M., "Richard Waller, S.R.S.: 'In the pursuit of nature'", *Notes and Records of the Royal Society* 38 (1984) 215–33.

Frohock R., *Buccaneers and Privateers: the Story of the English Sea Rover, 1675–1725* (Newark: 2012).

Guerrini A., "Perrault, Buffon and the Natural History of Animals", *Notes and Records of the Royal Society* 66 (2012) 393–409.

Guerrini A., "Tyson, Edward (1651–1708)", *Oxford Dictionary of National Biography* (Oxford: 2004), http://www.oxforddnb.com/view/article/27961, accessed 19 August 2015.

Gunther R.T., *Early Science in Oxford, vol. X: The Life and Work of Robert Hooke* (Oxford: 1935).

Henderson F., "Faithful Interpreters? Translation Theory and Practice at the Early Royal Society", *Notes and Records of the Royal Society* 67 (2013) 101–122.

Henderson F., "Making 'The Good Old Man' Speak English: the Reception of Antoni van Leeuwenhoek's Letters at the Royal Society, 1673–1723", in Cook H.J. – Dupré S. (eds.), *Translating Knowledge in the Early Modern Low Countries* (Zurich – Berlin: 2012) 243–268.

Henderson F. – Poole W., 'The Library Lists of Francis Lodwick FRS (1619–1694): An Introduction to Sloane MSS. 855 and 859, and a Searchable Transcript', *Electronic British Library Journal* (2009), art. 1, http://www.bl.uk/eblj/2009articles/pdf/ebljarticle12009.pdf, accessed 1 December 2016.

Hildyard D., "John Pell's Mathematical Papers and the Royal Society's English Atlas, 1678–82", *BSHM Bulletin: Journal of the British Society for the History of Mathematics* 29, 1 (2014) 18–31.

Inwood S., *The Man who Knew too Much: the Strange and Inventive Life of Robert Hooke 1635–1703* (London: 2002).

Jenner M.S.R., "Guidott, Thomas (1638?–1706)", *Oxford Dictionary of National Biography* (Oxford: 2004), http://www.oxforddnb.com/view/article/11719, accessed 19 August 2015.

Johns A., *The Nature of the Book: Print and Knowledge in the Making* (Chicago: 1998).

Kusukawa S., "Picturing Knowledge in the Early Royal Society: the Examples of Richard Waller and Henry Hunt", *Notes and Records of the Royal Society* 65 (2011) 273–294.

Lach D.F., "The Chinese Studies of Andreas Müller", *Journal of the American Oriental Society* 60 (1940) 564–575.

Lindenbaum P., "Authors and Publishers in the Late Seventeenth Century: New Evidence on their Relations", *Library* s6–17, 3 (1995) 250–269.

Lindenbaum P., "Authors and Publishers in the Late Seventeenth Century, II: Brabazon Aylmer and the Mysteries of the Trade", *Library* 3, 1 (2002) 32–57.

Lodwick Francis, *On Language, Theology, and Utopia*, ed. Henderson F. – Poole W. (Oxford: 2011).

Mandelbrote S., "Beaumont, John (c. 1640–1731)", *Oxford Dictionary of National Biography* (Oxford: 2004), http://www.oxforddnb.com/view/article/1876, accessed 19 August 2015.

Martens Friedrich, *Spitzbergische oder Groenlandische Reise beschreibung gethan im Jahr 1671* (Hamburg, Gottfried Schultze: 1675).

Moffat B., “Fraser, James (1645–1731)”, *Oxford Dictionary of National Biography* (Oxford: 2004), http://www.oxforddnb.com/view/article/73233, accessed 21 August 2015.

Oldenburg Henry, *The Correspondence of Henry Oldenburg*, ed. and trans. A.R. Hall – M. Boas Hall, 13 vols. (Madison, WI – London: 1965–1986).

Pearl J.H., “Geography and Authority in the Royal Society’s Instructions for Travelers”, in Hayden J.A. (ed.), *Travel Narratives, the New Science, and Literary Discourse, 1569–1750* (Farnham: 2012).

Plomer H.R., *A Dictionary of the Printers and Booksellers Who Were at Work in England, Scotland and Ireland from 1668 to 1725* (Oxford: 1922).

Poole W., “A Fragment of the Library of Theodore Haak (1605–1690)”, *Electronic British Library Journal* 6 (2007), http://www.bl.uk/eblj/2007articles/pdf/ebljarticle62007.pdf, accessed 19 August 2015.

Poole W., “Heterodoxy and Sinology: Isaac Vossius, Robert Hooke and the Early Royal Society’s Use of Sinology”, in Robertson J. – Mortimer S. (eds.), *The Intellectual Consequences of Religious Heterodoxy 1600–1750* (Leiden: 2012) 135–153.

Rivington C.A., “Early Printers to the Royal Society 1663–1708”, *Notes and Records of the Royal Society of London* 39 (1984) 1–27.

Robinson H.W. – Adams W. (eds.), *The Diary of Robert Hooke* (London: 1935).

Schilder G., “Development and Achievements of Dutch Northern and Arctic Cartography in the Sixteenth and Seventeenth Centuries”, *Arctic* 37 (1984) 493–514.

Taylor E.G.R., “‘The English Atlas’ of Moses Pitt, 1680–83”, *The Geographical Journal* 95, 4 (1940) 292–299.

CHAPTER 2

Networks and Translation within the Republic of Letters: The Case of Theodore Haak (1605–1690)*

Jan van de Kamp

A qui ou a quoy sert le Talent dans le Mouchoir? Il faut mieux d'avoir et de sçavoir moins, que d'en manquer la vraye jouissance, qui gist en la communication, en faisant le bien du mien, aux autres, et participant à ce contentement, que Dieu mesme poursuit, en maniere de dire, avec tant d'ardeur, et sans se lasser aucunement, de bien faire, voire au plus ingrate du monde.

Who or what is served by a talent wrapped in a handkerchief? It is better to have and to know less than to lack true joy, which consists in communicating, in doing good to others with my ability, and participating in this contentment which God Himself pursues, as it were, with such zeal and without in any way tiring of doing good, nay, not to the world's worst ingrate.

THEODORE HAAK TO MARIN MERSENNE, 6 August 1647[1]

Introduction

Among early modern European scholars there existed networks within which they exchanged data and insights. As these scholars lived before the time that science became increasingly specialized, they were polyhistors and wrote not only about natural philosophy, medicine and alchemy, but also on religion, history and politics. A famous correspondence network within this 'Republic of Letters' was the circle initiated by Samuel Hartlib.[2] An important participant

* I would like to express thanks to Christiaan Bremmer for his comments on this article and to Alexander Thomson for correcting the English.

1 Letter of Theodore Haak to Marin Mersenne, Aug. 6, 1647: Paris, Bibliothèque Nationale, Nouv. Acq. Francaises, 6206, fol. 167, as published in: Brown H., *Scientific Organizations in seventeenth century France, 1620–1680* (Baltimore: 1934) 269. Translation by Alexander Thomson.

2 On correspondence networks and the concept of the 'Republic of Letters' see: Berkvens-Stevelinck C. – Bots H. – Häseler J. (eds.), *Les grands intermédiaires culturels de la République*

 | DOI 10.1163/9789004349261_004

in the Hartlib Circle was Theodore Haak, who not only wrote his own texts but also translated many others.[3] As translator, he rendered scientific, political and, especially, theological texts into other languages.

Recent studies have provided us with more information about Haak's translations of theological writings, an aspect of his work which Pamela R. Barnett in her biography of Haak discussed only partially.[4] In order to attain a better understanding of these translations, one has to view them in connection with Haak's activities in the realms of science and politics. As this volume deals with scientific networks and correspondence, I will pose the question of what role the network in which Haak participated played in his translations of theological books.[5] In addition, I will try to find out in what manner Haak translated

---

des Lettres. Etudes de réseaux de correspondances du XVIe au XVIIIe siècles (Paris: 2005); Mulsow M., *Die unanständige Gelehrtenrepublik. Wissen, Libertinage und Kommunikation in der Frühen Neuzeit* (Stuttgart: 2007). On Hartlib, see: Turnbull G.H., *Hartlib, Dury and Comenius: Gleanings from Hartlib's Papers* (London: 1947); Greengrass M. – Leslie M. – Raylor T. (eds.), *Samuel Hartlib and universal reformation*, Studies in Intellectual Communication (Cambridge: 1994).

3 On the role of translations in the circulation of transfer and culture in Early Modern times in general, see: Pantin I., "The Role of Translations in European Scientific Exchanges in the Sixteenth and Seventeenth Centuries", in Burke P. – Po-chia Hsia R. (eds.), *Cultural Translation in Early Modern Europe* (Cambridge: 2007) 163–179; Stockhorst S. – Kiesant K. – Roloff H.-G. (eds.), *Cultural Transfer through Translation: The Circulation of Enlightened Thought in Europe by Means of Translation*, Internationale Forschungen zur allgemeinen und vergleichenden Literaturwissenschaft 131 (Amsterdam: 2010); Cook H.J. – Dupré S. (eds.), *Translating Knowledge in the Early Modern Low Countries* (Münster: 2012).

4 Barnett, *Theodore Haak* 13–18, 22, 71–75, 114–119. On Haak's theological translations, see: McKenzie E.C., *British devotional literature and the rise of German Pietism*, Ph.D. dissertation (University of St. Andrews: 1984) vol. 1, pp. 140.; Sträter U., *Sonthom, Bayly, Dyke und Hall. Studien zur Rezeption der englischen Erbauungsliteratur in Deutschland im 17. Jahrhundert* (Tübingen: 1987) 102–111; Damrau P, *The Reception of English Puritan Literature in Germany* (London: 2006) 96–133. In addition to the translations mentioned by Barnett, Haak was also the translator of: Scudder Henry, *Eines wahren Christen tägliche Wallfahrt oder eine trewhertzige Wegweisung und Anleytung, wie man zu einem gottseligen und gerühigen Leben und Wandel gelangen möge*, trans. T. Haak (Frankfurt am Main, Johann Friedrich Weiss: 1635), cf. McKenzie E.C., *A catalog of British devotional and religious books in German translation from the Reformation to 1750* (Berlin: 1997) 364, no. 1497. Barnett, *Theodore Haak* 14, n. 24 doubts the assertion by Martin Kempe that Haak was the translator of this book. Haak also was the translator of: Whitfield Henry, *Ermunter dich* (Amsterdam, Johann Jansson, 1638), cf. McKenzie, *British devotional literature* vol. 1, 143ff.; McKenzie, *A catalog* 433, no. 1769; *Bättkunst* (Basel 1639, publisher unknown, 1639), cf. McKenzie, *A catalog* 4, no. 4.

5 I have investigated the role of networks in regard to the production, distribution and reception of translations of devotional literature in my dissertation: Kamp J. van de, "auff bitte und einrahten etzlicher frommen Menschen ins hochteutsche ubersetzet": *Deutsche*

his source texts, and what a study of this process reveals about his practice as a translator.

This article is structured as follows. First, I will give an outline of Haak's life and work, within which his translating work will be discussed.[6] I will divide this biographical part into three subsections, which each deal with the contexts in which Haak was working at a given moment (although there are not always clear boundaries between these fields): Haak as a fundraiser for his oppressed fellow Christians in Germany; Haak's role in debating and translating natural philosophy; and Haak in his political offices. Second, I will give some examples of Haak's method in translating theological works, as a case study for his practice as a translator. Finally, I will draw a conclusion in which I seek to answer the question of the role of networks in Haak's translations. As the subject of translation in Royal Society circles is discussed elsewhere in this volume, this paper focusses not so much on his translation work there, but rather on his less-studied theological translations—both to show how he operated within a network of scholars, and as a window through which to view his translating practice more generally.

Haak as a Fundraiser for His Oppressed Fellow Christians in Germany

Haak, born on 25 July 1605 at Neuhausen near Worms, was through his mother Maria Tossanus a kinsman of the Huguenot families Tossanus and Spanheim, among whom were a couple of theologians and other learned men.[7] Already in his youth, Haak may have become acquainted with the irenic stance of Palatine Reformed theology, which strove for a reconciliation with Lutheranism.[8] Due

---

Übersetzungen englischer und niederländischer reformierter Erbauungsbücher 1667–1697 und die Rolle von Netzwerken, Ph.D. dissertation (Vrije Universiteit Amsterdam: 2011).

6 In this biographical part I draw particularly on Barnett, *Theodore Haak*, to which I have added data from the newer literature. See also Keller A.G., "Haak, Theodore (1605–1690)", in Harrison B. (ed.), *Oxford Dictionary of National Biography*, http://www.oxforddnb.com/view/article/11827, accessed 24th August 2015; Poole W., "A Fragment of the Library of Theodore Haak (1605–1690)", *Electronic British Library Journal* (2007) article 6, 1–37. See for this section, Barnett, *Theodore Haak* 9–33.

7 Rudolph, M., *Tossanus (Toussain de Beaumont). Beiträge zur Geschichte einer Hugenottenfamilie und der damit verbundenen Sippenkreise Couet du Vivier, Durant und Ferry* (Insingen: 2010). See also Barnett, *Theodore Haak* 9–33.

8 Benrath G.A., "*Die konfessionellen Unionsbestrebungen des Kurfürsten Karl Ludwig von der Pfalz († 1680)*", *Zeitschrift für Geschichte des Oberrheins* 116 (1968) 187–252; Hotson H., "Irenicism in the Confessional Age: The Holy Roman Empire, 1563–1648", in Louthan H.P. –

to the Thirty Years' War he could not study at the university of Heidelberg, for which reason he went abroad to England, a country which had strong ties with the Palatinate due to the marriage of the 'Winter King' Frederick V, Elector Palatine, to the British princess Elizabeth Stuart in 1613.[9]

Due to the unstable political and military situation in the Palatinate in the following years, Haak lived partly in England, partly on the continent. From 1625 to 1626, he studied at Oxford and Cambridge. Together with three fellow Germans, he followed a practical training for ministry led by the minister John White in Dorchester.[10] In 1626 he went to Cologne, where he joined an underground Reformed congregation. Every day while there, he read to the members an extract of a German translation he was producing of *The Mystery of Self-Deceiving* (1615) by Daniel Dyke. Reading the translation served as a substitution for having a minister.

From 1628 to 1631, Haak learned theology as well as mathematics from Thomas Allen in Oxford, at Gloucester Hall. Haak was ordained as deacon in the Church of England in 1631 by the well-known Bishop Joseph Hall. While Haak had not obtained a degree, he never obtained a benefice either, for he did not take full clerical orders.

In various ways, Haak tried to supply his countrymen with financial and spiritual goods. Around 1632, Haak was one of two men who were commissioned, among others via the London Dutch Reformed Church, to collect money in England for the Reformed ministers from the Palatinate who had been exiled due to the war.[11] From 1633 onwards, Haak travelled through Germany and the Netherlands. Haak's first translation, German into English, was published in 1633. The original was a lamentation sermon by his cousin Friedrich Schloer, minister at the German Church at The Hague, on the occasion of the deaths

---

Zachmann R.C. (eds.), *Conciliation and confession: the struggle for unity in the Age of Reform, 1415–1648* (Notre Dame, IN: 2004) 228–285, there 234–237, 258ff.

9 Wilson P.H., *The Thirty Years War: Europe's Tragedy* (Cambridge MA: 2009); Rüde M., *England und Kurpfalz im werdenden Mächteeuropa (1608–1632): Konfession—Dynastie—kulturelle Ausdrucksformen* (Stuttgart: 2007); Pursell B.C., *The Winter King: Frederick V of the Palatinate and the Coming of the Thirty Years' War* (Aldershot: 2003).

10 Grell O.P., *Dutch Calvinists in early Stuart London: The Dutch church in Austin Friars 1603–1642* (Leiden: 1989) 181. Grell mentions four students from the Palatinate who followed the training by White in 1626: 'Mr. Sleer, Mr. Fisher, Mr. Haake and Mr. Hopff', cf. Grell *Dutch Calvinists* 181. 'Sleer' is probably Frederike Schloer (see: n. 12), 'Fisher' and 'Hopff' cannot be identified.

11 Grell O.P., *Brethren in Christ: A Calvinist network in Reformation Europe* (Cambridge: 2011) 186, 201–214, 218, 226.

of both the King of Sweden, Gustavus Adolphus, and the King of Bohemia, Frederick V, in late 1632.[12]

About 1635, Haak came into contact with Samuel Hartlib and his friends and correspondence partners, all of whom shared an interest in natural philosophy and the digesting of all kinds of knowledge. Hartlib was in close touch with two men in particular: firstly, John Dury, a Scottish minister who strove after the reconciliation of the Protestant confessions.[13] In this context Dury regarded, among others, the popular English godly-living handbook *The practise of piety* (before 1612) by Lewis Bayly as a useful creed for an envisaged united church of Lutherans and Reformed.[14] The second was John Amos Comenius, bishop of the Moravian Brethren and advocate of pansophic knowledge.[15]

Hartlib himself strove for a more intensive manner of communication of all kinds of knowledge, which had, according to him, become corrupted after the Fall. One should not keep his talents hidden or secret, but should invest and share them. Hartlib called for the excerpting of books and the ordering of these excerpts in a synopsis, which would form a commonplace book on a range of topics.[16]

The three 'brothers', Hartlib, Dury and Comenius, all shared to some extent the expectation of an imminent kingdom of social, religious and political peace lasting for a thousand years (the theology of millenarianism). This would witness a substantial increase of knowledge that would lead to a restoration of the completeness of man's dominion over nature that had been lost at the Fall.

---

12 The title of the translation runs: *The Death of the Two Renowned Kings of Sweden and Bohemia* [...] *Publikely lamented in a Sermon, held before a Princely, Noble and Frequent Assembly, in the High-Dutch Congregation at The Hague* (London, I. D[awson]: 1633). The entry at Stationer's Hall adds 'translated out of the high Dutch by T:H:'. It contains a sermon on 1 Sam. 1:19–20. See also: Arber E., *A Transcript of the Registers of the Company of Stationers of London, 1554–1640* (London: 1877) vol. 4, entry for 18 Feb 1632/3, cf. Barnett, *Theodore Haak* 22.

13 See: Turnbull, *Hartlib, Dury and Comenius*; Trevor-Roper H.R., *Religion, the Reformation and Social Change and other essays* (London – Melbourne – Toronto: 1967) 237–293; Léchot P.-O., *Un christianisme "sans partialité": Irénisme et méthode chez John Dury (v. 1600–1680)* (Paris: 2011).

14 Together with Johann Arndt's *Vier Bücher vom wahren Christentum* (1605–10), Bayly's book was intended to constitute the Apostles' Creed, the Lord's Prayer and the Decalogue, the creeds of this united church, cf. Leube H., *Kalvinismus und Luthertum* (Leipzig: 1928) vol. 1, 237–238.

15 Blekastad M., *Comenius: Versuch eines Umrisses von Leben, Werk und Schicksal des Jan Amos Komenský* (Oslo: 1969).

16 Greengrass M, "Samuel Hartlib and the Commonwealth of Learning", in Barnard J. et al. (eds.), *The Cambridge History of the Book in Britain* (Cambridge: 2002) vol. IV, 1557–1695.

Hartlib and his brothers regarded the political upheaval of the English Civil War in the 1640s as an opportunity for a universal reformation.[17]

In addition to the financial aid that Haak gave his countrymen, he also sent spiritual aid by means of published translations: between 1635 and 1639, five translations of English Reformed devotional books by Haak were published. These writings were of a Puritan stance, i.e. they reflect the plea for an earnest and intensive form of Christianity.[18] As a key part of Haak's work as a translator, as well as his theological interests, these works merit some attention. The first was a translation of the work of Henry Scudder, which discussed several aspects of walking with God.[19] The second and third were translations of works by Daniel Dyke.[20] The books dealt respectively with deception in the spiritual realm and repentance and conversion. Both Dyke translations were afterwards combined into one volume and became bestsellers among the German translations of English devotional literature, being published 19 times. Two further

17 Webster C., *The Great Instauration: Science, Medicine and Reform, 1626–1660* (London: 1975). Webster was an exponent of the 'Merton thesis' and the results of his work should therefore be treated with caution. This article is not intended as a contribution to the 'Merton thesis', according to which the Puritan ethic (cf. the Weber thesis) was one of the main constitutive elements of the rise of an experimental form of science in the 17th century; cf. Merton R.K., "Science, Technology and Society in Seventeenth-Century England", *Osiris* 4 (1938) 360–632. This thesis has been justly questioned, see for example Cohen I.B. (ed.), *Puritanism and the Rise of Modern Science* (New Brunswick: 1990); Brooke J.H., *Science and Religion: Some Historical Perspectives* (Cambridge: 1991); Belluci D., *Science de la nature et Réformation* (Rome: 1998) 109–116.

18 Coffey J. – Lim P.C.H., "Introduction", in Coffey – Lim (eds.), *The Cambridge Companion to Puritanism* (Cambridge: 2008) 1–15.

19 Scudder, *Eines wahren Christen tägliche Wallfahrt*. This was Haak's translation of the English original: *The Christians daily walke in holy securitie and peace* (London, I. D [awson]: 1627). Cf. McKenzie, *British Devotional Literature* vol. 1, 140ff.; McKenzie, *A catalog* 364, nos. 1497–1500.

20 Firstly, Dyke Daniel, *Nosce Teipsum, das grosse Geheimnuß deß Selb-Betrugs*, trans. T. Haak (Frankfurt am Main, Johann Friedrich Weiss: 1636). A translation of the original *The mystery of selfe-deceiung* (1614). The title page states that the book was translated by one D.H.P. and from another source, we know that this likely stands for Dietrichus Haak Palatinus. See: McKenzie, *A Catalog* 168–173, nos. 698–716. Secondly, Dyke Daniel: *Eine sehr nothwendige vnd vberauß nützliche Betrachtung vnd Beschreibung der Wahren Buße*, trans. T. Haak (Frankfurt am Main, Johann Friedrich Weiss: 1637) See also: McKenzie, *A catalog* 174–177, nos. 719–737. See, generally: McKenzie, *British devotional literature* vol. 1, 205–212; Sträter, *Sonthom, Bayly, Dyke und Hall* 102–111; McKenzie, *A catalog* 168, no. 698; Damrau, *The Reception of English Puritan Literature* 96–133.

translations produced by Haak included German versions of a work by Henry Whitfield, and an unknown English book on the art of prayer.[21]

In keeping with the compiling and commonplacing activities of Hartlib and his circle, in about 1656 Haak played a role in an envisaged—but probably never accomplished—project for a systematic collection of fragments from English devotional literature.[22] This work had been planned since the 1630s by a number of theologians from the Palatinate and Wetterau, including several acquaintances of Haak's: the Palatine court chaplain Petrus Streithagen, the doctor of theology Paul Tossanus of Heidelberg (Haak's uncle), and the minister and natural philosopher Johann Moriaen of Frankfurt.[23] The writers of a request for support for this project made a plea for the distribution of the English writings, as they were a hidden treasure for foreigners:

> Talentum enim, quod a Deo singuli accepimus, fidei nostrae commissum est, ut illius dispensatores facti, non illud defodiamus, [...] sed ad Domini emolumentum impendamus [...], ne permittatis hoc talentum tam pretiosum ulterius abscondi et occultari ab Exocitorum minibus et oculis [...]

> For the Talent, which every one of us have received of God, is committed to our trust, that we being distributers thereof, should not hide it in the ground [...] but employ it to the advantage of our Lord [...] that ye would not suffer so precious a Talent to be hid and concealed any longer from the hands and eyes of Forreiners [...][24]

---

21 Whitfield Henry, *Ermunter dich, das ist: kurtze vnd einfältige Handleitung, wie sich ein jeder zu und in der übung der Gottseligkeit soll erwecken und aufmunteren*, trans. T. Haak (Amsterdam, Johann Janssonius: 1638). Title of the original: *Some helpes to stir up to Christian duties* (1634). See: McKenzie, *British devotional literature*, vol. 1, 143ff.; McKenzie, *A catalog* 433, no. 1769. *Bätt-Kunst oder einfältige und dabey sehr lehr- und trostreiche Betrachtung oder Erklärung des Gebätts des Herrn* [...] *aus dem Englischen übersetzt von D.H.P* (Basel: 1639). The translation is mentioned by McKenzie, *A catalog* 4, no. 4, which draws on Beck H., *Die religiöse Volkslitteratur der Evangelischen Kirche Deutschlands in einem Abriß ihrer Geschichte* (Gotha: 1891) 182. According to Beck, the Staats- und Universitätsbibliothek Carl von Ossietzky Hamburg holds a copy.

22 For the following paragraphs, see: Kamp J. van de, "Ein frühes reformiert-pietistisches Netzwerk in der Kurpfalz in der ersten Hälfte des 17. Jahrhunderts", *Archiv für Reformationsgeschichte* 103 (2012) 238–265.

23 Young J.T., *Faith, Alchemy and Natural Philosophy: Johann Moriaen, Reformed Intelligencer, and the Hartlib Circle* (Aldershot: 1998).

24 Quoted from the second edition: *An earnest plea for gospel-communion in the way of godliness, which is sued for by the protestant churches of Germanie, unto the churches of Great*

The proposed compilation would, the plea continued, make theologians strive more earnestly for peace and love than for controversy.[25] In 1656—three years after Streithagen's death—Haak shared the following information within Hartlib's network: 'The Collections out of so many Practical Divinity English Writers which Mr Streithagen brought together are with his wife in salvo and are going about to bee printed'.[26] Presumably, this expected publication was the book *Homo novus* [A new man] that appeared at Heidelberg in 1658 under Streithagen's name. It was published with an introduction by two of Streithagen's colleagues: Marcus Floccenius from Heidelberg and Paul Wirtz from Mannheim.[27] Probably, Streithagen was a driving force behind the compilation project and neither his colleagues or Haak had time enough to continue it.

Correspondence on Natural Philosophy

After years of wandering, it must have become clear to Haak that the political and military situation in the Palatinate would not allow him to return. In 1638, he settled definitively in London.[28] There, he had no permanent appointment, but he could afford this, because he was probably a man of independent means and high social standing.[29] In this new setting, Haak came into closer contact with Hartlib and his network, including the mathematician John Pell.

In 1639, Haak began to correspond with the French Father of the Minim Friars and scholar, Marin Mersenne, who was at the heart of a correspondence network dealing with natural philosophy. Most probably, Haak, with his good

---

Brittaine and Ireland, ed. Durie John (London, Richard Wodnothe: 1654), fol. A3r (Latin version), fol. B1r (English version).

25 *The Hartlib papers: A Complete Text and Image Database of the Papers of Samuel Hartlib (c. 1600–1662)* (Sheffield: 2002) [CD-Rom], Copy letters, Tossanus to archb. of Canterbury, & Hanoverian divines to English divines, in Latin, 20.11.1632 and 15.3.1633, 59/10/53A-60B; Durie (ed.), *An earnest plea*, fols. A3r–B3r.

26 The Hartlib Papers, *Ephemerides* (1656) part 1, Hartlib 29/5/63B.

27 I have used the following edition: Streithagen Peter, *Homo novus Das ist: Ein new, gelehrt und gottseliges Tractätlein, von deß Menschen Wider-Geburt, auß unterschiedlichen der berümbtesten Englischen Theologen ... zusammen getragen* (Heidelberg, Wilhelm Walter, William Fitzer, Abraham Lülß: 1670), fol.):(2r – fol. []:(4v].

28 See Barnett, *Theodore Haak* 34–50, 71–88, 120–157.

29 Martin Kempe described Haak as 'Nobilis Palatinus' (Palatine nobleman) and as 'Teutscher Edelmann' (German nobleman) in Kempe, *Charismatum Sacrorum Trias* 20, 660.

proficiency in French, was asked by Hartlib and his friends to make contact with Mersenne in order to bring together similar aims. Haak and Mersenne exchanged scientific instruments, books, news on scholarly discoveries, and questions on scientific matters. Topics included optical lenses, magnetism, mathematics and musical instruments. Due to Haak's workload, the correspondence came to an end in 1640, but was resumed in 1647 until Mersenne's death in 1648. In August 1647, Haak wrote to Mersenne the enthusiastic plea for the sharing of knowledge which is printed at the beginning of this contribution.

The question how religious differences were handled within these correspondence networks is revealed by an instruction of Joachim Hübner to Comenius. As Comenius also began to write to Mersenne, Hübner instructed him that he should write less about religion, but rather as a Christian to another Christian.[30]

The example of Mersenne, who organized meetings to discuss new discoveries in natural philosophy, may have been what inspired Haak to organize something similar. The result was the '1645 Group', which would meet in Gresham College London. Amongst other virtuosi and natural philosophers, Dr John Wilkins belonged to this group, who had been Chaplain to the Prince Elector Palatine Charles Louis, and who had an intense interest in mathematics and other sciences, as well as the physician Jonathan Goddard. The discussions of the group concerned topics like magnetics, astronomy and natural experiments. The meetings excluded politics as well as theology, as those topics were at that time too dangerously controversial.

Soon after the Restoration of the Stuart monarchy in 1660, the informal '1645 Group' was institutionalized into the Royal Society for the Improvement of Natural Knowledge. Haak was proposed as a candidate member of the society by Wilkins in 1661. He participated frequently in the meetings—for a time also serving in the Correspondence Committee and the Committee for Agriculture—and he was also active as correspondent and translator. Haak conducted an experiment with live vipers and a bottle of Malaga wine, and presented a sample of 'a kind of mastic made by ants in Franconia'.[31]

---

30 'multo minus de religione (quippe cujus causa optimus vir ille neminem odit), sed ut Christianus ad Christianum Christiane scribas' ('much less of religion (since that excellent man hates no-one on that account) but as a Christian to a Christian you are to write Christianly'), British Library, Sloane 639—Hübner J., *Epistolae ad Commenium*, Salmasium, Fabricium, Gronovium etc. (1635–1640), fol. 30 and 176 (b) (KK, I, no. 62), cf. Barnett, *Theodore Haak* 39–40.

31 Birch Thomas, *The History of the Royal Society* (London, [Samuel Richardson?], A. Millar: 1756), vol. 2, entry on 3.10.1667.

He corresponded, for example, with Abbess Elisabeth of the Palatinate and with John Winthrop, Governor of New England. Topics on which Haak corresponded were the breeding of oysters, experiments made by freezing coloured water in flasks, solar eclipses, geometrical problems, sugar refining and the generation of crystals in the Alps. Finally, he translated, among others, two Italian treatises on dyeing and, together with Henry Oldenburg, a German manuscript about amber.[32]

During the late 1660s and the 1670s, Haak seems to have been less active for the Society, for his name is absent from the *Journal Books* during this period. Nevertheless, he did some translation work, welcomed some continental visitors in England and introduced them to the Royal Society.[33] Two of those visitors were Martin Kempe and Heinrich Ludolf Benthem. Benthem, after his return home, wrote a book on the English church and education in which he also wrote on Haak and his work.[34] Its second edition contained a reference to a translation by Haak of a poem by his friend Georg Rudolf Weckherlin on *Psalm* 104: *The CIV Psalm according to the German Paraphrase of G.R.W. by T.H.* (1679).[35] No surviving copy of this translation has been found.

During the last years of his life, Haak was no longer a frequent attender of the meetings of the Society, but he was still active as correspondent and had a close friendship with the polymath Robert Hooke.[36] Topics covered in his correspondence now included magnetic experiments and phosphorus. In his own collection of curiosities, he possessed a 'strong loadstone' upon which he made some experiments.[37] In 1683 he gave two demonstrations for the Society

32 Birch, *The History of the Royal Society,* vol.1, entries on 14.5.1662 and 3.9.1662; vol.2, entry on 9.1.1668. See also: Barnett, *Theodore Haak* 125, 136–137.

33 In 1678, Haak translated some letters from the Dutch microscopist Anthonie van Leeuwenhoek to the Royal Society into English at the request of Robert Hooke: cf. Henderson F., "Making 'The Good Old Man' Speak English: The Reception of Antoni van Leeuwenhoek's Letters at the Royal Society, 1673–1723", in Cook H.J. – Dupré S. (eds.), *Translating Knowledge in the Early Modern Low Countries* (Zurich – Münster: 2013) 243–268, there 248.

34 Benthem Heinrich Ludolf, *Engeländischer Kirch- und Schulen-Staat* (Lüneburg, Lipper: 1694) 56–59.

35 Benthem Heinrich Ludolf, *Neu-eröffneter Engeländischer Kirch- und Schulen-Staat* (Hannover, Philipp Gottfried Saurmann's heirs: 1732) 116.

36 For translating in the 'Hooke circle' see the chapter by Felicity Henderson in this volume.

37 Birch, *The History of the Royal Society*, vol. 4, entry on 23.6.1686, see: Barnett, *Theodore Haak* 152.

in which he showed how a magnet which had apparently lost its powers could be recharged. He also made his own phosphorus lamp.[38]

Political Offices

According to Barnett Haak became distracted from his work on natural philosophy due to political or religious commissions during the Civil Wars and the Commonwealth.[39] He sided with Parliament, probably because his Reformed convictions were more compatible with the 'Puritan' position of Parliament than with the 'Anglican' position of King Charles I. In 1643/4, he was commissioned by Parliament to make a diplomatic journey to Denmark to resolve a quarrel about the seizing of ships both of the English Parliament and of Denmark. During this journey, Haak translated the *Solemn League and Covenant* (1642) and the *The Declaration of the Kingdoms of England and Scotland, joined in Arms for the Vindication and defence of their Religion, Liberties, and Laws, against the Popish, Prelatical, and Malignant party* (1643) into German. In both documents, England and Scotland declared that they had joined each other in the struggle for Reformed doctrine, a Presbyterian church government, independence of the church from the state, a plain liturgy and the reformation of manners.[40]

In Copenhagen, Haak took part in a discussion which sheds more light on the intentions which he may have had in mind with his translations of theological works. As we learn from a letter, he met a Dane, George Mosse, who had been arrested in London on account of the seized Danish ships. Mosse tried to harangue Haak and his companion, but Haak steered the discussion in the direction of spiritual topics. Mosse said that he esteemed the works of the German theologian Johann Arndt—probably his books on true Christianity—and that he had read Haak's translation *Nosce te ipsum*.[41]

In 1645, Haak was requested by the Westminster Assembly to render the Dutch States' Bible (*Statenvertaling*) (1637), including its annotations, into

38 Benthem Heinrich Ludolf, *Engeländischer Kirch- und Schulen-Staat* (Lüneburg, Johann Georg Lipper: 1694) 59–60.

39 See for this section particularly: Barnett, *Theodore Haak* 51–70, 89–119, 161–185.

40 Durston C. – Maltby J. (eds.), *Religion in revolutionary England* (Manchester – New York: 2006).

41 Berkshire Record Office, Reading: Trumbull Papers, Miscellaneous Correspondence XX—Haak's dispatches to Weckherlin from Denmark (1643–44) etc., 158 (16.12.1643).

English.[42] Haak may have been known to some of the deputies of the Assembly, such as John Dury, John White and Henry Scudder due to his previous translations of theological books.

After the Peace of Westphalia, Charles Louis, the son of the Winter King, who had held his court at London during the war, was installed as Elector Palatine at Heidelberg. He offered Haak the job of secretary, but Haak declined, wishing to stay in London. Instead, Haak became the unofficial London agent for the Elector in London: his task was to communicate on the main events and developments in England. In this function, upon request, he translated several political declarations into German, probably for those members of the staff of Charles Louis who were not acquainted enough with the English language: *Declarations of the Causes of the War with the Scotch* (1650) and probably also *Causes of the War with the Dutch* (1652). Haak also worked as correspondent for the English Parliament and as translator for the Secretary of State, John Thurloe. In 1651, he was asked to translate the proposition of an Oldenburg emissary from German into English, but declined.

In his secretarial function, Haak transferred the salary of his friend John Pell, for whom Haak had successfully negotiated a post as British Resident at Zurich, to Switzerland. Haak also sent books to Pell, among others theological books by William Gouge and James Duport. These books had been requested by the minister Johann Heinrich Hummel of Berne. Acting as couriers between England and Switzerland were, among others, Johann Zollikoffer, who had been adjunct to the German Church at Geneva. Both Hummel and Zollikofer had studied in England (among other countries), were in touch with Dury and his friends, and had translated English devotional books into German.[43]

In 1655 Haak asked Pell via Hartlib—who himself had discussed this with others—whether the chronicle on the Protestant Waldensians in Piedmont (*Waldenser Chronik*) was worth translating into English. The Waldensians had

---

42 On Haak's translation of the Dutch Bible and its annotations, cf. also: Bremmer C.C., "Historische achtergrond van de Haakbijbel", *StandVastig*, 49, 4 (2014) 18–19; 50, 1 (2015) 10–11; 50, 2 (2015) 12–13.

43 See on Hummel: Ryter B., "Hummel, Johann Heinrich", in *Historisches Lexikon der Schweiz*, version: 25.7.2005, URL: http://www.hls-dhs-dss.ch/textes/d/D10683.php, accessed on 24 August 2015; Larminie V., "Johann Heinrich Hummel, the Peningtons and the London godly community: Anglo-Swiss networks 1634–1674", *Journal for the History of Reformed Pietism* 2, 2 (2016) 1–26. On Zollikofer: Stückelberger H.-M., *Die appenzellische reformierte Pfarrerschaft seit dem Bestehen jeder reformierten Kirchgemeinde bis 1977* (Herisau: 1977) 30. See on their translations: McKenzie, *A catalog*, General Index under 'Hummel, Johann Heinrich' and 'Zollikoffer, Johannes'.

been massacred by soldiers of the Roman Catholic Duke of Savoy in that year. Haak probably did not translate this book.[44]

Due to the starting-up of the '1645 Group' as well as the scholarly and political correspondence, Haak's Bible translation project was delayed for several years. In 1655, Haak restarted the work with the help of two assistants whom he had engaged. Looking back in 1656, Haak wrote that the translation work had been 'put upon' him and that he wished not to undertake such a 'laborious work, at the publick desires, with so much losse of time and meanes to my self, instead of due encouragement or reward'.[45]

In 1657 Haak finally completed the Bible translation and it appeared under the title *The Dutch Annotations upon the whole Bible … by Theodore Haak Esq. London*.[46] The Bible translation also contained Haak's dedication to the Lord Protector, Oliver Cromwell. In the dedication of his translation to the Lord Protector, Haak recommended the annotations as a means to spread God's truth and asked Cromwell insistently to stimulate religious unity and concordance.[47]

Starting in the late 1660s, Haak spent much time in translating the first books of John Milton's poetic epos *Paradise Lost* (1655), the original of which was highly valued immediately after publication.[48] The epos traces the history of the Fall from war in heaven and the fall of Satan until the banishment of Adam and Eve from paradise.[49]

Haak must have remained in direct and indirect contact with Milton. After he had translated a first draft, he continued correcting it. In 1682 in the German city of Zerbst, a translation of Milton's epos appeared under the title *Das Verlustigte Paradeis* and under the name of Ernst Gottlieb von Berge. Berge from Haak had received a copy of his translation. He had followed Haak's translation almost unaltered and added his translation of the remaining books

44 Barnett, *Theodore Haak* 106; Poole, "A Fragment of the Library" 15–16.

45 In letters to John Pell, Aug. 1656, 16 June 1657: British Library, MS Additional 24850, fols. 6–7, and 12.

46 It was printed by Henry Hills for the following publishers: John Rothwell, Joshua Kirton and Richard Tomlins.

47 According to Kempe, *Charismatum sacrorum trias* 484 (cf. Barnett, *Theodore Haak* 14 n. 24), Haak also translated *The Old Pilgrim, being the History of the Bible*, but no copy has been found.

48 For this and the following paragraphs, see: Barnett, *Theodore Haak* 146–167.

49 Schwartz, L. (ed.), *The Cambridge Companion to Paradise Lost* (Cambridge: 2014).

FIGURE 2.1 *Sylvester Harding, portrait of Theodore Haak. Drawing, 151 × 122 mm. London, British Museum.*

of the original.[50] Haak must have handed a copy of a newer version of his translation to J.S. Fabricius, professor of Greek and history at Heidelberg.

In 1673, Haak made his will, in which he made bequests to, among others, the poor of his parish and the French and Dutch London Reformed Churches. Haak probably died on 5 May 1690. According to Anthony Wood, Haak had

50 From a comparison, it turns out that Haak used the 1667 edition of the original; von Berge, however, used the second edition of 1674 in combination with Haak's translation: Barnett, *Theodore Haak* 147–148.

left 3,000 proverbs translated from German into English and the same number translated from Spanish into German (?).[51] No traces of these proverbs remain.

Translation Method

In order to discover the manner in which Haak rendered his source texts into the target language, I have compared fragments from three translations. These were produced at the beginning, middle and the end of his career as translator: the sermon by Friedrich Schloer and the Dutch Bible.[52] I will compare my findings with the results of Barnett and Peter Damrau regarding the translation strategies that Haak used in his translations of the two books of Daniel Dyke and Milton's epos.

In the translation of Schloer's sermon from German into English, Haak changes the number of words.[53] He both augments and abbreviates the number of words, as the following two examples respectively demonstrate:

> [German, p. 3]: daß er dieselbige vns gnädiglich verzeihen, die hochbetrübte Königliche Fürstliche Häuser, vnd alle bekümmerte hertzen mit dem H. Geist trösten
>
> [English, fol. A3r-v]: to pray to GOD for a gracious pardon: and that thus reconciled, He would bee pleased with the dew of his blessed comforts to refresh, uphold, and cherish the Royall & Princesly Houses, and all the true hearted Friends & Dependants of those deceased *Worthies*

Haak adds 'to pray to GOD', 'and that thus reconciled, He would bee pleased with the dew of', 'refresh', 'cherish' and he specifies 'alle bekümmerte hertzen' [all anxious hearts] by 'all the true hearted Friends & Dependants of those deceased *Worthies*'.

51 Wood A.A., *Athenae Oxonienses, & Fasti Oxoniensis* (London: 1813) vol. 4, 278–280.

52 From the sermon by Schroer, nine pages were compared, and from the Dutch Bible three (larger-sized) pages. These sample pages are spread over the beginning, middle and end of the book. For the comparison, I have made use of the model by Chesterman A., *Memes of Translation* (Amsterdam – Philadelphia: 1997) 87–116.

53 Schloer Friedrich, *Klag- und Trawerpredigt, Uber den tödtlichen Abgang zweyer fürnemen Königen* (Leiden, Bonaventura Elzevier, Abraham I Elzevier, 1633); Schloer Friedrich, *The death of the tvvo renowned kings of Sweden and Bohemia*, transl. T. Haak (London, I. D[awson], Nicholas Bourne: 1633).

> [German, p. 23]: Ach wer wolte dann dieses leben lieb haben, welches thränen zu einem *prologo* vnnd eingang, den todt zu einem *epilogo* vnd außgang, vnd mühe vnd arbeit zur *tractation* vnd fortgang hat.

> [English p. 45–46]: who then would be in love with this life, whose Prologue is *weeping*, whose Act is *weeping*, whose Epilogue is *weeping*.

In the source text the contents of the prologue ('thränen': weeping), act ('mühe vnd arbeit': effort and work) and epilogue ('todt': dead) are different, but in the translation the content is the same: *weeping*. Haak also paraphrases certain phrases, for example reinforcing their propositions:

> [German, p. 3]: die zween fürnemste Patronen der Evangelischen Religion

> [English, fol. A2v]: two of the chiefest Patrones of true *Religion*

Haak replaces 'der Evangelische[n] Religion (Protestant religion) by "true *Religion*".

> [German, p. 15]: auch wegen der trefflichen Thaten

> [English, p. 27]: also for his matchlesse and imparalell Heroike Acts

Here Haak replaces 'trefflichen' [excellent] by 'matchlesse and imparalell Heroike'.

In addition, Haak changes tropes. In the following example, the source text does only contain the metaphor 'nehren' [to feed], but Haak adds a few, such as to root, plant, harbour, spoile and cumber:

> [German, p. 21]: So ist die sünde auch gut in vnserer *memori* vnd gedächtniß, offt daran zu gedencken, vnd was sie mit sich bringet: aber nicht gut in vnsern hertzen, dieselbige zu nehren vnd zu behertzigen.

> [English, p. 42]: so will sinne likewise be good, to be rooted in our memorie, to remember both it, and the grievous consequents of it, but not good to plant any in our hearts, to feed and harbour it there, to spoile and cumber the ground.

Haak has also a tendency to domesticate his source text. For this reason, he eliminates references to Germany as 'Vaterland Teutscher Nation' (our

Fatherland of the German Nation), and he translates quotes from foreign languages like Latin into English:

> [German, p. 3]: vnser geliebtes Vaterland Teutscher Nation
>
> [English, fol. A2r]: especially our own deare *Country* and *Nation*
>
> [German, p. 22]: *Ambrosius* tröstete die Vnterthanen nach dem todt *Theodosii* also: *Descedit, sed non totus: reliquit enim liberos, in quibus ipsum cernamus.*
>
> [English, p. 44]: Saint *Ambrose* comforted the people after the death of *Theodosius* in this manner: *Discedit, sed non totus; reliquit enim liberos, in quibus ipsum cernamus*; He is gone, but not wholly, for he left children, in whom we may see him still.

The tendency to domesticate is also found in regard to Bible verses quoted by Friedrich Schloer. Where the German version is largely identical to the King James Version (KJV), Haak follows the latter. Where it is not, he translates the Bible verse afresh out of the German.

> [German, p. 4]: [2 *Sam* 1,19–20] Und David klagt: Die Edelsten in Israel (oder, die Zierde in Israel) seind auff deiner höhe erschlagen: Wie sind die Helden gefallen. Sagets nicht an zu Gath, verkündigets nicht auff den Gassen zu Asklon, daß sich nicht frewen die Töchter der Philister, daß nicht frolocken die Töchter der vnbeschnittenen.
>
> [English, p. 1]: *And* David *lamented: The beautie of* Israel *is slaine upon thy high places; How are the Mightie fallen? Tell it not in* Gath, *publish it not in the streetes of* Askalon, *least the Daughters of the* Philistins *reioyce, least the Daugthers of the uncircumcised triumph.*
>
> KJV 1611: The beauty of Israel is slaine vpon thy high places: how are the mightie fallen! Tell it not in Gath, publish it not in the streetes of Askelon: lest the daughters of the Philistines reioyce, lest the daughters of the vncircumcised triumph.[54]

---

54 For an online full text version of the King James Bible, see http://www.kingjamesbibleonline.org, last accessed on 2 September 2016.

Haak follows the King James Version here.

> [German, p. 22]: [*Ps.* 146,3–5). Verlasset euch nit auf Fürsten sie sind menschen geist muß davon, vnnd er muß wider zu der erden werden, alsdann sind verlohren alle seine anschläge

> [English, p. 43]: *Put not your confidence in Princes, they are men and cannot helpe, their breath goeth forth, they returne to earth, in that very day their thoughts perish.*

> KJV 1611: Put not your trust in Princes: nor in the sonne of man, in whom there is no helpe. His breath goeth foorth, he returneth to his earth: in that very day his thoughts perish.

In the phrase 'sie sind menschen' the subject is plural, whereas in the King James Version it is singular ('the sonne of man'). Haak follows here the German source text: '*they are men*'.

Haak also tends to be more explicit in his translation, as the following example shows:

> [German, p. 3]: als wil es sich in alle weg gebüren, bey zeiten dem Allmächtigen in die ruthen zu fallen, durch waare unverfälschte busse und grössere sorgfältigkeit dem HErrn zu dienen, als wir bißhero gethan.

> [English, fol. A3r]: it is most requiste, that with unfained repentance, and effectuall resolutions of serving GOD more carefully than ever yet wee have done, we goe to meete the LORD, and to hold his hands from destroying us utterly.

In this translation, Haak adds the impending background for the exhortation to repentance and the resolution of serving God, namely 'to hold his hands from destroying us utterly'.

In addition, Haak sometimes adds information:

> [German, p. 15]: dardurch Sie [the kings of Bohemia and Sweden, JvdK] gesucht, das Evangelium außzubreiten.

> [English, p. 27–28]: to propagate the *Gospell* of *Christ*, whereby, all unhappie accidents notwithstanding.

In this case Haak adds 'whereby, all unhappie accidents notwithstanding', the accidents suffered by the kings.

The relation between author and reader is also changed by Haak. Catchwords like the complaint "Ach" (Oh!) are eliminated:

> [German, p. 4]: Ach! Jhr Geliebten

> [English, p. 2]: *Beloved*

Finally, Haak alters the voice of the linguistic act. In the following example, he has changed a question ('haben dann [...]?') into a statement ('I trow [...].'):

> [German, p. 21]: Wie? haben dann die Israeliter wöllen bottschafft nach Gath schicken vnd den Philistern die zeitung sagen lassen, daß sie jhren König Saul verlohren?

> [English, p. 40]: Why? I trow the *Israelites* would send no Messengers to *Gath*, to bring the newes to the *Philistines*, that they had lost *Saul* their King.

In a comparison of his second translation with the source text, the Dutch Bible together with its annotations, Haak has (as already analysed previously by Barnett), in comparison with his translation of theological books, only made minor changes.[55] In general, he stays very close to the source text. Where the Dutch translation of a specific verse is substantially identical to the King James Version he follows the latter; where it is not, he translates the verse straight from the Dutch.[56]

In addition to the translation strategies which Haak applies in his rendering of Schloer's sermon, he employs several other strategies in his translation of the annotations. First, he adds figures of speech.

> [Dutch, *Proverbia* fol. 1r]: met allerleye seer beweechlicke vermaningen tot onsen schuldigen plicht

---

55 For the Dutch State Bible (Statenvertaling), see the transcription of the 1637 edition overseen by Nicoline van der Sijs: http://www.bijbelsdigitaal.nl/statenvertaling-1637/, accessed on 2 September 2016. For Haak's translation: *The Dutch annotations upon the whole Bible* (London, Henry Hills, John Rothwell, Joshua Kirton, Richard Tomlins: 1657).

56 Barnett, *Theodore Haak* 117–118.

> [English, fol. 5A4r]: with all manner of effectuall exhortations and perswasions to the performance of our dutie

Haak adds assonance: 'exhortations'—'perswasions'—in the source text only 'vermaningen' [exhortations], as well as alliteration: 'perswasions'—'performance' (not in the source text).

How did Haak translate other writings? Barnett and Damrau give examples of the translation strategies regarding Daniel Dyke's books and John Milton's poem. In his translation of Dyke's books Haak, who is now translating from English into German, often adds synonyms and he enlarges passages, for example to explain or to reinforce an utterance. Haak also domesticates the source text. For example he omits the passage concerned with the late brother of the countess to whom the original was dedicated. Concerning Bible verses Haak either translates literally from the English or he follows Luther's Bible translation more or less. Finally, Haak sometimes paraphrases. He rewrites Dyke's specific rejection of Roman Catholic images and the mass into a general rejection of idolatry:

> [English]: O says one, I abhorre the Popish images, and the idolatry of the masse.[57]

> [German]: GER: Ich bin allem Götzendienst und Götzendienern von Hertzen Gram.[58]

> [I have a heartfelt abhorrence of all idolatry and idol-worship.]

In addition, Haak exchanges biblical characters for ordinary people and he omits the Church's misery and writes instead about the experience of the Thirty Year's War. Damrau's suggestion to interpret this as a form of secularization should be checked more thoroughly.[59]

Regarding Haak's translation of Milton's *Paradise lost*, we can observe that Haak renders his source text fairly literally, which, however, can lead to an un-

57 Dyke Daniel, "A treatise of repentance", in: Dyke Daniel, *Two treatises. The one, of repentance: the other, of Christs temptations, London, Edward Griffin, Ralph Mab* (London, Edward Griffin for Ralph Mab: 1616) (STC (2nd ed.), 7408.2) 64.

58 Dyke Daniel, "Die Wahre Buß", in Dyke Daniel, *Nosce te ipsum* (Frankfurt am Main, Johann Philipp Weiss: 1652) 646. See Barnett, *Theodore Haak* 32.

59 See for Haak's translation strategy regarding Dyke's writings: Barnett, *Theodore Haak* 16–18; Damrau, *The Reception of English Puritan Literature* 100–131.

common word order in the German as well as to a poor vocabulary, with some words recurring again and again.[60] Although Haak has a tendency to literal translation, in some cases he makes changes in unit length: he he has dissected one phrase into three separate phrases. The effect of this change is that the retardation and suspense that Milton had created in his epos is eliminated. Another kind of change that Haak applies concerns the structure of a constituent (a word or a group of words that function(s) as a single unit within a hierarchical structure): Haak changes a present participle ('moving') into the simple present ('stepft'), thus changing the tempo of the description: instead of the slow and steady action in the original, described by continuous tenses, and actions taking place partly simultaneously, events in the translation pass in quick succession.[61] In his translation, Haak also makes changes in hyponymy, turning a hyponym into a hyperonym: he summarizes a listing of Roman Catholic elements—'reliques, beads, / Indulgences, dispenses, pardons, bulls'—into the hyperonym 'heylthum' [relics].[62] Finally, whereas in Milton's original the characters, like Satan and his companion Beelzebub are depicted in shades of grey and with subtle references, Haak casts them in black and white.[63]

If one compares the translation strategies which Haak has applied, it appears that he translated the Dutch Bibel and Milton's poem quite literally. He made more changes, for example by adding synonyms or using paraphrase in his translations of theological works, for which reason his translations were longer than the respective source texts. He turns out to be most skilled in translation non-fiction literature and to be poorly skilled in literary translation.

Conclusion

We turn back to the question posed at the beginning of this article: what was the role of Haak's network in regard to his translations of various texts, ranging in subject matter from theology to natural philosophy and politics?

---

60 Milton John, *The Poetical Works of John Milton*, vol. 1: Paradise Lost, ed. H. Darbishire (Oxford: 1963); [Milton John], "Das verlustigte Paradeis", Kassel, Universitätsbibliothek—Landesbibliothek und Murhardsche Bibliothek, MS Poet. 4° 2.

61 Milton, *The Poetical Works of John Milton* vol. 1, book II, lines 674–676; [Milton], "Das verlustigte Paradeis", book II, lines 675–677. See Barnett, *Theodore Haak* 182–183.

62 Milton, *The Poetical Works of John Milton* vol. 1, book III, lines 490–492; [Milton], "Das verlustigte Paradeis", book III, lines 489–490. See Barnett, *Theodore Haak* 32–33.

63 See for Haak's translation strategy concerning Milton's epos: Barnett, *Theodore Haak* 168–186.

In some cases, it is clear that Haak rendered the texts at others' request, such as the Dutch Bible; the translations that he made in the context of his duties as political correspondent; and presumably also the texts on natural philosophy. The sermon by his cousin Schloer and the paraphrase of Psalm 104 by his friend Weckherlin may also have come into being by request: both men may have asked Haak to translate their publications into English. The background to his other translations, such as the theological works, the political declarations that Haak translated during his diplomatic mission to Denmark, and Milton's epos, is not clear. Haak may have been driven in these cases by his own interest.[64] However, this also will have applied to his other translations, although he complained about the workload of the Dutch Bible project.

His networks were thus responsible for certain requests to produce translations. This was not the only way in which his network played a role at the level of the production of translations: Haak co-produced a translation of a manuscript on amber together with Henry Oldenburg; he asked Pell if he considered the chronicle on the Vaudois people worth translating; he had direct and indirect contact with Milton; and he gave a copy of his translation to von Berge, who afterwards revised it. Third parties also played a role at the level of distribution and reception of Haak's translations: Haak alerted the participants in the Hartlib Circle to the publication of Streithagen's book, he gave a second copy of his rendering of *Paradise Lost* to Fabricius, and he read aloud parts of his Dyke translation to fellow members of the congregation at Cologne.

With his proficiency in Dutch, English, Italian, Latin and Spanish, Haak seems to have been exceptionally skilled during his times. In addition, he not only translated into his native language, but also into English. Also as a migrant and "go-between" between England and the continent, he was an excellent candidate for the role of translator. Peter Burke has pointed to the important

64 Haak's translation work may also have been connected with an interest in linguistics. In 1657, in a letter to Pell, he expressed the wish that Pell might help in the setting up of a proposal regarding 'universal character & language', British Library, MS Additional 24850—Original letters addressed to Dr. John Pell, English agent with the Swiss Protestant Cantons (1655–1658), fol. 12 (16 June 1657). In 1668, Haak belonged to the committee within the Royal Society which was tasked with reporting on John Wilkins's Essay toward a real character and a philosophical language, Stimson D., "Dr. Wilkins & the Royal Society", *Journal of Modern History* 3, 4 (1931) 557. Within the Royal Society, there were other translators who were working on the project of creating a universal language, like Francis Lodwick (1619–1694), cf. Lodwick F., *On Language, Theology, and Utopia*, ed. F. Henderson – W. Poole (Oxford: 2011). For seventeenth-century language projects, see Lewis R., *Language, Mind and Nature: Artificial Languages in England from Bacon to Locke* (Cambridge: 2007).

role of migrants in translating texts in early modern times.[65] Haak was able to profit from the international network in which he participated, which partly overlapped with other networks: the fund collectors for the exiled Palatine ministers, the group of Palatine men who made efforts for the production of a body of divinity out of English devotional literature, those who strove after a reconciliation of Lutherans and Reformed, and those who were interested in the developments of natural philosophy.

That Haak's translation work was, as it were, embedded within his network and correspondence is not astonishing, given his convictions on the need of communicating. In the quotation at the beginning of this article, Haak expressed—with an allusion to the parable of the talents which should not be hidden in a handkerchief (*Matth.* 25:14–30)—his joy at communicating his own goods to others, just as God does good with such an ardour to even the most ungrateful people of the world. In a letter to Johann Christian von Boyneburg, the chief court marshal of the Elector of Mainz, he describes the sole aim of the Royal Society as the honour of serving and doing good to humanity by doing one's utmost for 'the real, exacting and assiduous investigation of nature'.[66] Barnett is right in stating that the 'idea of communication provides the common motive behind all the major activities of his life and fits him equally for the roles of translator and of promoter of scientific discussion and experiment'.[67] The motive of not hiding but sharing one's talents is also found in Hartlib and the German theologians who longed for a systematic collection of fragments from English devotional works.

Haak's network consisted of men of different confessions. There seems to be a discrepancy between his correspondence with the Roman Catholic cleric Mersenne on the one hand and his translating works with a strongly Reformed theological stance and collecting money for Reformed ministers on the other.

---

65 Burke P., *Lost (and Found) in Translation: A Cultural History of Translators and Translating in Early Modern Europe* (Wassenaar: 2005) 10–12; Burke P., "The Renaissance-Translator as Go-Between", in Höfele A. – Koppenfels W. von, *Renaissance Go-Betweens: Cultural Exchange in Early Modern Europe* (Berlin: 2005) 17–31; Burke P., "Cultures of translation in early modern Europe", in Burke – Po-chia Hsia, *Cultural Translation in Early Modern Europe* 7–38.

66 'l'honneur de servir et faire du bien à tout le genre humain, s'adonnent et s'esvertuent, chacun à part soy et conjointement, à la reelle, severe et assidue perquisition de la nature', Haak to Johann Christian von Boyneburg, 18/28 Feb. 1663, published in Gruber Johann Daniel, *Commercii Epistolici Leibnitiani* (Hannover – Göttingen, brothers Schmidt: 1745) II, 1083 (20 Feb./ 1 March 1663), cf. Barnett, *Theodore Haak* 126.

67 Barnett, *Theodore Haak* 7.

The former was a domain where confessional differences were not supposed to be addressed. How does this square?

This combination may be explained by reference to the ideal of the humanistic Republic of Letters, which arose during the sixteenth century, to exchange knowledge across political and confessional boundaries. The topic of natural philosophy may have lent itself excellently to this correspondence, as it did not deal with systematic theology but with another division of theology, one which was regarded as common ground between Christians of all confessions and even between Christians and non-Christians: natural theology, the investigation of nature to find proofs for the existence and continuing work of God.[68] An indication for this from Haak's network is the instruction by Hübner to the Protestant Comenius that he should write to the Roman Catholic Mersenne less about religion, but rather as a Christian to another Christian.

The circumstance that Haak was an irenic man both in regard to political and religious matters may also have played an important role.[69] Good indications for this are his conversation with George Mosse at Copenhagen his book dedication to Cromwell as well as the reduction or general description of Roman Catholic elements in his translation. Haak may have been raised with this irenic spirit in the Palatinate and it may have been strengthened in the context of the devastating effects of religious strife in his home country.[70] He may have received additional impulses in this regard from Dury and others.

His talk with Mosse may demonstrate that Haak considered piety—as did Dury—a 'binder' between people of different political and confessional opinions: Mosse was probably a Lutheran from Denmark or Northern Germany. To a slight extent, Haak and Dury saw the doctrines of their own, Reformed, confession as merely relative to the importance of a godly life.

Nevertheless, Haak remained committed to the Reformed Church and its doctrines: later in life, he translated the Dutch States' Bible with its strong

68 Mandelbrote S., "Early Modern Natural Theologies", in Manning R.R – Brooke J.H. – Watts F. (eds.), *The Oxford Handbook of Natural Theology* (Oxford: 2013) 75–99. Cf. the theologian Johann Heinrich Alsted, who, deeply influenced by the pansophic thoughts of Comenius, wrote *Theologia naturalis* (1615). Cf. Hotson H., *Johann Heinrich Alsted 1588–1638: Between Renaissance, Reformation, and Universal Reform* (Oxford: 2000).

69 See for example the fact that Haak participated in the political and diplomatic circles of the Commonwealth and after the Restoration joined the Royal Society, which was established by a royal charter. Barnett interprets this not as the conduct of a chameleon, but she assumes rather that Haak only had major qualms with political changes if they affected his moral and spiritual ideas. She assumes that this was not the case with the Commonwealth and the Restoration: see Barnett, *Theodore Haak* 122.

70 Barnett, *Theodore Haak* 32–33.

Reformed stance, and in his will he bequeathed a substantial portion of his goods to the French and Dutch Reformed Churches in London.

Finally, regarding Haak's translation method, it has turned out that he translated the Dutch Bible more literally than theological works and that he was more skilled in translating non-fiction writings than literary work.

Selective Bibliography

Barnett P.R., *Theodore Haak, F.R.S. (1605–1690). The first German translator of Paradise Lost* (The Hague: 1962).

Berkvens-Stevelinck C. – Bots H. – Häseler J. (eds.), *Les grands intermédiaires culturels de la République des Lettres: Etudes de réseaux de correspondances du XVIe au XVIIIe siècles* (Paris: 2005).

Burke P. – Po-chia Hsia R. (eds.), *Cultural Translation in Early Modern Europe* (Cambridge: 2007).

Greengrass M. – Leslie M. – Raylor T. (eds.), *Samuel Hartlib and Universal Reformation*, Studies in intellectual communication (Cambridge: 1994).

Greyerz K. von – Kaufmann T. – Siebenhüner K. – Zaugg R. (eds.), *Religion und Naturwissenschaften im 16. und 17. Jahrhundert*, Schriften des Vereins für Reformationsgeschichte 21 (Gütersloh: 2010).

Kamp J. van de, "auff bitte und einrahten etzlicher frommen Menschen ins hochteutsche ubersetzet": *Deutsche Übersetzungen englischer und niederländischer reformierter Erbauungsbücher 1667–1697 und die Rolle von Netzwerken*, Ph.D. dissertation (Vrije Universiteit Amsterdam: 2011).

Kamp J. van de, "Ein frühes reformiert-pietistisches Netzwerk in der Kurpfalz in der ersten Hälfte des 17. Jahrhunderts", *Archiv für Reformationsgeschichte* 103 (2012) 238–265.

McKenzie E.C., *A Catalog of British Devotional and Religious Books in German Translation from the Reformation to 1750* (Berlin: 1997).

McKenzie E.C., *British Devotional Literature and the Rise of German Pietism*, Ph.D. dissertation (University of St. Andrews: 1984).

Pantin I., "The Role of Translations in European Scientific Exchanges in the Sixteenth and Seventeenth Centuries", in Burke P., Po-chia Hsia R. (eds.), *Cultural Translation in Early Modern Europe* (Cambridge: 2007) 163–179.

Poole W., "A Fragment of the Library of Theodore Haak (1605–1690)", *Electronic British Library Journal* (2007) article 6, 1–37.

Turnbull G.H., *Hartlib, Dury and Comenius: Gleanings from Hartlib's papers* (London: 1947). Webster C., *The Great Instauration: Science, Medicine and Reform, 1626–1660* (London: 1975).

Wilson P.H., *The Thirty Years War: Europe's Tragedy* (Cambridge, MA: 2009).

CHAPTER 3

What Difference does a Translation Make? The *Traité des vernis* (1723) in the Career of Charles Dufay

Michael Bycroft

Charles Dufay (1698–1739) is known to historians as an energetic member of the Paris Académie des sciences, as a methodical investigator of light and electricity, and as the author of a set of regulations that Louis XV imposed on French textile dyers in 1737.[1] All these aspects of Dufay's short but active career were connected to a project that is almost entirely *un*known to historians, namely his translation of a work by the Italian naturalist Filippo Buonanni, *Trattato sopra la vernice*. Dufay's translation was published as *Traité des vernis* in 1723, the year in which he entered the Académie; a second French edition appeared seven years later.[2]

Buonanni's text was an influential exploration of methods for imitating Chinese varnish with materials available in Europe (section 1). Dufay's translation was a major project that led him to replicate, augment and correct many of Buonanni's procedures. These edits were the result of wide and careful reading, face-to-face contact with artisans, and personal experience in the laboratory (section 2). The translation helps to clear up the mystery of Dufay's entry into the Académie (section 3). It also explains some of the methods and materials that he deployed, with considerable success, in the experimental research he carried out as an academician (section 4). Finally, it helps to explain his

1 The standard accounts of Dufay's life and work are Brunet P., "L'oeuvre scientifique de Charles François du Fay", *Petrus Nonius* 3, 2 (1940) 77–95, and Heilbron J., "Dufay, Charles – François de Cisternay", in Gillispie C.C. (ed.), *Dictionary of Scientific Biography*, 17 vols. (New York: 1970–1980) vol. 4, 214–217.

2 Buonanni Filippo, *Trattato sopra la vernice detta comunemente Cinese* (Rome, Giorgio Placho: 1720); Buonanni Filippo, *Traité des vernis*, trans. Charles Dufay, 1st ed. (Paris, Laurent d'Houry: 1723), 2nd ed. (Paris, Laurent d'Houry: 1733). The full title of the two French editions is *Traité des vernis, Où l'on donne la manière d'en composer un qui ressemble parfaitement à celui de la Chine, & plusieurs autres qui concernent la Peinture, la Dorure, la gravure à l'eau forte, &c.* The two French editions are identical apart from the "Avis au lecteur" and a few minor typographical differences. Further references to the *Traité des vernis* in these footnotes are to the second edition unless otherwise indicated.

 | DOI 10.1163/9789004349261_005

ongoing interest in scientific translation, including the translation of numbers as well as words (section 5).

This episode shows that the translation of artisanal knowledge from one language to another was a form of research in its own right. Moreover, it was a form of research that led naturally into the systematic search for empirical laws that preoccupied experimenters at the Académie early in the eighteenth century. Above all, the *Traité des vernis* shows that translations made a difference not only to the texts translated but also to the fortunes of the translators.

1 Buonanni's *Trattato* and its French Translator

When Buonanni's *Trattato sopra la vernice* appeared in 1720, it was the most detailed account of Chinese varnish hitherto published by a European author. Buonanni was an aging professor of mathematics at the Jesuit college in Rome, a position he had held since the death of his teacher, the prolific polyhistor Athanasius Kircher, in 1680. The book emerged from the Jesuit tradition of publishing descriptions of Eastern geography, languages, customs and technologies. According to Buonanni, the earliest text in this tradition that dealt with Chinese varnish was Father Martino Martini's *Novus Atlas Sinensis*, first published in 1655.[3] Kircher himself published a recipe for Chinese lacquer, based on the report of an Augustinian monk, in his 1667 *China Illustrata*.[4] Martini had described how the Chinese applied their lacquer to writing desks, chests, tables, wooden roofs and floors, and other domestic items. The effect they sought was a mirror-like sheen that could be enhanced by mixing pigments with the lacquer or by attaching golden decorations to its surface once the lacquer was dry.[5] Europeans valued lacquers for their appearance, their hardness, and their resistance to heat, dust and humidity; they applied them not only to domestic furniture but also to watch boxes, paintings, and even shells. The European demand for oriental lacquers, like the contemporaneous demand for oriental porcelain, had given rise to many imitations, of uneven quality, over the course of the seventeenth century.[6]

---

3 Martini Martino, *Novus Atlas Sinensis* (Amsterdam, Johannes Blaeu: 1655) 115.

4 Kircher Athanasius, *China Illustrata* (Amsterdam, Jan Janszoon van Waesberge and Eliza Weyerstraet: 1667) 220–222.

5 Buonanni, *Traité des vernis* 1–2, 154–56, 170. Cf. Perugini F., "Filippo Buonanni and the Treatise", in Buonanni Filippo, *Techniques of Chinese Lacquer: The Classic Eighteenth-Century Treatise on Asian Varnish*, trans. F. Perugini (Los Angeles: 2009) x.

6 Buonanni surveys these imitations at *Traité des vernis* chaps. 2, 3, 4, 5, 12, 18.

Buonanni's main aim in the *Trattato* was to reveal the definitive method for imitating Chinese varnish. He believed that his recipe excelled all the others in the ease of its manufacture and in its resemblance to the original. The only difference between his lacquer and the Chinese one, he boasted, was that the former could be made with materials readily available in Italy. In modern-day terms, his recipe was a mixture of rosin, linseed oil, turpentine, and gum copal. Rosin and linseed oil were the main ingredients; Buonanni added gum copal to harden the varnish and turpentine to thin it out. Buonanni arrived at this recipe with the help of earlier recipes published by Martini, Kircher and many other European writers. He was also indebted to Cosimo III, Grand Duke of Tuscany, who supplied him with samples of *chiaram* and *girgili*, the main ingredients in the Chinese recipe. One chapter of the book was devoted to a long series of trial-and-error experiments in which Buonanni sought domestic substitutes for these two ingredients.[7] The other 19 chapters dealt with other kinds of varnish, such as Japanese varnish and a heat-proof varnish, and with other parts of the varnishing process, from the heating of linseed oil to the application of golden arabesques to a varnished surface. The *Trattato sopra la vernice* went on to become the most influential work on Chinese lacquer in eighteenth-century Europe. By 1770 it had gone through at least seven editions, including translations into Dutch and Spanish as well as French.[8]

The French translation of 1723 came at a key period in the development of French lacquer. French artisans had been imitating Oriental lacquers already in the seventeenth century, but French production lagged behind that in Amsterdam, London and Berlin, and it was only after 1710 that French lacquer work came into its own. The art developed rapidly in Paris from that date onwards. In 1713, three artisans were granted a royal privilege to manufacture imitation lacquer for a period of twenty years.[9] By all accounts their work was of high quality, but it was overshadowed by the output of Guillaume Martin and his three brothers, whose name was synonymous with fine French lacquer work by the end of the century, as it still is today. As early as 1711, Guillaume Martin was signing contracts with artisans specialized in the engraving of lacquered objects, presumably to decorate objects that he had lacquered in his workshop in the Faubourg Saint-Antoine, the artisans' quarter in Paris. Within a decade, 'vernis Martin' was sufficiently well known to consumers in Paris

7 Ibidem chap. 14, esp. 142–143, cf. 58, 72–74, 157, 167.

8 Perugini, "Filippo Buonanni and the Treatise" ix–xii.

9 Kopplin M., "Naissances des laques françaises dans le contexte européen du XVIIe siècle", in Forray-Carlier A. – Kopplin M. (eds.), *Les secrets de la laque française: Le vernis Martin* (Paris: Les Arts Décoratifs, 2014) 11–16, esp. 15–16.

to feature on a shopkeeper's counter in the *Enseigne de Gersaint*, a work by the painter Antoine Watteau designed to decorate the shopfront of his friend Edme-François Gersaint. In 1725, Louis XV granted the Martin brothers the position of *vernisseurs du roi*; in the same decade the king's cousin, Louis-Henri, Duke of Bourbon, set up a workshop-laboratory in his castle at Chantilly to perfect the art of imitating Oriental varnish and porcelain [Fig. 3.1].[10] We do not know whether the *Traité des vernis* was of any use to artisans, such as those in Louis-Henri's workshop, who wished to discover or perfect the techniques

FIGURE 3.1 *Wooden plate decorated with black lacquer, gold, and silver, with highlights in red lacquer. Made in Japan in the second half of the seventeenth century, and designed for the European market, this plate was among the items confiscated by French revolutionaries from the collections of the Princes of Condé in 1793. One of those princes, Louis-Henri de Bourbon, experimented with lacquer recipes at his castle at Chantilly, a residence that Dufay visited at least once in the 1730s. Unknown artist. Diameter: 50cm, height: 9.8 cm. Paris, Musée Guimet, inv.* MR *385.*

10 Forray-Carlier A., "Les débuts des Martin", in Forray-Carlier – Kopplin, *Vernis Martin* 53–59, esp. 54, 55, 57. Louis-Henri's workshop is mentioned in a caption at Forray-Carlier – Kopplin, *Vernis Martin* 28.

pioneered by Guillaume Martin and his brothers.[11] But the technical and commercial success of lacquer in Paris in the 1710s and 1720s adds extra interest to the question of who translated the work from Italian to French.

Neither of the French editions name the translator. Nearly all writers who consider the question attribute the translation to the lawyer, naturalist, and author Antoine-Joseph Dezallier d'Argenville.[12] The origins of this attribution are unclear. They may be connected to the fact that Argenville traveled to Italy in the 1710s, that he went on to publish best-selling works on natural history and the history of art, and that he was acquainted with lacquer through his study of shell collections in France, England and Holland (shells decorated with lacquer were all the rage among rococo collectors).[13] Argenville may have written the preface to the second French edition of 1730, which refers to the translator in the third person. But he did not translate Buonanni's text, and the preface tells us who did:

> Un Cavalier François, qui l'année passée eut l'honneur d'accompagner Mgr le Cardinal de Rohan dans son voyage à Rome, en se satisfaisant dans cette Ville sur le goût naturel qu'il a pour les Arts, par la fréquentation de toutes les personnes qui y excellent en quelque genre que ce soit, lia connoissance avec le R.P. Buonanni, Jésuite, si connu par tant d'ouvrages curieux qu'il a mis au jour [...].

> A French cavalryman, who last year had the honour of accompanying M. Cardinal de Rohan on his voyage to Rome, in satisfying in that city his innate taste for the arts, by meeting everyone there who excels in any art

11 Nor do we know much about what those techniques were. Even the question of whether the Martins' lacquer contained gum copal, a key ingredient in Buonanni's recipe, is impossible to answer at present. Forray-Carlier Anne, "L'engouement pour le vernis martin: décor intérieur et ameublement", in Forray-Carlier – Kopplin, *Vernis Martin* 71–77, esp. 72–73.

12 The book is attributed to Dezallier at Pinault-Sørensen M., "Dezallier d'Argenville, l'*Encyclopédie* et la *Conchyliologie*", *Recherches sur Diderot et sur l'Encyclopédie* 24, 1 (1998) 101–148, esp. 103; Guichard C., "Vernis", in Lafont A. (ed.), *1740, un abrégé du monde: savoirs et collections autour de Dezallier d'Argenville* (Paris: 2012) 249; and Perugini, "Buonanni and the Treatise" x. The only secondary source I have found that credits Dufay with the translation is St. le Tourneur, "Du Fay, Charles-François de Cisternay", in Balteau J. – Rastoul A. – Prévost M. (eds.), *Dictionnaire de biographie française* (Paris: 1929-), vol. 2, 1387.

13 Dezallier toured Italy in 1713–1716: Laissus Yves, "Argenville, Antoine – Joseph Dezallier", in Gillispie, *Dictionary of Scientific Biography*, vol. 1, 243–244, esp. 243.

whatsoever, made the acquaintance of R.P. Buonanni, a Jesuit very well known for the many intriguing works he has written […].[14]

The unnamed '*cavalier*' must have been Charles-François de Cisternay Dufay. Born in 1698, Dufay entered the French army as a young man and was a lieutenant in the Regiment of Picardy until 1723, the year the first French translation of *Traité des vernis* appeared.[15] Charles Dufay's father, Charles-Jérôme, was a good friend of the Cardinal Armand-Gaston de Rohan, the bishop of Strasbourg and a leading churchman in the courts of Louis XIV and Louis XV. The Cardinal went to Rome in March 1721 to elect a new pope following the death of Clement XI.[16] Charles-François traveled with the Cardinal, 'dont il étoit fort connu & fort gouté' [who he knew well and liked very much].[17] The Cardinal, and presumably Dufay, stayed in Rome for ten months, plenty of time for Dufay to meet Buonanni and to 'satisfy his innate taste for the arts'. Although Dezallier did go to Rome, he went there well before Buonanni's book was published and he did not go with the Cardinal.[18] Corroborating evidence for Dufay's authorship can be found in his biographical dossier at the Académie des sciences, which includes a copy of the *Traité des vernis* in Dufay's hand, and in the inventory of a library belonging to one of Dufay's acquaintances, wherein Dufay is cited as the translator of the work.[19] It is worth adding that Dufay was, certainly in the 1730s and perhaps earlier as well, a customer of Edme-François Gersaint and a visitor to Louis-Henri's castle at Chantilly.[20]

14 "Avis au lecteur", in *Traité des vernis* iii.

15 Desmaze C., *Infanterie française: le régiment de Picardie* (Paris, 1888) 85, 113. Cited in Heilbron J., *Electricity in the 17th and 18th Centuries: a Study of Early Modern Physics* (Berkeley: 1979) 250.

16 Claude M., *Le siècle des Rohan: une dynastie de cardinaux en Alsace au XVIIIe siècle* (Strasbourg: 2006) 7; Heilbron, *Electricity* 251. Cf. "Rohan, Armand Gaston de", in Michaud L.-G. (ed.), *Biographie universelle*, 45 vols. (Paris: 1843–18??), vol. 36, 336–337.

17 Fontenelle Bernard le Bovier de, "Eloge de M. Dufay", *Histoire de l'Académie Royale des Sciences* (1739) 73–83, esp. 75.

18 See note 13 above.

19 "Vernis", Archives de l'Académie Royale des Sciences, Paris, Dufay dossier, in folder entitled "Donner à l'abbé Nollet". *Catalogue des livres et estampes de la bibliotheque de feu Monsieur Pajot, comte d'Onsenbray* (Paris: 1756) item no. 3742. I am grateful to Madeleine Pinault-Sørensen for drawing my attention to the latter citation.

20 Bycroft M., *Physics and Natural History in the Eighteenth Century: the Case of Charles Dufay*, PhD dissertation (University of Cambridge: 2013) 114–116.

2 Translating the *Trattato*

Dufay's translation was a major project. He did not merely render Buonanni's text into the French language. He also tested nearly all the recipes in the book, a large task given the many recipes Buonanni had gathered from friends, books, and his own experience.[21] Traces of these replications can be found in the 86 footnotes that Dufay inserted into his 200-page translation. Most of the footnotes are only one or two lines long. But some are more substantial, such as those containing a description of how to transfer gold leaf onto a varnished surface, a recipe for 'a curious way of painting figures very easily without knowing how to draw', and an account of a beautiful red varnish that Dufay saw in the hands of the physician to the queen of Poland.[22] Some of these notes contained recipes that Buonanni had overlooked, or advice that Dufay had drawn from his own experience. For example, where Buonanni's text described a recipe containing lye, Dufay's note told readers how to make this ingredient for themselves.[23] On some occasions Dufay even corrected Buonanni when his own trials belied the Italian's advice. It is unwise to distill turpentine, Dufay counselled, because it easily catches fire and its flames are hard to extinguish. In the case of the 'curious way of painting figures', Dufay criticized Buonanni's method and suggested a better one.[24] Overall, Dufay augmented Buonanni's recipes on nine occasions and corrected them on five other occasions.

Dufay's additions to the text were as often literary as they were experimental. Fourteen of his footnotes improved upon Buonanni's rather threadbare citations of his sources. He added the dates and titles of texts that Buonanni had mentioned only in passing. He supplied page numbers for citations that either had no page numbers in Buonanni's text or whose numbers referred to an edition different from the one Dufay consulted.[25] He wrote out quotations from Latin works that Buonanni had translated into Italian, and he listed ingredients and recipes that he had found in Buonanni's sources, that the Italian had omitted, and that Dufay considered useful to the reader.[26] In four footnotes he took issue with the way Buonanni used his sources, disputing the Italian's translation of French and Latin terms and pointing out discrepancies between

21 Dufay, "Avis au lecteur" iii.

22 Notes at *Traité des vernis* 159–160, 186, 138.

23 Note at ibidem 112, cf. 123, 138, 178.

24 Notes at ibidem 93, 186, cf. 94, 98.

25 Notes at ibidem 76, 120, 124, 182, 183, 198, 199.

26 E.g. notes at ibidem 76, 120.

the procedures and ingredients described by Buonanni and the ones described by his sources.[27]

These additions show that Dufay read at least five of the 28 sources that Buonanni cited in his *Trattato*. These were seventeenth-century pharmacopoeia by the druggist Pierre Pomet and the chemist Nicolas Lémery; a 1685 work by the Jesuit father Jean Zahn, entitled *Fundamentum Tertium Practico-mechanicum Syntagma*; Abraham Bosse's 1645 etching manual, *Traité de manières de graver en taille-douce*; and the 1557 French edition of Alessio Piemontese's famous book of secrets.[28] Dufay was particularly sensitive to the details of the latter two works. He warned readers that Buonanni had made unannounced changes to Bosse's recipe for a heat-resistant varnish, changes that included replacing walnut oil with linseed oil and omitting Bosse's practice of *mixing* Greek pitch and resin before adding the oil to them.[29] In the case of Piemontese, Dufay inserted the kind of precaution that could make the difference between the success and failure of a recipe. For example, an ingredient for a pigment needed to be cleaned repeatedly *with a fresh portion of hot water each time*—Piemontese included the italicized detail, Buonanni omitted it, and Dufay restored it in a footnote.[30]

Dufay's third source of information, in addition to his reading and experimenting, was his experience of artisanal practices during his tour of Italy with the Cardinal. We know from the preface of the *Traité des vernis* that Dufay frequented artists of various kinds while in Rome. We may speculate that Buonanni was acquainted with artisans in the city, or at least with those who made lacquers, and that he introduced Dufay to some of them. Dufay's footnotes provide a vivid account, though a frustratingly incomplete one, of the world of painters and practical men that he encountered in 'cette Capitale du Monde' [that capital of the world], as Fontenelle described the Rome of his age.[31] When he inquired about the meaning of '*acqua di rasa*', Dufay received the same answer from 'Tous les Peintres Italiens que j'ai consultés' [All the Italian painters I consulted].[32] He observed lacquer on Roman streetlamps, recorded the address of an artisan known for his golden arabesques, and recalled with pleasure the elegant black finish on locks of Roman watch boxes.[33]

27 Notes at ibidem 11, 94, 121, 124.

28 Notes at ibidem 11, 22, 33, 45, 76, 90, 94, 120, 121, 124, 127, 128, 137, 182–184, 198–199.

29 Note at ibidem 121n1, cf. 124.

30 Note at ibidem 183, cf. 94n2, 184, 189.

31 Fontenelle, "Eloge de Du Fay" 75.

32 Note at Buonanni, *Traité des vernis* 22.

33 Notes at ibidem 78, 108, 119.

Perhaps it was also in Rome that he learned the painters' name (*or couleur*) for the mordant they used to attach gold leaf to a varnished surface. Finally, Dufay described the test that painters used to determine whether a batch of linseed oil was hot enough to be mixed with the other ingredients in a lacquer (a feather dipped in the oil should crackle and turned brown).[34]

Dufay sometimes had trouble finding French equivalents for Buonanni's Italian terms. He was candid about this problem in his preface to the *Traité*. There he explained that he had been obliged to 'ajouter quelques notes pour donner des éclaircissements sur plusieurs noms propres de drogues, qui sont differens en France, ou même qui ne s'y rencontrent pas communément, à cause du peu d'usage qu'on en fait' [added some notes to shed light on several names of substances that are different in France [than in Italy], or that are rarely found here because we make little use of them].[35] The case of Greek pitch illustrates the kind of hermeneutic challenge he faced. Buonanni's recipe for Chinese varnish included something that Buonanni called '*pece Grece*'. In the main text of his translation, Dufay rendered this as '*poix Grecque*'. But this was a literal translation that meant nothing to French speakers. So Dufay clarified in a footnote that *poix Grecque* was what the French called '*colophenne*'. Dufay made this identification by comparing the properties of *pece Grece* with those of *colophenne*. For example, both substances were used by archers in Italy and France to treat the hairs of their bows.[36] In other cases Dufay based his identifications on Buonanni's descriptions of the substances he used and on information (procured in Rome) about their price and origin. For example, he learned that Buonanni's '*terebintho*' came from Cyprus and cost 8 jules in Rome, data which suggested to him the translation '*térébenthine de Chypre*'.[37]

Acqua di rasa was an even thornier term. The literal translation was '*eau de résine*', but this was no more meaningful to French artisans than '*poix Grecque*'. Dufay explained in a footnote that all the painters he consulted in Italy had assured him that '*acqua di rasa*' was just another name for *oglio di spigo*. The latter term was naturally translated into French as '*huile d'aspic*'. Dufay could not rest here, however, because French authorities disagreed about the identity of *huile d'aspic*: for Pomet it was an essential oil of lavender, and for others a spirit of turpentine.[38] Another complication was that Buonanni sometimes used the term '*oglio di spigo*', as if it were something different from *acqua di rasa*.

34 Notes at ibidem 82, 108.

35 "Avis au lecteur", in Buonanni, *Traité des vernis*, 1st ed.

36 Buonanni, *Traité des vernis* 18n1, cf. 94n2.

37 Ibidem 137–138, cf. 15, 34.

38 Ibidem 22.

Buonanni also referred to another substance, *oglio di rasa*, without saying how it differed from *acqua di rasa*.[39] Dufay's solution was to hedge his bets. In the main text he translated both '*oglio di spigo*' and '*acqua di rasa*' as '*huile d'aspic*'. However for each occurrence of '*huile d'aspic*' he added a footnote indicating the word that Buonanni had used. He did so, he explained, so that 'le Lecteur soit en état de l'entendre autrement, s'il sçait mieux que moi la signification de ce mot' [the reader is free to interpret [the text] differently, if he knows better than me the meaning of the word].[40] Translating Buonanni's texts required not only a knowledge of two languages, Italian and French, but also familiarity with the substances to which Buonanni referred—their origin, price, uses and their counterparts in the works of French naturalists.

3 Dufay's Entry into the Académie des sciences

A project as substantial as the translation of *Traité des vernis* could not fail to affect Dufay's development as a scientist. In the short term, it eased his entry into the Paris Académie des sciences, then the most prestigious scientific institution in France and, arguably, in all of Europe. This appointment was the determining event in Dufay's career, for several reasons. Firstly, the Académie brought him into regular contact with France's leading scientists, some of whom became his friends and collaborators. Examples are the naturalists René Réaumur and Henri-Louis Duhamel du Monceau, the chemists Jean Hellot and Claude-Joseph Geoffroy, the natural philosopher Jean-Jacques d'Ortous de Mairan, and the astronomer Charles-Marie de la Condamine.[41] Secondly, the Académie offered two high-profile forums—the annual *Histoire et Mémoires* of the Académie and their twice-annual public meetings—that brought Dufay and his research to the attention of the wider public.[42] Finally, the institution was a stepping stone to three further appointments that shaped Dufay's research and strengthened his influence in scientific and industrial circles in

39 Ibidem 88, 162.

40 "Avis au lecteur", in Buonanni, *Traité des vernis*, 1st ed.

41 On Dufay's relationship with Hellot and Geoffroy see Bycroft M., "Wonders in the Academy: the Value of Strange Facts in the Experimental Research of Charles Dufay", *Historical Studies in the Natural Sciences* 43, 3 (2012) 334–370, esp. 361. On Duhamel du Monceau see Dufay, "Observations sur la Sensitive", *Mémoires de l'Académie Royale des Sciences* (1736) 87–111, esp. 87. See also Bycroft, *Physics and Natural History* 83–107 (Réaumur), 95 (Duhamel), 112 (Geoffroy), 99 (de Mairan), 97 (la Condamine).

42 Bycroft, *Physics and Natural History* 17, 186–189.

France. In 1731 he acquired the title of *Inspecteur-général des teintures* and was given the task of developing tests of the quality of textile dyes; the following year he became *Intendant of the Jardin du Roi*, an appointment that placed him at the centre of network of naturalists and collectors in Paris and London and in French territories such as Guyana; and in 1733 and 1738 he served as the Académie's *Directeur*.[43] Without the Académie it is doubtful that Dufay would have had anything resembling a 'scientific career' at all, let alone a career as distinguished as the one he forged for himself.

The *Traité des vernis* was a factor in Dufay's appointment to the Académie because he had few other scientific accomplishments to his name and because Chinese varnish was of special interest to Réaumur, a powerful figure at the Académie and the main supporter of Dufay's candidacy. Dufay's appointment in May 1723 was based partly on two papers—on salt of lime and capillary action—that he submitted to the Académie in the preceding April. Réaumur and the chemist Etienne-François Geoffroy read these papers on behalf of the Académie and reviewed them positively.[44] Aside from these papers, Dufay's scientific CV was thin. It did not include a medical degree, a background in pharmacy, or participation in chemistry courses at the Jardin du Roi, the three routes by which the majority of the Académie's chemists made their way into that body in the first century of its existence.[45] Nor had Dufay undergone an informal apprenticeship under an existing member of the Académie, as had two other youngsters appointed in the early 1720s, Pierre Maupertuis and Jean Pitot.[46] Moreover, neither of the two papers that Dufay submitted to the Académie was practical in orientation. They showed his theoretical imagination and his skill as an experimenter, but they contained no hint of his interest in describing and perfecting the arts. Dufay's translation of the *Traité des vernis* was a perfect complement to his research on capillary action and the salt of lime.

43 Bycroft, *Physics and Natural History* 14–15 (overview), 85–89 (textile dyes), 93–94 (Jardin du Roi).

44 Procès-verbaux de l'Académie des Sciences 1723, 10 Apr (fol. 67r), 28 Apr (fol. 79r), 30 Apr (fol. 81r). The full text of the two papers can be found at Dufay, "Sur le sel de chaux", *Mémoires de l'Académie Royale des Sciences* (1724) 88–93, and Procès-verbaux 1724, 12 Feb (fol. 60r).

45 Sturdy D., *Science and Social Status: The Members of the Académie des Sciences, 1666–1750* (Woodbridge: 1995) 262–63, 400–401, 430–432.

46 Terrall M., *The Man Who Flattened the Earth: Maupertuis and the Sciences in the Enlightenment.* (Chicago: 2002); Mayr O., "Pitot, Henri", in Gillispie, *Dictionary of Scientific Biography* vol. 11, 4–5.

Practical applications mattered for Dufay's appointment because they mattered to Réaumur. By the early 1720s the naturalist had cemented his position as the leading practitioner of the empirical sciences at the Académie and as the mainstay of its Baconian programme of utilitarian research. Since at least 1711 he had led the *Déscription des arts et métiers*, the Académie's attempt to describe and codify the arts and industries practised in France.[47] In 1722 he published the first volume in that project, a treatise on steel and cast iron.[48] In the 1710s he led a survey of France's mineral resources, and early in the 1720s he wrote an impassioned plea for increased state funding for the Académie and for its useful research.[49] Some of the projects he mentioned in that plea—especially those on gold refining—bear a striking resemblance to research that Dufay carried out in 1727 and 1728. Dufay followed through on two other projects conceived by Réaumur: the creation of new workshops for manufacturing steel and iron, traces of which can be found in the procès-verbaux of the Académie's meetings in the 1720s;[50] and the *Déscription des arts et métiers*, to which Dufay's study of textile dyes was intended as a contribution.[51] It seems likely that Réaumur had this kind of project in mind when he threw his weight behind the young Dufay in 1723.

The *Traité des vernis* was not simply a description of an art, but of a *decorative* art, and moreover an art with Chinese origins. Réaumur had always paid particular attention to the decorative arts, from his first practical undertaking

---

47 Demeulenaere-Douyère C. – Sturdy D., *L'Enquête du Régent, 1716–1718: sciences, techniques, et politique dans la France pré-industrielle* (Turnhout: 2008) 26.

48 Réaumur R., *L'Art de convertir le fer forgé en acier, et l'art d'adoucir le fer fondu* (Paris, Michel Brunet: 1722). English quotes from this text are from Réaumur René, *Réaumur's Memoires on Steel and Iron*, ed. C.S. Smith, trans. A.G. Sisco (Chicago: 1956). My page references use Réaumur's pagination.

49 Demeulenaere-Douyère – Sturdy, *L'Enquête du Régent*, passim. Réaumur René, "Reflexions sur l'utilité dont l'Académie des sciences pourroit être au royaume, si le royaume luy donnoit les secours dont elle a besoin", transcribed in Maindron E., *L'Académie des sciences: histoire de l'Académie, fondation de l'Institut national* (Paris: 1888) 103–110. Internal evidence allowed Maindron to date this document between 1716 and 1727. A further date suggests that it was written in 1722 or earlier: Réaumur refers (107) to the Conseil de commerce, a body that was replaced by the Bureau du commerce in June 1722: Bonnassieux P. – Lelong E., *Conseil de commerce et Bureau du commerce 1700–1791: inventaire analytique des procès-verbaux* (Paris: 1900) xii.

50 Procès-verbaux de l'Académie Royale des Sciences, 23 Dec 1724 (fol. 371v); 2 Jun 1725 (fol. 131r); 18 Jul 1725 (fol. 165r); 18 Aug 1725 (fol.219v); 3 Apr 1726 (fol. 129r); 3 Mar 1728 (fol. 87v); 30 Jul 1728 (fol. 291r). Cf. Bycroft, *Physics and Natural History* 91–92.

51 Dufay Charles, "Observations physiques sur le mélange de quelques couleurs dans la teinture", *Mémoires de l'Académie Royale des Sciences* (1737) 253–268, esp. 254.

for the Académie, which concerned the feasibility of making clothes from spiders' silk, to his book on iron and steel, where he declared that 'the production of more beautiful work, without sacrifice of quality and at lower cost, is the route to progress which we must endeavour to guide the arts'.[52] When Dufay began his translation, Réaumur was in the middle of a project on Chinese porcelain that had much in common with Buonanni's on varnish. In 1722 Réaumur acquired samples of the main ingredients in Chinese porcelain (*kaolin* and *petuntse*), the equivalents of Buonanni's *chiaram* and *girgili*. These samples allowed Réaumur to narrow down his search for French substitutes for the Chinese materials, a search that resembled Buonanni's hunt for Italian versions of *chiaram* and *girgili*.[53] Réaumur's investigation was in full flow in 1723, when he took on the engineer Henri Pitot as a laboratory assistant to help him with the many trials that were necessary to find an adequate (let alone perfect) recipe for hard-paste porcelain.[54] It is hard to imagine a topic better suited to Réaumur's research interests in 1723 than the domestic production of Chinese lacquer. Dufay had finished his translation by the end of March;[55] two weeks later Réaumur put forward Dufay's name for the vacant position of adjoint chemist at the Académie.[56] These two events were no doubt connected.

4 Material and Methodological Echoes of the *Traité des vernis*

Dufay's translation of the *Trattato* was not just a canny career move but also a source of material and methodological intelligence that Dufay drew upon in the course of his research as an academician. Consider the practice of consulting past authors. The importance of this practice was a lesson that Dufay learned twice over in translating the *Trattato*, once as he admired the numerous citations in Buonanni's text and again as he corrected and augmented those citations. It is no surprise, then, that Dufay's experimental papers, and especially his important ones on static electricity and on the luminescence of barom-

52 Réaumur René, "Examen de la soye des araignées", *Mémoires de l'Académie Royale des Sciences* (1710) 386–408. Réaumur, *Steel and Iron* 529.

53 Réaumur René, "Idée générale des differentes manières de faire la porcelaine; & quelques sont les véritables matiéres de celle de la Chine", *Mémoires de l'Académie Royale des Sciences* (1727) 185–203, on 192–201.

54 Fouchy Jean-Paul Grandjean de, "Eloge de M. Pitot", *Histoire de l'Académie Royale des Sciences* (1771) 143–157, on 146–147.

55 Buonnani, *Traité des vernis*, 1st ed., "Approbation."

56 Procès-verbaux de l'Académie Royale des Sciences, 10 Apr 1723 (fol. 67r).

eters and diamonds, are exemplary in their 'full command of earlier writings'.[57] The same papers have impressed historians for their 'clear prescriptions for producing the phenomena under study'.[58] Dufay gave long lists of the conditions necessary for the success of his experiments on light and electricity, and he criticized past authors for 'quelques circonstances obmises dans le récit de ces faits' [some circumstances omitted in the reporting of these facts].[59] These circumstances were 'toutes si nécessaires, que l'obmission de quelques-unes en diminuë considérablement, ou même en empêche absolument le succès' [all so crucial [to experiments], that the omission of any one of them reduces their success or completely prevents it].[60] These remarks echo passages that Dufay had met in the *Trattato sopra la vernice*. Buonanni's Chinese varnish only succeeded when he followed it exactly, and he gave detailed instructions so that others would succeed invariably.[61] 'Mais il ne dit ni la dose des ingrédiens', Buonanni complained of one author, 'ni la manière de les unir ensemble: ainsi il nous suffit de les avoir indiqués, passons présentement à d'autres plus surs & plus expérimentés' [but he reports neither the dose of the ingredients nor the manner of uniting them: so it is enough for us to mention [his recipes],

57 Heilbron, *Electricity* 251. Dufay Charles, "L'histoire de l'électricité", *Mémoires de l'Académie Royale des Sciences* (1733) 23–35. Cf. Dufay Charles, "Mémoire sur les baromètres lumineux", *Mémoires de l'Académie Royale des Sciences* (1723) 295–306, on 295–298; Dufay Charles, "Recherches sur la lumière des diamants et de plusieurs autres matières", *Mémoires de l'Académie Royale des Sciences* (1735) 347–372, on 347–352.

58 Heilbron, *Electricity* 251.

59 Dufay Charles, "Sixième mémoire sur l'électricité: Quel rapport il y a entre l'électricité et la faculté de rendre de la lumière [...]", *Mémoires de l'Académie Royale des Sciences* (1734) 503–526, esp. 503.

60 idem, "Quatrième mémoire sur l'électricité: L'Attraction et la répulsion des corps électriques", *Mémoires de l'Académie Royale des Sciences* (1733) 457–476, esp. 474. Cf. Fontenelle's comments on Dufay's method at "Sur le sel de la chaux", *Histoire de l'Académie Royale des Sciences* (1724) 39–40, esp. 39; idem, "Sur l'électricité", *Histoire de l'Académie Royale des Sciences* (1737) 1–6, esp. 6. Cf. Dufay's precautions at "Troisième mémoire sur l'électricité: Des corps qui sont les plus vivement attirés par les matières électriques, et de ceux qui sont les plus propres à transmettre l'électricité", *Mémoires de l'Académie Royale des Sciences* (1733) 233–254, esp. 245–248; idem, "Attraction et répulsion" 465–466, 471–474; idem, "Cinquième mémoire sur l'électricité: Des nouvelles découvertes sur cette matière", *Mémoires de l'Académie Royale des Sciences* (1734) 341–362, esp. 348–49; idem, "Lumière des diamants", 353–354; idem, "Septième mémoire sur l'électricité: Quelques additions aux mémoires précédants", *Mémoires de l'Académie Royale des Sciences* (1737) 86–100, esp. 92.

61 Buonanni, *Traité des vernis* 145–156.

and to hasten to others that are surer and more thoroughly tested].[62] Dufay must have learnt these lessons from Buonanni's text and from the problems he encountered when replicating Buonanni's experiments. Replication is a third theme of Dufay's papers that harks back to the *Trattato*. We have seen that Dufay repeated many of the recipes in Buonanni's text in the course of his translation. Later, as an academician, Dufay extolled the practice of repeating other people's experiments in order to check the veracity, accuracy and completeness of their written reports. And he practiced what he preached, replicating Robert Boyle's experiments on the luminosity of diamonds, Stephen Gray's on electricity, and Duhamel's on sensitive plants.[63]

A final methodological echo of the *Trattato* is Dufay's habit of systematically varying the materials in an experiment. For Buonanni, and hence for his translator, the key to imitating Chinese porcelain was to find the right ingredients. But Buonanni had precious little theory to guide his choice of materials, and as a result much of his research was a long process of trial and error in which patience and vigilance were key virtues. 'La meilleure composition', he wrote, 'étoit celle que le hasard m'avoit fait découvrir parmi une infinité d'expériences & de différentes combinaisons que j'ai essayées' [The best composition was that which chance picked out from the infinity of experiments and combinations that I tried].[64] A few years after translating the *Trattato*, Dufay applied the same procedure in his search for wood that gave acid-resistant glass bottles and for pigments that could be used to colour marble.[65] In both of these cases—as in Réaumur's work on porcelain—the procedure helped to identify substances with useful properties and to perfect existing recipes by replacing the usual ingredients with new ones.

---

62 Buonanni, *Traité des vernis* 32. Buonanni's phrasing was 'ma non assegna la quantità degl'ingredienti nè il modo di unirli assieme, onde basti averli quì indicati, sacendo passaggio a cose più esperimentate': Buonanni, *Trattato sopra la vernice* (2nd ed.) 23.

63 For Gray see: Dufay, "Nouvelles découvertes" 345; idem, "Huitième mémoire sur l'électricité", *Mémoires de l'Académie Royale des Sciences* (1737) 307–325, esp. 312, 317. For Boyle: idem, "L'électricité et la lumière" 514, 521. For Duhamel: idem, "Observations sur la Sensitive", *Mémoires de l'Académie Royale des Sciences* (1736) 87–111, esp. 87.

64 *Traité des vernis* 167, cf. 131. Buonanni's phrasing was 'ho esperimentato essere migliore la composizione, che il caso mi sece scoprire nelle quasi infinite esperienze, e conbinazioni tentate': Buonanni, *Trattato sopra la vernice* (2nd edn.) 114.

65 Dufay Charles, "Expériences sur la dissolubilité de plusieurs sortes de verres", *Mémoires de l'Académie Royale des Sciences* (1727) 32–39; idem, "Mémoire sur la teinture et la dissolution de plusieurs especes de pierres", *Mémoires de l'Académie Royale des Sciences* (1728) 50–67; "Second mémoire sur la teinture des pierres", *Mémoires de l'Académie Royale des Sciences* (1732) 169–181.

Dufay used the same procedure in his research on light and electricity.[66] Some time in the 1720s he began a study of precious and semi-precious stones in which he applied a large number of chemical and physical operations to each of the stones in his collection. The main outcome of this project was a series of surprising generalizations: nearly all bodies become electric when heated then rubbed; nearly all of them can be made to glow in the dark after being exposed to light; all precious stones glow when rubbed in the dark; and so on. Material-driven experimentation also lent itself to a particular kind of experimental law. These were laws of the form: all materials of *this* kind behave one way, while all materials of *that* kind behave a different way. An example is Dufay's discovery that the phenomenon of double refraction—an optical effect in which objects seen through a refracting medium appear twice—is absent in all transparent stones with faces that meet at right angles, and present in all other transparent stones. Laws of this kind were not incidental to Dufay's discoveries in experimental physics. Indeed, they formed the backbone of his pioneering papers on static electricity.

Dufay's translation of the *Trattato* also had a more concrete legacy. Replicating Buonanni's recipes meant acquiring Buonanni's materials, and some of these materials reappeared in Dufay's later experiments. Pigments are the most obvious example. Buonanni's book included three chapters on coloured lacquers, and the recipes therein were a thorough introduction to natural dyestuffs. When he tested these recipes, Dufay handled many pigments—gamboge, cochineal, cinnabar, saffron, and others—that he later used to colour stones and textiles.[67] There were also continuities in the way Dufay selected and manipulated these pigments. As noted above, Dufay sometimes used the price and origin of substances to identify the French equivalents of the substances Buonanni mentioned. In a later paper on coloured marble, Dufay distinguished between dragon's blood from the Canary Islands and that from St Lawrence Island, and recommended the former as the more vivid pigment. In the same paper he described how to draw portraits on a piece of carnelian by covering the stone with colcothar (iron III oxide) and removing parts of this pigment with a pin (the parts of the stone covered in colcothar turned white when heated). Shading could be achieved by varying the thickness of the colcothar.[68] These procedures recall a footnote in the *Traité des vernis* where Dufay described how to remove gold from a lacquered surface

---

66 This paragraph summarizes material in Bycroft, *Physics and Natural History* chapters 3 and 4.

67 Buonanni, *Traité des vernis* chapters 5, 20, and 22. Cf. Dufay Charles, "Couleurs dans la teinture" 255–260.

68 Dufay, "Seconde mémoire sur la teinture" 171; 177–179.

with a pin, and another footnote in which he advised aspiring painters on how to prepare a piece of glass for painting by transferring the ink of an engraving onto its reverse side.[69]

Gums and resins are another set of recurring materials in Dufay's career. They appeared in most of Buonanni's lacquer recipes, and one of the chapters of the *Trattato sopra la vernice* is simply an enumeration of all the known varieties.[70] Gums and resins served Dufay well in his work on electricity, since aside from glass they were the materials most apt to attract and repel light objects when rubbed. It is fitting that gum copal, an ingredient in Buonanni's recipe for Chinese varnish, made a material contribution to what was arguably Dufay's single most important legacy for experimental physics. This was his discovery of the distinction between 'resinous' and 'vitreous' electricity, an influential precursor to Benjamin Franklin's distinction between 'positive' and 'negative' electricity. The distinction between the 'two electricities' also

FIGURE 3.2
An instrument for studying the transmission of pressure in liquids. Made of metal; glass; and wood decorated with gold on black lacquer, with red lacquer trim. The gold landscapes on the base are particularly sinisant *in their style and subject-matter. From the collection of Jean-Antoine Nollet, a natural philosopher and instrument-maker who assisted Dufay in his experiments on electricity in the early 1730s. Unknown artist. Height: 89cm, width: 53cm, depth: 28cm. Paris, Conservatoire national des arts et métiers, inv. 04409.*

69 The painter would then use the outline and shading provided by the ink as a guide as he applied colour to the front side of the glass. See below for the full text of the footnote on gilding. Buonanni, *Traité des vernis* 178, 186.

70 Ibidem, chapter 2, esp. 6–7.

intrigued Jean-Antoine Nollet, a student of Dufay's who (incidentally) used lacquer and gold leaf to decorate the scientific instruments he manufactured for fashionable clients such as Voltaire and the King of France [Fig. 3.2]. Dufay discovered the 'two electricities' when he approached a piece of rubbed copal to a piece of gold leaf suspended above a rubbed glass tube. The levitation of the gold leaf was due to the electrical repulsion exerted upon it by the glass tube, which Dufay had electrified by rubbing. Dufay thought the copal would behave like the glass tube, repelling the gold leaf. To his astonishment, it did the opposite. He concluded that there are two modes of electrification, one exemplified by the glass tube and the other by the copal. Dufay called the former mode 'vitreous' and the latter 'resinous', in honour of the materials that led him to the unexpected result.

The gold leaf in this experiment is another echo of Dufay's research on Chinese varnish. Buonanni referred to gold leaf in two chapters dedicated to the application of gold leaf to a lacquered surface.[71] In a long footnote, Dufay

FIGURE 3.3 *Wooden box decorated with red lacquer and engraved gold (exterior) and black lacquer sprinkled with gold particles (interior). Probably made in Paris, c. 1700–1720. The gold is finely shaded: consider for example the folds in the robes of the figure at the centre of the lid. In a footnote to the* Traité des vernis, *Dufay described how to apply gold leaf to a red lacquered surface, and how to shade the gold thus applied. Unknown artist. Height: 10cm, width: 29cm, depth: 22cm. London, Victoria and Albert Museum.*

71 Buonanni, *Traité des vernis*, chapters 5 and 17, esp. 42, 159–160.

described how to apply gold leaf to a ground of red lacquer [Fig. 3.3]. He went into considerable detail about this delicate and multi-staged operation:

> Lorsque le Vernis est sec & bien poli, on prend un blanc d'oeuf que l'on bat bien avec un petit bâton, on le laisse ensuite reposer, & on en met avec un pinceau ou avec le doigt dans les endroits que l'on veut dorer, ensuite on y applique des morceaux de feuilles d'or à peu près de la grandeur de la figure que l'on veut dorer, cela sèche dans le moment: lorsque cela est sec, on prend une épingle dont la pointe est un peu émoussée; & qui est emmanchée dans un petit bâton, & on s'en sert pour enlever l'or, suivant le dessein que l'on veut; les traits peuvent être aussi fins que l'on souhaite, jusques à faire même les ombres par des hachûres comme dans les Estampes: lorsqu'il y a des endroits un peu grands où il faut enlever l'or, comme aux bordes de l'ouvrage, on se sert d'un petit bâton applati par le bout, que l'on mouille, & avec lequel on enlève facilement l'or: après l'ouvrage est fini, on le couvre d'une ou deux couches de Vernis composé seulement de gomme-lacque dissoute dans l'esprit de vin, cela conserve parfaitement l'or.

> Once the lacquer is dry and well polished, take an egg white well beaten with a little stick, leave [the egg white] to sit, and apply it with a brush or finger to the places you want to gild. Then apply pieces of gold leaf of roughly the same size as the shape you want to gild, this dries in an instant. When it is dry, take a pin with a slightly blunted point, hafted to a little stick, and use this to remove the gold according to the desired shape; the lines can be as fine as you like, even to the point of forming shadows by hatching as in engravings. Where there are large areas of gold that need to be removed, as on the edges of the work, use a little stick with a flattened end, which you wet, and with which you can easily remove the gold. When the work is complete, cover it with one or two layers of lacquer made simply of shellac dissolved in spirit of wine, this preserves the gold perfectly.[72]

The level of detail in this description suggests that Dufay had observed artisans applying gold leaf to lacquer. Probably he had also performed the operation himself. In any case, his familiarity with gold leaf helps to explain why he was the first natural philosopher to use that material systematically as a detector of

72 Ibidem 179–180.

electricity.[73] The same expertise served him well when, some time in 1725, he set out to build a spherical mirror made of plaster and gold leaf.[74]

In the case of electricity, it is not clear whether Dufay could have drawn any theoretical conclusions from the gilding procedure he described. If anything, the procedure suggests that gold leaf is *attracted* to rubbed objects (such as the blunt pin), rather than being *repelled* by them as Dufay maintained in his third paper on electricity.[75] Nevertheless, Dufay's study of electricity required that he slice gold leaf into fragments, move those fragments to the right places in his experiments, and drop them onto electrified objects at the right times. All of these operations would have been facilitated by the gilding skills he picked up from lacquer artists in Rome or Paris.

5 Translation after the *Traité*

Dufay's translation of the *Trattato* made a difference to the methods and materials he used in his later research, but it did not make this difference *because he translated it*. The *Trattato* would have had much the same effect on Dufay's methods and materials if Buonanni had written it in French rather than Italian, and if Dufay had simply read this French version and repeated the experiments therein. But there are other echoes of the *Traité* in Dufay's experimental career that *did* depend on the fact that it was a translation.

Most obviously, Dufay continued to translate experimental findings between vernacular languages, especially French, English, and Italian. For example, his research on electricity required that he read the latest findings on the topic published in English and Italian, and that he summarize these findings in French in the papers he published in the Académie's *Mémoires*. The Italian text was a translation, from the original English, of a treatise on experimental

73 Both Boyle and Newton had used gold leaf in electrical experiments, but neither had done so systematically. See Boyle Robert, *Experiments and Notes about the Mechanical Origine or Production of Electricity* (London, Flesher: 1675), in Hunter M. – Davis E. (eds.), *The Works of Robert Boyle* (London: 1999) vol. 8, 510–523, esp. 516; Guerlac H., "Newton's Optical Aether: His Draft of a Proposed Addition to His *Opticks*", *Notes and Records of the Royal Society of London* 22, 1 (1967) 45–57, esp. 48.

74 Procès-verbaux de l'Académie Royale des Sciences, 14 Nov 1725 (242v); Dufay Charles, "Sur quelques expériences de catoptrique", *Mémoires de l'Académie Royale des Sciences* (1726) 165–175, esp. 165–166.

75 Dufay, "Attraction et repulsion" 458–464.

philosophy by the Englishman Francis Hauksbee.[76] The principal texts in English were ten articles by Stephen Gray published in the *Philosophical Transactions of the Royal Society of London* between 1721 and 1736.[77] Dufay published French summaries of eight of these papers, wrote down full translations of at least three of them, published a full translation of a ninth paper by Gray, and referred to a tenth paper, which he had apparently read, in his last publication on electricity.[78] Also significant were two letters, written in English, that Dufay received from the experimenter Granville Wheler in 1736 and 1737; in the 1737 volume of the *Mémoires* Dufay published a comprehensive translation of one of these letters and a summary of the other.[79] The fact that the three manuscript translations of Gray's papers are in Dufay's hand suggests that he translated them himself; he was probably also the author of the two translated texts (Gray's paper and Wheler's letter) that he incorporated into his published papers on electricity. Dufay shared the burden of French-English translation with his correspondents across the English Channel. He wrote to the naturalist Hans Sloane in French and received replies in the same language, and he sent a French summary of his major electrical discoveries to the Duke of Richmond, who had the text translated into English for publication in the *Philosophical Transactions*.[80] This two-way traffic between the English and French languages was essential to the dramatic advances in the science of electricity that took place in the 1720s and 1730s. Dufay's main discoveries in the field were the result of his efforts to replicate, systematize and extend the findings of Hauksbee and Gray; Gray repaid the compliment by building on Dufay's findings after he

76 The Italian text was a 1716 translation of Francis Hauksbee's *Physico-Mechanical Experiments on Various Subjects*, first published in London in 1709. Dufay, "Histoire de l'électricité" 28; idem, "Corps attirés" 235.

77 For translation between the Royal Society and the Académie des sciences, see also the chapter by Meghan C. Doherty in this volume.

78 Gray papers summarized in Dufay, "Histoire de l'électricité" 31–35; idem, "Nouvelles découvertes" 341–348; idem, "Quelques additions" 86–90; and idem, "Huitième mémoire sur l'électricité" 307–11. Gray papers translated in full in Archives de l'Académie Royale des Sciences, Dufay dossier, folder entitled "Pieces manuscrites" (3 papers), and in Dufay, "Huitième mémoire sur l'électricité" 310–311, where Dufay also refers to the 10th paper by Gray on p. 323.

79 Wheler to Cromwell Mortimer, 21 November 1737, and Wheler to Dufay, 9 March 1738, both in "Pieces manuscrites". Cf. Dufay, "Huitième mémoire sur l'électricité" 318–21 (translation), and 322–323 (summary).

80 London, British Library, Sloane Mss. 4050, 4053, 4054, 4055, 4056, 4058, 4068. Dufay C. and T.S., "A Letter Concerning Electricity", *Philosophical Transactions* 38 (1733–1734) 258–266.

read the English summary of them in the spring of 1734.[81] Dufay's work on the *Traité des vernis* helps to explain the energy and care with which he deployed translated material in his research as an academician.

Dufay did not just translate *words* in the years after the *Traité des vernis*. He also translated *numbers*—that is to say, he expressed French units of weight and distance in terms of English units, and vice versa. He did so as part of a collaborative project undertaken by the Paris Académie and the Royal Society of London in the 1730s. The aim was to equip each institution with sets of brass rods and lead balls that embodied the main units of length and weight in both countries. In addition, the *Philosophical Transactions* published a list of ratios of the French and English units. These ratios would allow for 'the better comparing together the Success of Experiments made in England and in France'.[82] In other words, they were an aid to replication: by converting yards into *toises*, or pounds into *marcs*, an experimenter in England could find out whether he had successfully replicated an experiment that had been carried out in France.[83]

The conversion from yards to *toises* was a translation in the sense that a term in one language (English) was expressed in another language (French). Of course, these terms represent quantities rather than things. However the analogy between these two kinds of conversion would not have been lost on Dufay. In the footnotes of *Traité des vernis*, he had not only recorded the French words for Italian materials but also the French versions of Italian quantities: he tells us repeatedly that one Italian pound is equivalent to 12 French ounces; and that one Italian florin is equivalent to 2 French *livres* and 14 French *sous*.[84] Moreover, two of Dufay's closest collaborators at the Académie used the translation of words as a metaphor for the standardization of units. Consider René Réaumur's complaint about the vagaries of thermometers: 'Les manieres dont ils s'expriment, s'il est permis de parler de la sorte, étant toutes différentes, on n'entend que la langue d'un qu'on a suivi pendant plusieurs années' [They all express themselves, if you will, in different ways, and we only understand

---

81 Dufay's debts to Gray and Hauksbee are spelled out in Heilbron, *Electricity* 250–260, and Bycroft, *Physics and Natural History* chap. 4. Gray mentioned his debt to Dufay in his "Experiments and Observations upon the Light that is Produced by Communicating Electrical Attraction to Animal or Inanimate Bodies", *Philosophical Transactions* 39 (1735) 16–24, esp. 16–17.

82 "An Account of the Proportions of English and French Measures and Weights, from the Standards of the Same, Kept at the Royal Society", *Philosophical Transactions* 42 (1742) 185–188, esp. 185.

83 Ibidem, passim. Cf. Daumas M., *Scientific Instruments of the Seventeenth and Eighteenth Centuries and Their Makers* (London, Batsford: 1972) 131.

84 Notes at Buonanni, *Traité des vernis* 44, 51, 89.

the language of a thermometer when we have known it for several years].[85] A few years later, the astronomer Charles-Marie de la Condamine wrote that scientific academies would 'speak the same language' if they all used the same unit of length. He used the same metaphor to defend a standardized currency against the criticism that it would put money-changers out of work. 'Si tous les hommes parloient la même langue, l'office d'interprète deviendroit inutile. Conclurra-t-on de là que la diversité des langues est avantageuse à la société ? [...] à parler, pour ainsi dire, la même langue' [If everyone spoke the same language, interpreters would be useless. Shall we then conclude that the diversity of languages is advantageous to society?].[86] For Dufay, as for Réaumur and Condamine, the conversion of units was a special case of the translation of languages. Dufay's work on English and French units is another echo, albeit a distant one, of his translation of the *Traité des vernis*.

Conclusion

Fontenelle implied that Dufay was chosen to examine textile dyes for the Bureau du commerce because he had displayed his experimental skill in the research on chemistry and physics that he had carried out between his entry into the Académie in 1723 and his appointment to the Bureau du commerce in 1731.[87] Fontenelle thereby suggested that experimental research came first for Dufay, and the decorative arts second. Subsequent biographers of Dufay have not revised this order of events. The *Traité des vernis* shows that this is a misleading picture of Dufay's development as a scientist. Dufay's experimental skills emerged at the same time as, and under the influence of, his engagement with the technology of the decorative arts. His project on textile dyes was not an accidental by-product of his virtuosity in the laboratory but an extension of skills that he had developed while translating Buonanni's treatise on Chinese varnish.

The basic aim of this project was to translate Buonanni's Italian prose into French. But Dufay was a diligent translator, committed to clarity, correctness

85 Réaumur René, "Regles pour construire des thermometres dont les degrés soient comparables, et qui donnent des idées d'un chaud ou d'un froid qui puissent être rapportés à des mesures connues", *Mémoires de l'Académie Royale des Sciences* (1730) 452–507, esp. 453.

86 Condamine Charles-Marie de la, "Nouveau projet d'une mesure invariable propre à servir de mesure commune à toutes les nations", *Mémoires de l'Académie Royale des Sciences* (1747) 489–514, esp. 493 and 512 respectively.

87 Fontenelle, "Eloge de Du Fay" 76.

and completeness. He repeated Buonanni's experiments, read his sources, consulted Italian artisans, and converted Italian units into French ones. As a result, the translation was not just an exercise in linguistics but a wide-ranging project that involved experiment, scholarship, metrology, travel, and on-the-spot observations of craft practices. It was a substantial contribution to applied chemistry, one that impressed René Réaumur and ensured Dufay a position at the Académie at the young age of twenty-five.

The translation also supplied Dufay with a set of methodological precepts, a familiarity with certain key materials, and a facility as a translator. These by-products of the *Traité des vernis* help to explain some key features of Dufay's later research on topics such as light and electricity. This research was characterized by thorough accounts of earlier authors, clear descriptions of experimental procedures, scrupulous replications of past experiments, systematic variation in the materials used, skilled use of gold, gums and dyestuffs, close attention to texts written in English and Italian, and concern for the interconversion of French and English units. All these features can be found in abundance in the *Traité des vernis*, either in Buonanni's text, in Dufay's activities as the translator, or both. Perhaps Dufay would have acquired the same set of skills and resources, in some other way, if he had never translated a treatise by Buonanni, or by anyone else for that matter. But the fact is that he *did* translate Buonanni's treatise. And we cannot deny that his scientific activities prior to his entry into the Académie—of which none were more substantial than the translation—made a difference to his activities after that date. It is surely no coincidence that Dufay's translation shares many important features with the experimental research he carried out as an academician. The former undoubtedly made a difference to the latter.

Selective Bibliography

Brunet P., "L'oeuvre scientifique de Charles François du Fay", *Petrus Nonius* 3 (1940) 77–95.

Buonanni Filippo, *Trattato sopra la vernice detta comunemente Cinese* (Rome, Giorgio Placho: 1720).

Buonanni Filippo, *Traité des vernis*, trans. Charles Dufay, 1st ed. (Paris, Laurent d'Houry: 1723), 2nd ed. (Paris, Laurent d'Houry: 1733).

Buonanni Filippo, *Techniques of Chinese Lacquer: The Classic Eighteenth-Century Treatise on Asian Varnish*, trans. F. Perugini (Los Angeles: 2009).

Bycroft M., "Physics and Natural History in the Eighteenth Century: the Case of Charles Dufay", Ph.D. dissertation (University of Cambridge: 2013).

Demeulenaere-Douyère C. – Sturdy D. (eds.), *L'Enquête du Régent, 1716–1718: sciences, techniques, et politique dans la France pré-industrielle* (Turnhout: 2008).

Dufay dossier, Archives de l'Académie Royale des Sciences.

Dufay Charles, "Expériences sur la dissolubilité de plusieurs sortes de verres", *Mémoires de l'Académie Royale des Sciences* (1727) 32–39.

Dufay Charles, "Mémoire sur la teinture et la dissolution de plusieurs especes de pierres", *Mémoires de l'Académie Royale des Sciences* (1728) 50–67.

Dufay Charles, "Second mémoire sur la teinture des pierres", *Mémoires de l'Académie Royale des Sciences* (1732) 169–181.

Dufay Charles, "Premier mémoire sur l'électricité: L'histoire de l'électricité," *Mémoires de l'Académie Royale des Sciences* (1733) 23–35.

Dufay Charles, "Troisième mémoire sur l'électricité: Des corps qui sont les plus vivement attirés par les matières électriques, et de ceux qui sont les plus propres à transmettre l'électricité", *Mémoires de l'Académie Royale des Sciences* (1733) 233–254.

Dufay Charles, "Quatrième mémoire sur l'électricité: L'Attraction et la répulsion des corps électriques", *Mémoires de l'Académie Royale des Sciences* (1733) 457–476.

Dufay Charles, "Cinquième mémoire sur l'électricité: Des nouvelles découvertes sur cette matière", *Mémoires de l'Académie Royale des Sciences* (1734) 341–362.

Dufay Charles, "Sixième mémoire sur l'électricité: Quel rapport il y a entre l'électricité et la faculté de rendre de la lumière […]", *Mémoires de l'Académie Royale des Sciences* (1734) 503–526.

Dufay Charles, "Recherches sur la lumière des diamants et de plusieurs autres matières," *Mémoires de l'Académie Royale des Sciences* (1735) 347–372.

Dufay Charles, "Septième mémoire sur l'électricité: Quelques additions aux mémoires précédants", *Mémoires de l'Académie Royale des Sciences* (1737) 86–100.

Dufay Charles, "Observations physiques sur le mélange de quelques couleurs dans la teinture," *Mémoires de l'Académie Royale des Sciences* (1737) 253–268.

Dufay Charles – T.S., "A Letter Concerning Electricity", *Philosophical Transactions* 38 (1733–1734) 258–266.

Fontenelle, Bernard le Bovier de, "Eloge de M. Dufay", *Histoire de l'Académie Royale des Sciences* (1739) 73–83.

Forray-Carlier A. – Kopplin M. (eds.), *Les secrets de la laque française: Le vernis Martin* (Paris: 2014).

Heilbron J., "Dufay, Charles-François de Cisternai," in Gillispie C.C. (ed.), *Dictionary of Scientific Biography*, 17 vols. (New York: 1970–1980), vol. 4, 214–217.

Heilbron J., *Electricity in the 17th and 18th Centuries: a Study of Early Modern Physics* (Berkeley, CA – London: 1979).

Procès-verbaux de l'Académie Royale des Sciences.

Réaumur R., *L'Art de convertir le fer forgé en acier, et l'art d'adoucir le fer fondu* (Paris, Michel Brunet: 1722).

Torlais J., *Un esprit encyclopédique en dehors de L'Encyclopédie: Réaumur, d'après des documents inédits* (Paris: 1961).

CHAPTER 4

'Ordinary Skill in Cutts': Visual Translation in Early Modern Learned Journals*

Meghan C. Doherty

In the final article of the 30 March 1665 issue of the *Journal des Sçavans*, the editor took notice of a new English enterprise: the *Philosophical Transactions*, marking the first in a series of references to the journal that at the time was affiliated with, but not officially produced by, the Royal Society of London.[1] Denis de Sallo (1626–1669), first editor of the *Journal des Sçavans*, lauded the state of 'la belle Philosophie' in England and praised the constant flow of new information and high-quality research coming out of the Royal Society.[2] Although Henry Oldenburg (1619–1677), founding editor of the *Philosophical Transactions*, would go to great lengths to separate his journal from the Royal Society of London, it was initially received as being directly related to the Society, which as Sallo put it, had the honour of having the King of England for its founder.[3] The linguistic difficulty presented by the English journal, as

* I would like to thank Sietske Fransen and Niall Hodson for inviting me to participate in the panel they organized at the Renaissance Society of America conference in 2015, and to contribute to this volume.

1 "Philosophical Transactions. A Londres, chez Jean Martin & James Allistry, Imprimeurs de la Societé Royale, & se trouve à Paris chez Jean Cusson, ruë S. Jacques", *Journal des Sçavans* (30 March 1665) 156. For an overview of the early history of both journals see: McCutheon R.P., "The 'Journal Des Scavans' and the 'Philosophical Transactions of the Royal Society'", *Studies in Philology* 21, 4 (1924) 626–628; Brown H., *Scientific Organizations in Seventeenth Century France* (New York: 1967) 185–207; Brown, "History and the Learned Journal", *Journal of the History of Ideas* 33, 3 (1972) 365–78; Andrade E.N. da C., "The Birth and Early Days of the *Philosophical Transactions*", *Notes and Records of the Royal Society* 20, 1 (1965) 9–22; Kronick D.A., *A History of Scientific and Technical Periodicals: The Origins and Development of the Scientific and Technical Press, 1665–1790*, 2nd ed. (Metuchen, N.J.: 1976); and Kronick D.A., "Notes on the Printing History of the Early *Philosophical Transactions*", *Libraries and Culture* 25, 2 (1990) 243–268. Throughout this chapter, the spelling and grammar of the original sources have been maintained.

2 "Philosophical Transactions" 156.

3 Oldenburg H., "To the Royal Society", *Philosophical Transactions* 1, 1 (6 March 1665) n.p.. 'Elle a l'honneur d'avoir le Roy d'Angleterre pour Fondateur', in "Philosophical Transactions" 156. On the perception among the French that the Royal Society was deeply underwritten by the

 | DOI 10.1163/9789004349261_006

most articles were written in English, compelled Sallo to find a translator who could provide his readers with access to all that was happening in England.[4] Oldenburg did not need to rely on a separate translator as he was already maintaining a multilingual correspondence before embarking on this publication project.[5] But linguistic translation was not the only matter to hand: the movement of images presented an additional and complex set of epistemological demands on the circulation of natural knowledge. This chapter examines this impulse to translate, looking specifically at illustrated articles moving back and forth between the two journals and finds that both text and image required specific literacies.[6]

This paper examines the cross-pollination between the *Philosophical Transactions* and the *Journal des Sçavans* by focussing specifically on images and authors' views of their potential as an open form of communication.[7] Studying how these articles were illustrated and how the illustrations related to the texts lends insight into the role of images within the larger missions of these periodicals. I argue that these cases increase our understanding of the communication of knowledge by foregrounding the possibilities that images could overcome linguistic barriers while paradoxically requiring their own form of literacy. This essay has two parts, first a brief survey of the illustrated articles that were translated between the two journals which demonstrates the range of knowledge that was transmitted visually across the journals and the problems inherent in trying to provide access to information published in vernacular languages. Following this survey, the second section offers a close study of the translation of an article in which Robert Hooke described 'both how to sound depths of the sea without a Line, and to fetch up water from any depth of the same' and the series of articles that responded to it. Hooke's article is significant because its translation was followed by a response in the *Journal des Sçavans* and then a rebuttal by Hooke in the *Philosophical Transactions*, which hinged on readers' lack of fluency with regards to the twinned subjects of

---

Crown, see Stroup A., "Royal Funding of the Parisian Académie Royale des Sciences during the 1690s", *Transactions of the American Philosophical Society* 77, 4 (1987) 4.

4 "Philosophical Transactions" 156.

5 See also the Introduction to this volume.

6 For the sake of space and clarity, this essay focuses solely on articles that were translated between the English and French journals. However, there was a great deal of borrowing between the Italian and German learned journals that were being published in the second half of the seventeenth century. This broader phenomenon is the subject of my current research.

7 For the role of images in the translation of scientific knowledge across language, see also the chapter by Charles van den Heuvel and Joyce van Leeuwen in this volume.

linguistic skills and visual literacy. This series of articles, published in the early years of two of the primary learned journals of the mid-seventeenth century, foreshadowed two of the key functions of the scientific periodical: acting first as a forum for priority debates, and second as a platform for methodological debates.[8] These crucial pursuits in natural knowledge occurred through the processes of translation, both visual and linguistic, as new concepts and discoveries passed between the two journals.

Although it was not an official publication of the Royal Society, the *Philosophical Transactions* was understood in France as being directly tied to the Royal Society. This perception emerged because the information contained in many of the early issues was intricately tied to the activities of the Royal Society. While the project was personally financed by Henry Oldenburg, the first issue of the journal was dedicated to the Royal Society.[9] Indeed, Oldenburg positioned his work as editor as being subservient to the Society's mission:

> In these Rude Collections, which are onely the Gleanings of my *private* diversions in broken hours, it may appear, that many Minds and Hands are in many places industriously employed, under Your Countenance and by Your Example, in the pursuit of those Excellent Ends, which belong to Your Heroical Undertakings.[10]

He sets up the *Transactions* as being a record of the works of 'many Minds and Hands' in diverse locations, all of whom are working towards improving the collective knowledge of nature. Oldenburg's vision for the journal paralleled that of the Society's, which Thomas Sprat in his *History of the Royal Society*

8 For other literature on the development of scholarly journals see for instance: Broman T., "Criticism and the Circulation of News: The Scholarly Press in the late Seventeenth Century", *History of Science* 51 (2013) 1–26; Bazerman C., *Shaping Written Knowledge: The Genre and Activity of the Experimental Article* (Madison, WI: 1988); Atkinson D., "The *Philosophical Transactions of the Royal Society of London*, 1675–1975: A Sociohistorical Discourse Analysis", *Language in Society* 25 (1996) 333–371.

9 This distinction has been noted by many scholars writing on journals in general and the *Philosophical Transactions* in particular. See for instance: Johns A., "Miscellaneous Methods: Authors, Societies and Journals in Early Modern England", *The British Journal for the History of Science* 33, 2 (2000) 166. In his article on the early history of the *Philosophical Transactions* Noah Moxham investigates how the journal survived outside of an institutional context: Moxham N., "Fit for Print: Developing an Institutional Model of Scientific Periodical Publishing in England, 1665–ca.1714", *Notes and Records of the Royal Society* 69 (2015) 241–260.

10 Oldenburg, "To the Royal Society" n.p.

described as 'to bring out experimental knowledge, from the *retreats*, in which it had long hid it self, to take its part in the *Triumphs* of that universal Jubilee [i.e. the Restoration]'.[11] Oldenburg then saw his publication of the selections of the correspondence of members of the Society with foreign scholars as part of the project of bringing experimental knowledge to the forefront.

Furthermore in the introduction to the first issue of the *Philosophical Transactions*, Oldenburg made explicit the importance of communication and the printing press to the improvement of experimental knowledge as he aligned his work with that of Francis Bacon (1561–1626) who was regarded as the intellectual founder of the Royal Society:

> Whereas there is nothing more necessary for promoting the improvement of Philosophical Matters, than the communicating to such, as apply their Studies and Endeavours that way, such things as are discovered or put in practise by others; it is therefore thought fit to employ the *Press*, as the most proper way to gratifie those, whose engagement in such Studies, and delight in the advancement of Learning and profitable Discoveries, doth entitle them to the knowledge of what this Kingdom, or other parts of the World, do, from time to time, afford, as well of the progress of the Studies, Labours, and attempts of the Curious and learning in things of this kind, as of their compleat Discoveries and performances.[12]

The allusions in this passage to Bacon's *Novum Organum* (London: 1620) and *Of Proficience and Advancement of Learning Divine and Human* (London: 1605) were not incidental as Oldenburg was an acolyte of Bacon and closely adhered to his principles, as adopted by the Royal Society, in his exchanges with his correspondents.[13] Bacon stressed the importance of the printing press as one of the three great inventions, along with gunpowder and the compass, that 'changed the face and condition of things all over the globe [...] so that no empire or sect or star seems to have exercised a greater power and influence on

11 Sprat Thomas, *The History of the Royal Society of London for the Improving of Natural Knowledge* (London, Printed by T.R. for J. Martyn and J. Allstrey: 1667) 59.

12 Oldenburg H., "The Introduction", *Philosophical Transactions* 1, 1 (6 March 1665) 1–2.

13 John Henry in his review of *The Correspondence of Henry Oldenburg* documents numerous instances in which Oldenburg rejected speculation and theorization on the part of his correspondents and instead encouraged them to report on facts they observed about nature. Henry J., "Review: The Origins of Modern Science: Henry Oldenburg's Contribution", *The British Journal of the History of Science* 21, 1 (1988) 103–110.

human affairs than those mechanical things'.[14] The press then figured largely in the prehistory of the Royal Society through Bacon's writings and Oldenburg saw his efforts as fulfilling part of the Society's mission.

Whereas Oldenburg turned to the press to promote 'the improvement of Philosophical Matters', by publishing much of his correspondence, Denis de Sallo positioned the *Journal des Sçavans* first and foremost as a news outlet, particularly news of books published in Paris and throughout Europe.[15] As a result, 'Philosophical Matters' were third on de Sallo's outline for the content of the *Journal*, as priority was given to book reviews and obituaries.[16] The material published in the *Journal* was more wide-ranging than the *Philosophical Transactions* as historical and religious books would be reviewed alongside natural historical and natural philosophical ones. Another important distinction to make between the *Journal des Sçavans* and the *Philosophical Transactions* is that the Académie Royale des Sciences was not founded until nearly two years after the *Journal des Sçavans* began.[17] Where Oldenburg had to work to distinguish his 'Rude Collections' from the royally sanctioned society whose first charter of 1662 included specific privileges related to printing, de Sallo began his project as a personal venture without the insulation a learned society provided against censure.[18] This lack of protection quickly became a problem for de Sallo as the *Journal* was suppressed. Because there was not a complete overlap in the goals of both journals, it follows that not everything that was published in one was of interest to the other. This chapter is concerned with the areas of common interest by exploring what was translated and how those translations were received.

---

14 Bacon F., *The New Organon*, ed. L. Jardine – M. Silverthorne, (Cambridge: 2000) 100 (Book I, CXXIX).

15 "L'Imprimeur au Lecteur", *Journal des Sçavans* (5 January 1665) n.p.

16 "L'Imprimeur au Lecteur" n.p.

17 The Académie was founded in December 1666. Stroup A., *A Company of Scientists: Botany, Patronage, and Community at the Seventeenth-Century Parisian Royal Academy of Sciences* (Berkeley: 1990) 3.

18 Margery Purver discusses the terms and impact of the Charter in the context of the founding of the Society: Purver M., *The Royal Society: Concept and Creation* (Cambridge, MA: 1967) 136–142. The privilege to print the *Journal* was signed on 8 August 1664 and registered on 30 December 1664. Brown, *Scientific Organizations* 187–188.

Translations Across the Channel

Although scholars in France complained about the inaccessibility of the information contained in the English articles in the *Philosophical Transactions* and Henry Oldenburg complained about attempts to create pirated full Latin translations of his journal, I am not going to discuss the attempts to create complete Latin translations of the *Philosophical Transactions*.[19] Instead, my focus is on the illustrated articles the editors of each journal selected to translate and republish.[20] For the purpose of this essay, I examine the illustrated articles published from the journals' inceptions in 1665 through the year of Oldenburg's death, 1677. These were tumultuous years for both journals as both publications were interrupted at points whether due to political disruptions or natural disasters, such as fire and plague. Despite these interruptions, issues of both journals kept appearing and the editors kept translating illustrated content from each other.

In 1665 there were no illustrations in the *Journal des Sçavans* and it only ran for three months before being suppressed at the request of the papal nuncio.[21] Publication resumed in January 1666 under the editorship of Jean Gallois (1632–1707). There were thirteen illustrated articles that year compared to ten in the *Philosophical Transactions*.[22] In 1666 roughly a third of the illustrated content in each was translated from the other journal (four went to the *Journal des Sçavans* and three went in the other direction). I would argue that there was a high frequency of translation in these first years as both journals were still establishing their sources of content: as the editors sought to collect articles for their journals the other periodical provided ready content.

One significant difference in the journals' strategies for including images was in how images were inserted in the text. In the *Journal des Sçavans*, images

19 For a detailed account of French translations of the *Philosophical Transactions* and attempts to create a Latin translation see: Turner A., "An Interrupted Story: French Translations from *Philosophical Transactions* in the Seventeenth and Eighteenth Centuries", *Notes and Records of the Royal Society* 62 (2008) 341–354.

20 For an overview of a broader sample of articles from these journals as well as the *Mémoires de l'Académie Royale des Sciences* see: Gross A.G. – Harmon J.E. – Reidy M.S., "Argument and 17th-Century Science: A Rhetorical Analysis with Sociological Implications", *Social Studies of Science* 30, 3 (2000) 371–396.

21 Turner, "An Interrupted Story" 342.

22 There were four illustrated articles in 1665 in the *Philosophical Transactions*. For the sake of counting, I am considering the articles published in one year of the Gregorian calendar (i.e. January-December) to be a unit. For a more detailed discussion of the illustrated articles published in both journals in 1666 see: Doherty M.C. "Giving Light to Narrative: The Use of Images in Early Modern Journals", *Nuncius* 30, 3 (2015) 543–569.

were printed alongside the text of the article [Fig. 4.1]. That is, each page with an illustration went through the press twice: first, through a common press to have the text printed and second, most likely in a different workshop, through an intaglio press for the image.[23] This strategy had additional fiscal costs, but potential intellectual benefits. Although the additional labour added an expense, this process meant that there was a literal closeness between the text and the image that allowed a reader to have easy access to the image while reading the text. In the *Philosophical Transactions*, however, the majority of the images were combined onto intaglio printed sheets that were included with the issue of the journal [Fig. 4.2]. While in some cases this led to somewhat jarring juxtapositions, in others it served to bring disparate articles into direct conversation with one another. The plate of images in issue 14 of the *Philosophical Transactions* included material for four different articles, all of which were related to observational astronomy.[24] A different closeness is achieved in this instance, instead of being close to their respective texts, the images are now close to one another, which in this case allows for a direct visual comparison of how Robert Hooke understood Mars to look along with how Jean Dominique Cassini (1625–1712) understood it. When Hooke's article was translated and printed in the *Journal des Sçavans*, the image was close to Hooke's text, but distant from both Cassini's observation, which had been published two months earlier, as well as from Hooke's observations of Jupiter and Saturn, which came in the next issue of the *Journal des Sçavans* [Figs 4.3 and 4.4].[25] This example shows both the intertwined nature of these journals in their early years and

23 For a detailed discussion of these processes, see: Gaskell R., "Printing House and Engraving Shop. A Mysterious Collaboration", *The Book Collector* 53 (2004) 213–251.

24 Hooke Robert, "The Particulars. Of Those Observations of the Planet Mars, Formerly Intimated to Have Been Made at London in the Months of February and March A. 1665/6", *Philosophical Transactions* 1, 14 (2 July 1666) 239–242; Cassini Jean Dominique, "Observations Made in Italy, Confirming the Former, and Withall Fixing the Period of the Revolution of Mars", *Philosophical Transactions* 1, 14 (2 July 1666) 242–245; Hooke Robert, "Some Observations Lately Made at London Concerning the Planet Jupiter", *Philosophical Transactions* 1, 14 (2 July 1666) 245–246; Hooke Robert, "A Late Observation about Saturn made by the Same", *Philosophical Transactions* 1, 14 (2 July 1666) 246–247.

25 "*Martis circa axem proprium revolubilis Observationes. Bononiae a Johanne. Dominico Cassino habitae.* Romae, 1666", *Journal des Sçavans* (31 May 1666) 259–262; "Extrait du Journal d'Angleterre, Contenant les particularitez de quelques observations qui ont esté faites à Londres au mois de Fevrier & de Mars dernier touchant la planette de Mars", *Journal des Sçavans* (23 August 1666) 403–406; "Extrait du Journal d'Angleterre, Contenant quelques nouvelles Observations faites à Londres touchant la Planete de Jupiter", *Journal des Sçavans* (30 August 1666) 416–418.

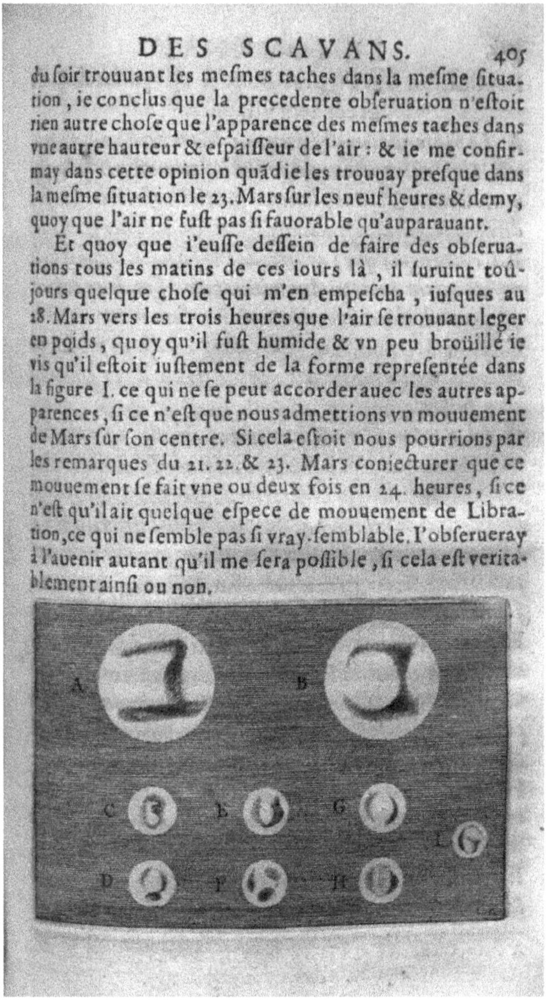

DES SCAVANS. 405

du soir trouuant les mesmes taches dans la mesme situation, ie conclus que la precedente obseruation n'estoit rien autre chose que l'apparence des mesmes taches dans vne autre hauteur & espaisseur de l'air : & ie me confirmay dans cette opinion quãd ie les trouuay presque dans la mesme situation le 23. Mars sur les neuf heures & demy, quoy que l'air ne fust pas si fauorable qu'auparauant.

Et quoy que i'eusse dessein de faire des obseruations tous les matins de ces iours là, il suruint toûiours quelque chose qui m'en empescha, iusques au 28. Mars vers les trois heures que l'air se trouuant leger en poids, quoy qu'il fust humide & vn peu broüillé ie vis qu'il estoit iustement de la forme representée dans la figure I. ce qui ne se peut accorder auec les autres apparences, si ce n'est que nous admettions vn mouuement de Mars sur son centre. Si cela estoit nous pourrions par les remarques du 21. 22. & 23. Mars coniecturer que ce mouuement se fait vne ou deux fois en 24. heures, si ce n'est qu'il ait quelque espece de mouuement de Libration, ce qui ne semble pas si vray-semblable. I'obseruray à l'auenir autant qu'il me sera possible, si cela est veritablement ainsi ou non.

FIGURE 4.1 *"Extrait du Journal d'Angleterre, Contenant les particularitez de quelques observations qui ont esté faites à Londres au mois de Fevrier & de Mars dernier touchant la planette de Mars". Engraving, page 227 mm × 157 mm (plate 65 mm × 101 mm), in* Journal des Sçavans *(23 August 1666) 405. Evanston, IL, Northwestern University Library.*

the ramifications of their different editorial (and perhaps fiscal) decisions regarding images.

This early level of exchange was not maintained consistently during the years under consideration. In part this was due to a lack of content. The years between 1665 and 1677 were fraught for the *Journal des Sçavans*. After being

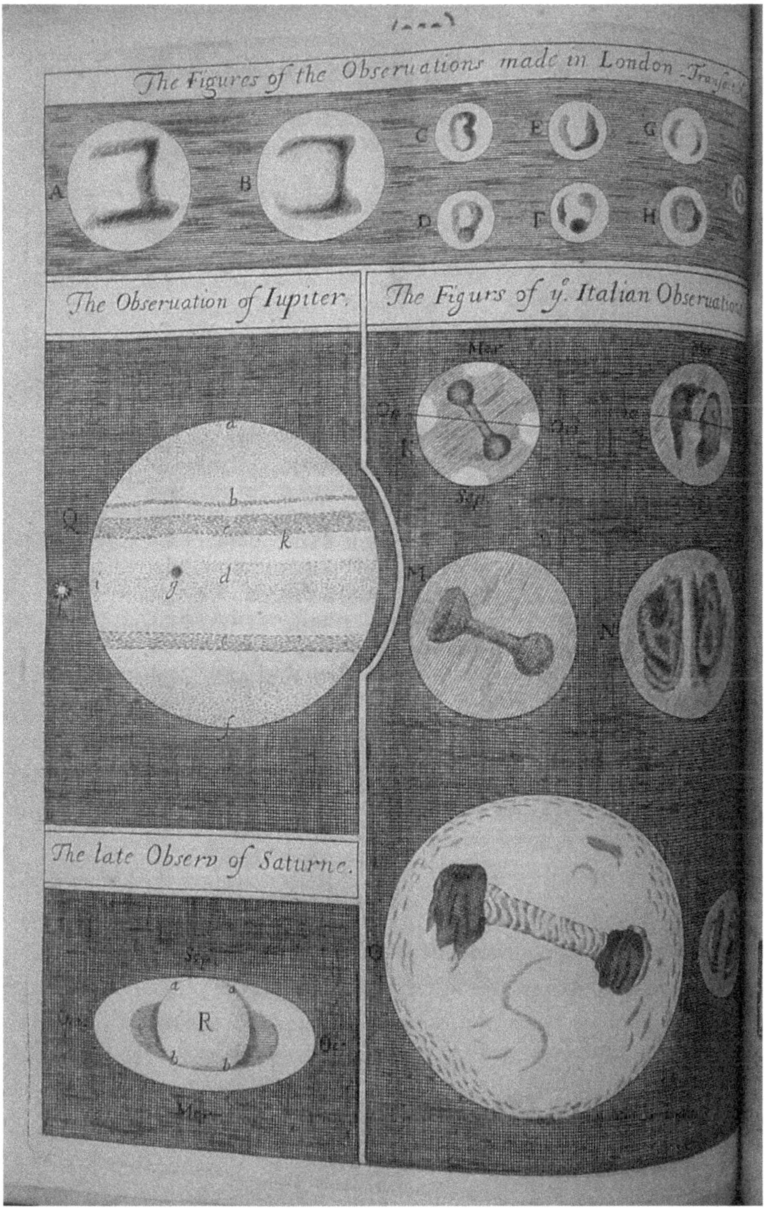

FIGURE 4.2 *"The Figures of the Observation made in London". Engraving, 209 × 160mm in* Philosophical Transactions *1, 14 (2 July 1666) facing 231. London, Royal Society Library.*

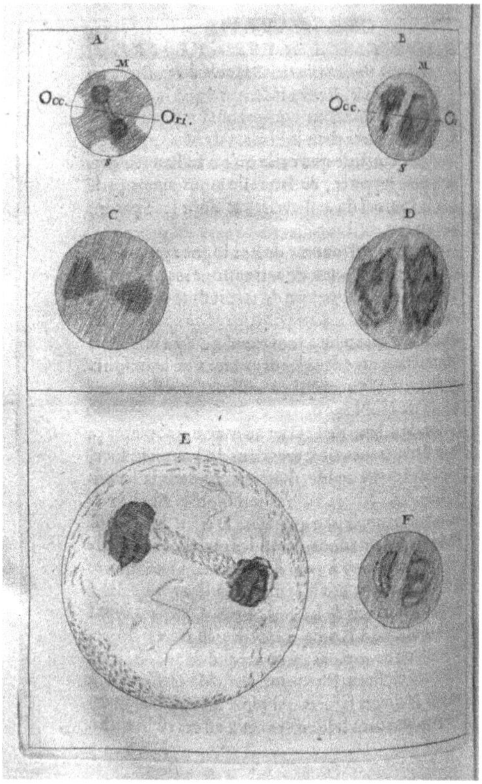

FIGURE 4.3
"Martis circa axem proprium revolubilis Observationes. Bononiae a Johanne Dominico Cassino habitae. *Romae, 1666". Engraving, page 227 mm × 155 mm (plate 192 mm × 128 mm), in* Journal des Sçavans *(31 May 1666) 260. Evanston, IL, Northwestern University Library.*

suppressed in March 1665, the *Journal* took on a less critical stance and as Henri Justel put it in a letter to Henry Oldenburg: 'You will see in the *Journal* how the ecclesiastics have their way, but we may say nothing to it'.[26] Harcourt Brown argued that the *Journal* should be considered dormant for the years 1669 to 1674.[27] In spite of this there were still a number of illustrated articles that moved from the *Journal des Sçavans* to the *Philosophical Transactions* during this time.[28]

Like so much of what was published in both periodicals, the illustrated articles defy neat categorization. However, they cluster around key international figures in the pursuit of natural knowledge (Huygens, Cassini, Hevelius) and key topics of concern (new instruments, observations of natural phenomena, reports of experiments and experiences). These topics align neatly with the

26 Quoted in Brown, *Scientific Organizations* 198.

27 Brown, *Scientific Organizations* 199.

28 In 1668, 1671, and 1672 one illustrated article was translated into the *Philosophical Transactions* and two were translated in 1669.

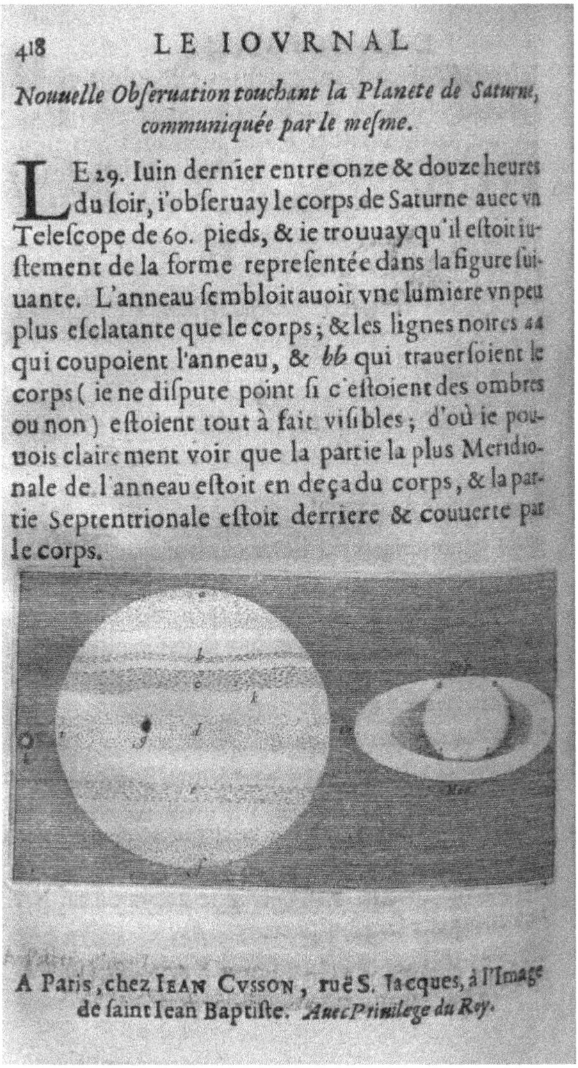
418 LE IOVRNAL

Nouuelle Obſeruation touchant la Planete de Saturne, communiquée par le meſme.

LE 29. Iuin dernier entre onze & douze heures du ſoir, i'obſeruay le corps de Saturne auec vn Teleſcope de 60. pieds, & ie trouuay qu'il eſtoit iuſtement de la forme repreſentée dans la figure ſuiuante. L'anneau ſembloit auoir vne lumiere vn peu plus eſclatante que le corps; & les lignes noires *aa* qui coupoient l'anneau, & *bb* qui trauerſoient le corps (ie ne diſpute point ſi c'eſtoient des ombres ou non) eſtoient tout à fait viſibles; d'où ie pouuois clairement voir que la partie la plus Meridionale de l'anneau eſtoit en deçà du corps, & la partie Septentrionale eſtoit derriere & couuerte par le corps.

A Paris, chez IEAN CVSSON, ruë S. Iacques, à l'Image de ſaint Iean Baptiſte. *Auec Priuilege du Roy.*

FIGURE 4.4 *"Extrait du Journal d'Angleterre, Contenant quelques nouvelles Observations faites à Londres touchant la Planete de Jupiter". Engraving, page 227 mm × 159 mm (plate 60 mm × 107 mm), in* Journal des Sçavans *(30 August 1666) 418. Evanston, IL, Northwestern University Library.*

content areas outlined by de Sallo in the first issues of the *Journal des Sçavans* as well as being the object of Oldenburg's considerations.[29]

29 "L'Imprimeur au Lecteur" n.p.

Direct translation by (or at the behest of) the editor was not the only means by which images appeared in both journals. Johannes Hevelius's simultaneous submission of his observations of the new star in the breast of Cygnus to both journals meant that his image and report appeared in both journals within a matter of weeks instead of months as was the case with translated articles. The text of Hevelius's report appeared in issue 19 of the *Philosophical Transactions* on 19 November 1666 and the image two issues later on 21 January [Fig. 4.5].[30] The text and image appeared in the *Journal des Sçavans* on 13 December 1666 [Fig. 4.6].[31] Hevelius (1611–1687) wrote to both Oldenburg and Antoine-François Payen (c. 1610-?) on 29 October reporting on his observations from 24 September.[32] Instead of sending his observations to one correspondent and eventually seeing his report in both publications, he wrote simultaneously to both Paris and London, presumably including drawings with both that looked like the one held in the archives of the Royal Society [Fig. 4.7]. The article as it appeared in the *Journal des Sçavans* included an editorial section at the beginning that provided some historical context for Hevelius's observations and stressed its fidelity to its source, stating that the following extract of Hevelius's letter had been translated word for word from Latin with the figure he sent at the end.[33] Aside from the framing provided in the French version of the article, the text of the two articles is identical in the details provided about Hevelius's observations and his methods. In both the *Journal des Sçavans* and the *Philosophical Transactions*, Hevelius's drawing has been translated into print using type. The apparent magnitude of the stars, which Hevelius described in the text, was conveyed using different size sorts of asterisks. Where most of the images that ended up in both journals were there because of decisions made by the editors, in this case the image appeared in both because of the

30 "Promiscuous Inquiries, chiefly about Cold, formerly sent and recommended to Monsieur Hevelius; together with his Answer return'd to some of them", *Philosophical Transactions* 1, 19 (19 November 1666) 344–352; "The Figure of the Stars in the Constellation of Cygnus; together with the New Star in it, discover'd some years since, and very lately seen by M. Hevelius again", *Philosophical Transactions* 1, 21 (21 January 1666/7) 372.

31 "Extrait d'une lettre de Monsieur Hevelius ecrite a Monsieur Payen Advocat au Parlement, touchant l'Observation de la nouvelle etoile decouuerte dans la constellation du Cygne", *Journal des Sçavans* (13 December 1666) 486–490.

32 "Extrait d'une lettre de Monsieur Hevelius" 487; for Hevelius's letter see: Archives of the Royal Society, Early Letters EL/H2/12. This letter has been published in Oldenburg's correspondence: *The Correspondence of Henry Oldenburg*, ed. and trans. A.R. Hall – M. Boas Hall, 13 vols. (Madison, WI – London: 1965–1986), vol. 3, 248–259.

33 'Voicy l'extrait de sa Lettre traduit mot à mot du Latin, avec la figure telle qu'elle est au bas de l'observation', in "Extrait d'une lettre de Monsieur Hevelius" 487.

author's decision. Working in Gdansk, at the periphery of the European intellectual community, Hevelius sought to be included in both periodicals on his own terms, not relying on one editor to copy his work from the other journal.

Although material moved intermittently between the two journals throughout the period under consideration, there was a sustained interest in what was being published on either side of the Channel. While larger organizational problems plagued the *Journal des Sçavans* during the period, a lack of capable English translators added to the difficulties faced by the editors of the *Journal* in providing their readers access to what was printed in the *Philosophical Transactions*.[34] The case study of an article by Robert Hooke that follows examines the issues that arose when articles were translated across both languages and intellectual contexts.

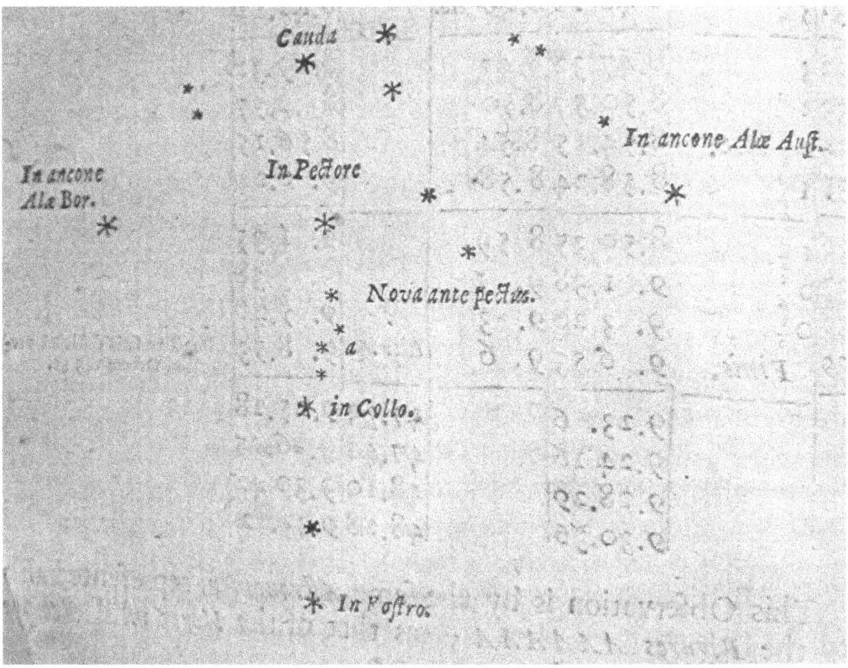

FIGURE 4.5 *Johannes Hevelius, "The Figure of the Stars in the Constellation of Cygnus". Typeset, page 223 × 160 mm (figure ca. 80 × 110 mm), in* Philosophical Transactions *1, 21 (21 January 1666/7) 372. London, Royal Society Library.*

34 Brown discusses how the translations organized by Colbert focussed on 'books of travel and history'. Although partial translation of the *Philosophical Transactions* is extant from this period, Brown describes it as 'rather roughly done'. Brown, *Scientific Organizations* 204.

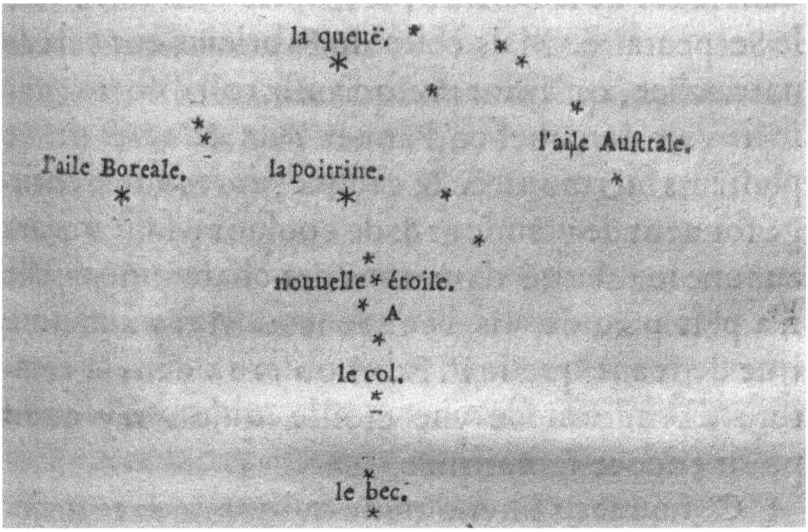

FIGURE 4.6 *Johannes Hevelius, "L'Observation de la nouvelle etoile decouverte dans la constellation du Cygne". Typeset, page 225 × 160 mm (figure 60 × 85 mm), in* Journal des Sçavans *(13 December 1666) 489. Evanston, IL, Northwestern University Library. Detail.*

'Ordinary Skill in Cutts and the English Language'

This overview of the cross-pollination of visual materials in the *Journal des Sçavans* and the *Philosophical Transactions* serves as a context for a more in depth examination of a single case of translation. This case study focusses on a series of articles related to Robert Hooke's report on two instruments that he claimed would improve the data mariners could collect while at sea, which was published in the 12 February 1666 issue of the *Philosophical Transactions* as 'An Appendix to the Directions for Seamen, bound for far Voyages'.[35] In this article, Hooke described 'both how to sound depths of the sea *without a Line*, and to fetch up water from any depth of the same' [Fig. 4.8].[36] This article was a follow-up to one in the previous issue of the journal, 'Designs for Sea-men, bound for far Voyages'.[37] The articles directed at mariners from issues eight

35 Hooke Robert, "An Appendix to the Directions for Seamen, bound for far Voyages", *Philosophical Transactions* 1, 9 (12 February 1666) 147–149.

36 Hooke, "An Appendix" 147.

37 "Directions for Sea-men, bound for far Voyages", *Philosophical Transactions* 1, 8 (8 January 1666) 140–143.

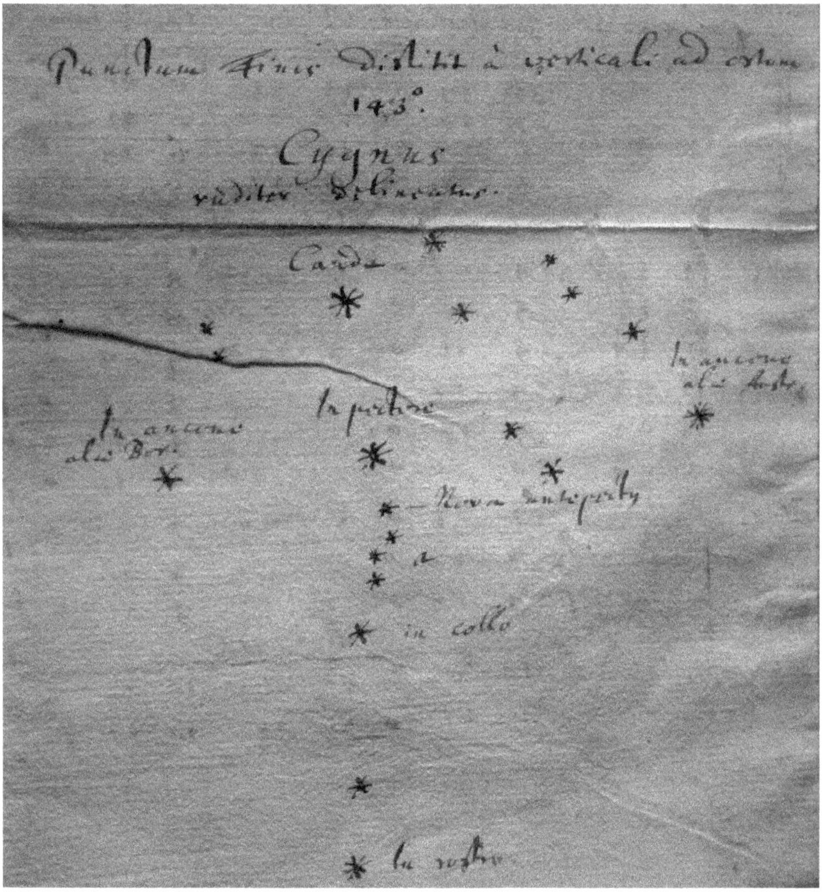

FIGURE 4.7 *Johannes Hevelius, "Cygnus" (29 October, 1666). Ink drawing, page 335 × 223 mm (image ca. 110 × 100 mm). London, Royal Society Library, Early Letters, EL/H2/12.*

and nine were later combined and expanded in issue 24.[38] These articles included lists of questions, instructions for data collection, and descriptions of how to use particular instruments. Taken together they sought to guide mariners in accumulating information that would 'be of good use, both *Naval* and *Philosophical'* as the Fellows of the Royal Society sought to document weather, tidal, and current patterns around the world as part of their aim of supporting

38 "Directions for Observations and Experiments to be made by Masters of Ships, Pilots, and other fit Persons in their Sea-Voyages", *Philosophical Transactions* 2, 24 (8 April 1667) 433–448.

the greater, national good while also developing a more detailed explanation of variations in these phenomena.[39]

A translation of the first article appeared in the 19 April issue of the *Journal* and one of the 'Appendix' in the 3 May issue [Fig. 4.9].[40] Both of these translations maintained the basic structure of the English articles as well as the dis-

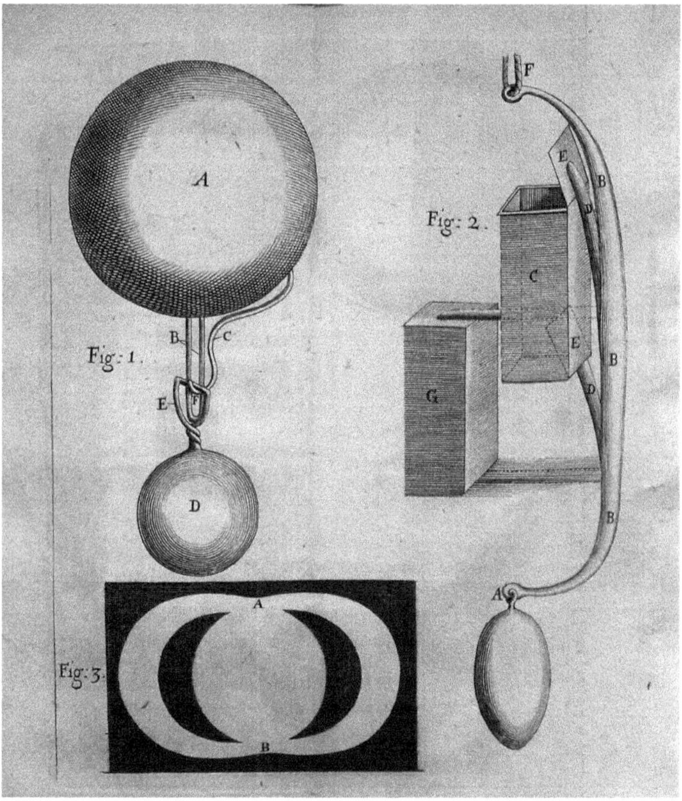

FIGURE 4.8 *Robert Hooke, "Instruments to sound depths of the sea without a Line, and to fetch up water from any depth of the same". Engraving, page 215 × 203 mm (plate 195 × 163 mm), in* Philosophical Transactions *1, 9 (12 February, 1666) facing 147. London, Royal Society Library.*

39 Hooke, "An Appendix" 147.

40 "Extrait du Journal d'Angleterre, contenant des instructions pour ceux qui ont à faire de longs voyages sur mer", *Journal des Sçavans* (19 April 1666) 193–196; "Extrait du Journal d'Angleterre, touchant une nouvelle maniere de sounder la profondeur de la mer sans corde, & de reconnoistre la nature de l'eau qui est au fond de la mer", *Journal des Sçavans* (3 May 1666) 217–120 [i.e. 220].

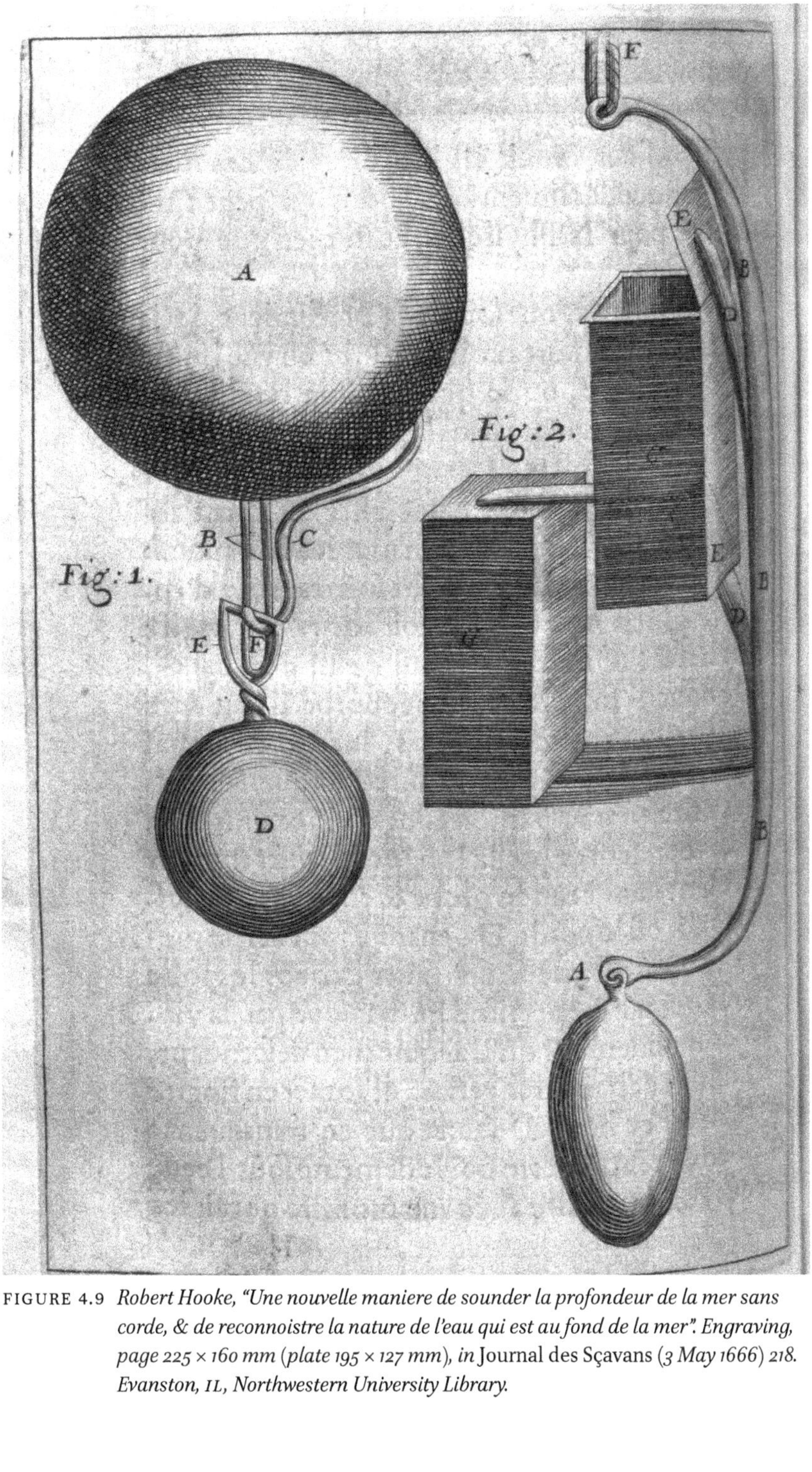

FIGURE 4.9 *Robert Hooke, "Une nouvelle maniere de sounder la profondeur de la mer sans corde, & de reconnoistre la nature de l'eau qui est au fond de la mer". Engraving, page 225 × 160 mm (plate 195 × 127 mm), in* Journal des Sçavans *(3 May 1666) 218. Evanston, IL, Northwestern University Library.*

tant editorial tone. In the 24 May issue, there was an extract of a letter from Pierre Petit to Jean Gallois in which Petit took issue with Hooke's descriptions of the two instruments.[41] On 6 June, a rebuttal appeared in the *Philosophical Transactions*.[42] This particular series of articles is significant in part because of the sustained nature of the published discussion of Hooke's description of his instrument. In addition, this series indicates both the role of the periodical in the intellectual life of mid seventeenth-century Europe and most significantly the central role images had in communicating knowledge.

Although Petit implied that Hooke had done so in the first article, Hooke did not make a claim for the novelty of his instruments, instead providing a detailed description of how to construct his version of the instrument along with an account of the experiments he had conducted with it. Petit wrote that this was not a new invention and that many others had written on how to sound the depths.[43] Hooke instead insisted that his particular method for sounding was an improvement, rather than a novelty: 'Whereas it is further excepted, That this way of Sounding Depths is no new Invention; The answer is ready, that neither is it pretended to be so, in the often quoted *Tract*; it being only intimated there, that the manner of performing it, as it is in that place represented and described, is new'.[44] Hooke argued that he had not made a claim for the priority of inventing the general instrument for sounding the depths, but rather that he had made specific methodological improvements to how to perform that task. He also objected to the assumption that he had no prior knowledge of French versions of similar instruments: 'Mean while, that way, which the *French* Author recommends for this purpose as more simple, *Videl. a Brass Pump with double Valves*, is not at all unknown in *England*, nor his [has] bin left untried there; but was found inconvenient [...]'.[45] He then goes on to detail the ways in which he found the French version to be lacking, finishing the paragraph by again asserting the superiority of his version: 'Whereas by the way proposed in *Num. 9*. both is [are] perform'd with great ease and security'.[46] This exchange highlights the way in which journals were viewed by readers

41 Petit Pierre, "Extrait d'une Lettre Escritte par M. Petit Intendant des Fortifications, à M. Galloys P. touchant la profondeur de la Mer, la nature de l'Eau qui est au fond de la Mer, & quelques autres curiositez", *Journal des Sçavans* (24 May 1666) 247–250.

42 [Hooke Robert], "Some Considerations Touching a Letter in the *Journal des Scavans* of May 24. 1666", *Philosophical Transactions* 1, 13 (4 June 1666) 228–230.

43 'Au reste ce n'est pas une invention nouvelle. Car plusiers ont escrit, que pour sonder la mer [...]', in Petit, "Extrait d'une Lettre" 249.

44 Hooke, "Some Considerations" 230.

45 Hooke, "Some Considerations" 229.

46 Hooke, "Some Considerations" 229.

and authors as a space for airing contentions about who knew what when and what they did with that knowledge.

One of the fundamental disagreements found in these pages is about the basic method for pursuing knowledge about the natural world. Hooke summarized the problem neatly in his response to Petit's charges about determining the salinity of sea water: stating that it '[...] is a matter, much better to be found out by *Trial*, than *Discourse*'.[47] Petit insisted on the necessity of obtaining a complete mathematical understanding of the rate at which the instrument would descend and ascend in the water before using the instrument.[48] Hooke set the tone of his rebuttal in methodological terms when he asserted that Petit claimed that the instrument could not be used before his concerns were answered: '[...] proposing also some Difficulties, relating to that Subject, and esteemed by him necessary to be satisfied, before any use could be made of the said Instruments'.[49] While Petit required mathematical proofs, Hooke asserted the value of his experimental data:

> Whereas the *French* Author is of opinion, that 'tis unknown, how much time a Heavy Body requires to sink in water, according to a certain depth; he may please to take notice, that that hath been made out in *England* by frequent Experiments; by which, several Depths, found by this Method of sounding *without* a Line, were examin'd by trying them over again in *the same* place *with* a Line, after the common way.[50]

Hooke grounded his claims by describing his own experimental process. Petit, however, wanted Hooke to address the trials he and Marin Mersenne (1588–1648) had conducted to determine whether arrows and other projectiles, shot perpendicularly, would descend proportionally to the rate at which they ascended.[51] Hooke, on the other hand, just wanted people to use the instrument 'with the help of some *Tables*' to correlate the time it took for the instrument to resurface to determine the depth.[52] Instead of insisting on a complete mathematical understanding of the rate of acceleration, Hooke wanted to pass on knowledge of a useful instrument. This difference in emphasis was in large

47 Hooke, "Some Considerations" 230.

48 Petit, "Extrait d'une Lettre" 248.

49 '[...] il reste encore bien des choses à sçavoir avant que de tirer quelque utilité de cette invention', in Petit, "Extrait d'une Lettre" 249; Hooke, "Some Considerations" 228.

50 Hooke, "Some Considerations" 229.

51 Petit, "Extrait d'une Lettre" 248.

52 Hooke, "An Appendix to the Directions for Seamen" 148.

part connected to differences in audience: Petit was concerned with a community of scholars and Hooke with seamen who were familiar with using charts and tables to make calculations. At issue here then is an ideological distinction between providing a non-specialist audience with a useful instrument, which provided data the Royal Society was interested in, and providing a scholarly audience with a theoretical explanation of how an instrument worked.

While the intellectual disagreements found in these articles highlight their place in larger debates about the role of the journal, issues of translation strained the communicative reach of the journals. Hooke was characteristically dismissive of Petit's linguistic skills and argued for the clarity of his own prose. I would note however that I share some of Petit's difficulty in understanding particular aspects of the functionality of the instrument for sounding depths without a line, especially the manner by which the ball returned to the surface.[53] Hooke began his response graciously, but his tone quickly changed: 'The Author of the French *Journal des Scavans* found good, to insert them both [the article and the figure] in his *Journal* of *May 3.* but in another of *May* 24. intimates, that the said *Schemes* and their *Descriptions* are not very clear and intelligible (he means, that they were not well understood by *French* Readers) [...]'.[54] By presenting the text as belonging to the images, Hooke's phrasing here prioritizes the image, the Scheme, as the primary vehicle for communicating his ideas. He went on to address the linguistic difficulties: '*First*, That *Englishmen* and such others, as are well versed in the *English* tongue, find no difficulty in understanding the descriptions of these *Engines*, nor in apprehending their structure, exhibited by the *Figures* [...]'.[55] The basic tenet of Hooke's argument is that if Petit had read the original article in English and had enough facility with the language, he would not have misunderstood Hooke. At its core, this is an argument against linguistic translation.

Instead of appreciating the value of linguistic translation for the circulation of knowledge, Hooke seemed to argue that images could overcome the linguistic barriers implied by journals published in vernacular languages, rather than Latin. He wrote in his rebuttal that 'both the *Figure* and the annexed *Description* thereof are so plain and clear, that tis some wonder here, that any difficulty of understanding them is pretended by any, that hath but ordinary skill in *Cutts* and the *English* language'.[56] Petit did not agree and argued that the engraved

53 I want to thank Jim Bennet for discussing this plate with me and helping me to better understand how the instruments might work.

54 Hooke, "Some Considerations" 228.

55 The emphasis is in the original. Hooke, "Some Considerations" 229.

56 Hooke, "Some Considerations" 229.

figure, called a cut because of how the image had to be cut into the copper, did not intelligibly represent the shape and placement of the spring and how it released the ball from the plumb.[57] It is important to note at this point that there was an erratum published in issue 10 for the image: 'In *Fig.* I. of *Num.* 9 of these Tracts, the Graver hath placed the bended *end* of the *springing Wire* C F, above the *Wire-staple* B, between it and the *Ring E*, of the *Weight* D; whereas *that* end should have been so expressed, as to pass *under* the *Wire-staple*, betwixt its two Wires, into the said *Ring*'.[58] In the drawing that accompanied his report, Hooke clearly outlined the different pieces of bent wire with brown ink before lightly shading the wires with ink wash [Fig. 4.10]. With his light source coming from the right, Hooke added shading to the springing wire CF to indicate the shadow cast by the right hand (outer) part of the wire staple B. When the articles were republished in issue number 24, the image was corrected [Fig. 4.11]. Despite the error in the version of the image that was available in Paris, Hooke placed a great deal of explanatory weight on this figure and claimed that it was as clear as his prose.

Coupled as it was with his comments about his readers' fluency in English, Hooke's phrase, 'that hath but ordinary skill in *Cutts*', sets up the ability to clearly understand the engraved line as another skill necessary for participation in seventeenth century intellectual life. While Petit seemed to argue for mathematics as a necessary language for understanding physical problems, Hooke argued here for the necessity of visual literacy. Although Hooke, like so many of his peers, complained at times about the errors made by engravers and the problems inherent in printing images, he consistently included drawings in his reports to the Royal Society and included printed images in his published works and Hooke took an interest in his own visual education.[59] With the phrase, 'skill in Cutts', he implied that others within this community of scholars should also cultivate their visual acuity. In Hooke's view, the visual was another language to be learned by scholars in order to fully participate in intellectual debates in the seventeenth century.

---

57 This usage of cut is noted in the Oxford English Dictionary as now referring only to image cut into wood, but in the seventeenth century was used to refer to the use of both wood or copper. "Cut", *Oxford English Dictionary Online*, n.2, definition 22,a, http://www.oed.com, accessed 20 November, 2015. Petit, "Extrait d'une Lettre" 247.

58 "Note", *Philosophical Transactions* 1, 10 (12 March 1666) [178].

59 Doherty M.C., "Discovering the 'True Form:' Hooke's Micrographia and the Visual Vocabulary of Engraved Portraits", *Notes and Records of the Royal Society of London* 66 (2012) 211–234.

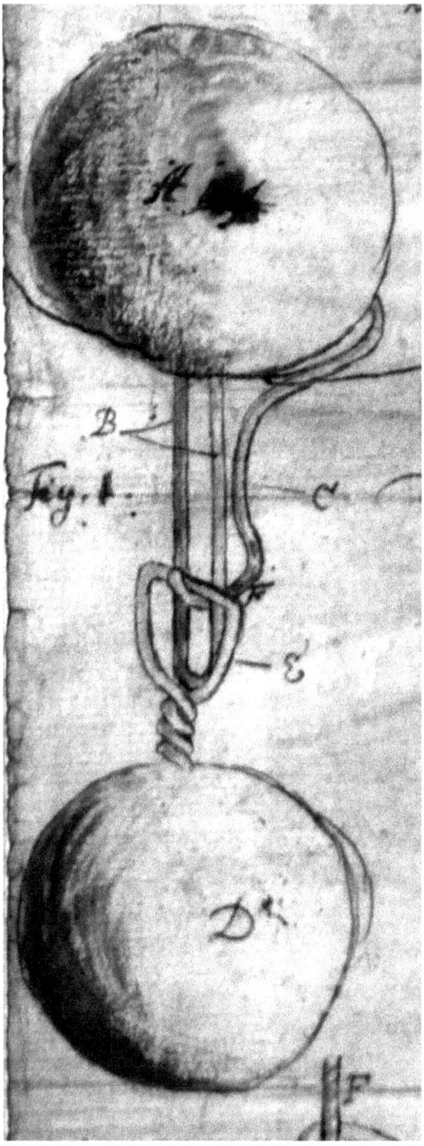

FIGURE 4.10
Robert Hooke, "The way preferred to sound the depths of the sea" (30 September 1663). Ink and wash drawing, page 294 × 189 mm (image 121 × 45 mm). London, Royal Society Library, Classified Papers, Cl.P./20/23.

In the middle of the seventeenth century, as Latin waned as the shared scholarly language, philosophers sought new ways to communicate across an increasing array of linguistic boundaries.[60] The rise to dominance of English as the

60 For an account of this quest in the English context see: Lewis R., *Language, Mind and Nature: Artificial Languages in England from Bacon to Locke* (Cambridge: 2007).

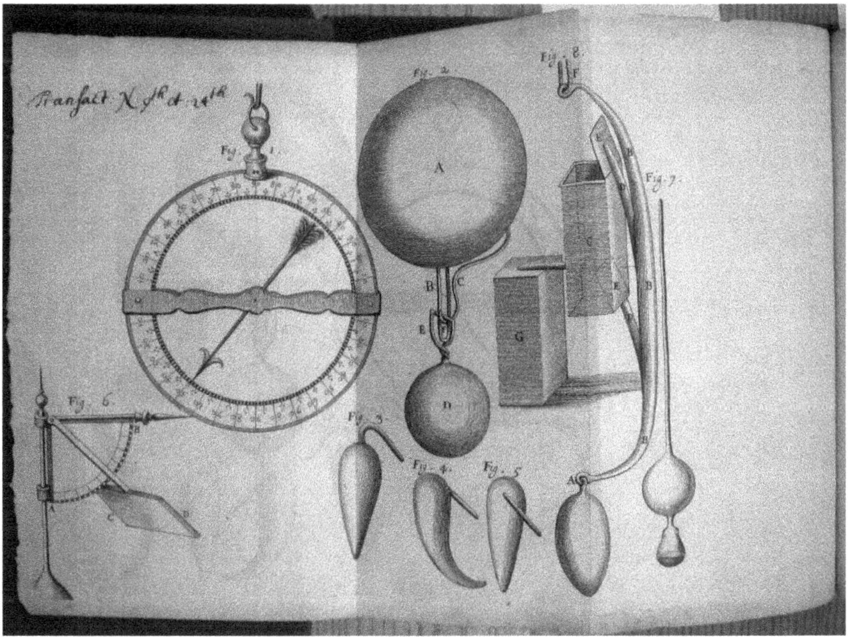

FIGURE 4.11 *Robert Hooke, "Directions For Observations and Experiments to be made". Engraving, in* Philosophical Transactions *2, 24 (8 April 1667) facing 433. London, Royal Society Library.*

language of science did not happen overnight.[61] Instead, editors of the *Journal des Sçavans* struggled to translate English articles into French while at the same moment Fellows of the Royal Society were deeply involved in projects to create a universal language. Some of them, like Hooke and John Evelyn, advocated in particular for an image-based answer. Evelyn, another Fellow of the Royal Society, in his history of engraving, claimed images could transcend language: 'For *picture*, is a kind of *Universal Language*, how diverse soever the tongues and vocal expressions of the several *Nations* which speak them may appear'.[62] Despite this lofty claim, images were just one part of the communication of knowledge in this period. As this overview of illustrated and translated articles and the close examination of one instance of translation has shown, authors, editors, and readers struggled with both linguistic and visual literacies. Hooke's

61 For a thoughtful account of this rise see: Gordin M.D., *Scientific Babel: How Science was Done Before and After Global English* (Chicago: 2015).

62 Evelyn John, *Scupltura: or the History, and Art of Chalcography and Engraving in Copper* (London, Printed by J.C. for G. Beedle, and T. Collins, at the Middle-Temple Gate, and J. Crook in St. Paul's Church-yard: 1662) 140.

assertion that readers needed twinned literacies, in English and cuts, did not match the reality of his readers' abilities. What the images allowed was not a perfect translation, but rather a bridge across a linguistic divide.

Selective Bibliography

Archives of the Royal Society

Early Letters EL/H2/12.

Sources pre-1800

"L'Imprimeur au Lecteur", *Journal des Sçavans* (5 January 1665) n.p.

"Philosophical Transactions. A Londres, chez Jean Martin & James Allistry, Imprimeurs de la Societé Royale, & se trouve à Paris chez Jean Cusson, ruë S. Jacques", *Journal des Sçavans* (30 March 1665) 156.

"Directions for Sea-men, bound for far Voyages", *Philosophical Transactions* 1, 8 (8 January 1666) 140–143.

"*Martis circa axem proprium revolubilis Observationes. Bononiæ à Jo. Dominico Cassino habitæ*. Romæ, 1666", *Journal des Sçavans* (31 May 1666) 259–262.

"Extrait du Journal d'Angleterre, Contenant les particularitez de quelques observations qui ont esté faites à Londres au mois de Fevrier & de Mars dernier touchant la planette de Mars", *Journal des Sçavans* (23 August 1666) 403–406.

"Extrait du Journal d'Angleterre, Contenant quelques nouvelles Observations faites à Londres touchant la Planete de Jupiter", *Journal des Sçavans* (30 August 1666) 416–418.

"The Figure of the Stars in the Constellation of Cygnus; together with the New Star in it, discover'd some years since, and very lately seen by M. Hevelius again", *Philosophical Transactions* 1, 21 (21 January 1666/7) 372.

"Directions for Observations and Experiments to be made by Masters of Ships, Pilots, and other fit Persons in their Sea-Voyages", *Philosophical Transactions* 2, 24 (8 April 1667) 433–448.

"Extrait du Journal d'Angleterre, contenant des instructions pour ceux qui ont à faire de longs voyages sur mer", *Journal des Sçavans* (19 April 1666) 193–196.

"Extrait du Journal d'Angleterre, touchant une nouvelle maniere de sounder la profondeur de la mer sans corde, & de reconnoistre la nature de l'eau qui est au fond de la mer", *Journal des Sçavans* (3 May 1666) 217–120 [i.e. 220].

Bacon Francis, *The New Organon*, ed. L. Jardine – M. Silverthorne (Cambridge: 2000). Cassini Jean Dominique, "Observations Made in Italy, Confirming the Former, and Withall Fixing the Period of the Revolution of Mars", *Philosophical Transactions* 1, 14 (2 July 1666) 242–45.

Evelyn John, *Scupltura: or the History, and Art of Chalcography and Engraving in Copper* (London, printed by J.C. for G. Beedle, and T. Collins: 1662).

Hevelius Johannes, "Promiscuous Inquiries, chiefly about Cold, formerly sent and recommended to Monsieur Hevelius; together with his Answer return'd to some of them", *Philosophical Transactions* 1, 19 (19 November 1666) 344–352.

Hevelius Johannes, "Extrait d'une lettre de Monsieur Hevelius ecrite a Monsieur Payen Advocat au Parlement, touchant l'Observation de la nouvelle etoile decouverte dans la constellation du Cygne", *Journal des Sçavans* (13 December 1666) 486–490.

Hooke Robert, "An Appendix to the Directions for Seamen, bound for far Voyages", *Philosophical Transactions* 1, 9 (12 February 1666) 147–149.

[Hooke Robert], "Some Considerations Touching a Letter in the *Journal des Scavans* of May 24. 1666", *Philosophical Transactions* 1, 13 (4 June 1666) 228–230.

Hooke Robert, "The Particulars. Of Those Observations of the Planet Mars, Formerly Intimated to Have Been Made at London in the Months of February and March A. 1665/6", *Philosophical Transactions* 1, 14 (2 July 1666) 239–242.

Hooke Robert, "Some Observations Lately Made at London Concerning the Planet Jupiter", *Philosophical Transactions* 1, 14 (2 July 1666) 245–246.

Hooke Robert, "A Late Observation about Saturn made by the Same", *Philosophical Transactions* 1, 14 (2 July 1666) 246–247.

Oldenburg Henry, "The Introduction", *Philosophical Transactions* 1, 1 (6 March 1665) 1–2.

Oldenburg Henry, "To the Royal Society", *Philosophical Transactions* 1, 1 (6 March 1665) n.p.

Oldenburg Henry, *The Correspondence of Henry Oldenburg*, ed. and trans. A.R. Hall – M. Boas Hall, 13 vols. (Madison, WI – London, 1965–1986).

Petit Pierre, "Extrait d'une Lettre Escritte par M. Petit Intendant des Fortifications, à M/ Galloys P. touchant la profondeur de la Mer, la nature de l'Eau qui est au fond de la Mer, & quelques autres curiositez", *Journal des Sçavans* (24 May 1666) 247–250.

Sprat Thomas, *The History of the Royal Society of London for the Improving of Natural Knowledge* (London, Printed by T.R. for J. Martyn and J. Allstrey: 1667).

Sources post-1800

Andrade E.N. da C., "The Birth and Early Days of the *Philosophical Transactions*", *Notes and Records of the Royal Society* 20, 1 (1965) 9–22.

Atkinson D., "The *Philosophical Transactions of the Royal Society of London*, 1675–1975: A Sociohistorical Discourse Analysis", *Language in Society* 25 (1996) 333–371.

Bazerman C., *Shaping Written Knowledge: The Genre and Activity of the Experimental Article* (Madison, WI: 1988).

Broman T., "Criticism and the Circulation of News: The Scholarly Press in the late Seventeenth Century", *History of Science* 51 (2013) 1–26.

Brown H., *Scientific Organizations in Seventeenth Century France* (New York: 1967).

Brown H., "History and the Learned Journal", *Journal of the History of Ideas* 33, 3 (1972) 365–378.

Doherty M.C., "Discovering the 'True Form:' Hooke's Micrographia and the Visual Vocabulary of Engraved Portraits", *Notes and Records of the Royal Society of London* 66 (2012) 211–234.

Doherty, M.C., "Giving Light to Narrative: The Use of Images in Early Modern Journals," *Nuncius: Journal of the Material and Visual History of Science* 30, 3 (2015) 543–569.

Gaskell R., "Printing House and Engraving Shop. A Mysterious Collaboration", *The Book Collector* 53 (2004) 213–251.

Gordin M.D., *Scientific Babel: How Science was Done Before and After Global English* (Chicago: 2015).

Gross A.G. – Harmon J.E. – Reidy M.S., "Argument and 17th-Century Science: A Rhetorical Analysis with Sociological Implications", *Social Studies of Science* 30, 3 (2000) 371–396.

Johns A., "Miscellaneous Methods: Authors, Societies and Journals in Early Modern England", *The British Journal for the History of Science* 33, 2 (2000) 159–186.

Kronick D.A., *A History of Scientific & Technical Periodicals: The Origins and Development of the Scientific and Technical Press, 1665–1790*, 2nd ed. (Metuchen, N.J.: 1976).

Kronick D.A., "Notes on the Printing History of the Early *Philosophical Transactions*", *Libraries and Culture* 25, 2 (1990) 243–268.

Lewis R., *Language, Mind and Nature: Artificial Languages in England from Bacon to Locke* (Cambridge: 2007).

McCutheon R.P., "The 'Journal Des Scavans' and the 'Philosophical Transactions of the Royal Society'", *Studies in Philology* 21, 4 (1924) 626–628.

Moxham N., "Fit for Print: Developing an Institutional Model of Scientific Periodical Publishing in England, 1665–ca.1714", *Notes and Records of the Royal Society* 69 (2015) 241–260.

Purver M., *The Royal Society: Concept and Creation* (Cambridge, MA: 1967). Stroup A., "Royal Funding of the Parisian Académie Royale des Sciences During the 1690s" *Transactions of the American Philosophical Society* 77, 4 (1987) 1–167.

Stroup A., *A Company of Scientists: Botany, Patronage, and Community at the Seventeenth-Century Parisian Royal Academy of Sciences* (Berkeley: 1990).

Turner A., "An Interrupted Story: French Translations from *Philosophical Transactions* in the Seventeenth and Eighteenth Centuries", *Notes and Records of the Royal Society* 62 (2008) 341–354.

PART 2

Translating Practical Knowledge

CHAPTER 5

'As the Author Intended': Transformations of the Unpublished Writings and Drawings of Simon Stevin (1548–1620)

Charles van den Heuvel

Introduction

In the years 1605 and 1608, the Flemish mathematician Simon Stevin (1548–1620) published five books in two volumes with the title *Wisconstighe Ghedachtenissen* [Mathematical Memoirs] that can be seen as a compilation of his private lessons to Prince Maurice of Orange on mathematics, natural sciences and military arts at the court in The Hague.[1] In the fifth book *Vande Ghemengde Stoffen* [Miscellanea], Stevin explained in a note that he had not been able to finish several treatises announced in the table of contents (on Arithmetic, Book keeping, Architecture, Music Theory, Military Arts and other topics) in time for the printer and planned, therefore, to publish these at a later moment.[2] By the time Stevin died in 1620, only a few fragments of these announced treatises had appeared in other publications.[3] Fortunately, several scholars made tables of content and transcripts (Isaac Beeckman), published some of the material (Hendrik Stevin) or corresponded about the content of

1 Stevin Simon, *Wisconstighe Ghedachtenissen, inhoudende t'ghene daer hem in gheoeffent heeft den Doorluchtichsten Hooghgheboren Vorst ende Heeren, Maurits, Prince van Oraengien* [...] (Leiden, Jan Bouwensz.: 1605–1608). For the life and work of Simon Stevin the Dutch biography Dijksterhuis E.J., *Simon Stevin* (The Hague: 1943) is still the most accurate. For a more recent biography in English see Devreese J.T. – Vandenberghe G., *The Wonderful World of Simon Stevin: Magic is No Magic* (Southampton: 2007). For translations of titles in this chapter I have made use of the series *The Principal Works of Simon Stevin*, ed. E. Crone – E.J. Dijksterhuis – R.J. Forbes – M.G.J. Minnaert – A. Pannekoek, 5 vols. (Amsterdam: 1955–1966).

2 Stevin S., *Wisconstighe Ghedachtenissen*, Vijfde Stuck [Part Five] 107.

3 For a bibliography and reconstruction of Simon Stevin's unpublished works see, Heuvel C. van den, *'De Huysbou': A reconstruction of an unfinished treatise on architecture, town planning and civil engineering by Simon Stevin* (Amsterdam: 2005) and Heuvel C. van den, "Le traité incomplet de l'Art Militaire et l'ínstruction pour une école des ingénieurs de Simon Stevin", in *Simon Stevin (1548–1620) L'emergence de la nouvelle science*, Exhibition Catalogue, Bibliothèque Royale Albert I, Bruxelles, 17-09-2004/30-10-2004 (Turnhout: 2004) 101–111.

 | DOI 10.1163/9789004349261_007

the manuscripts (Constantijn Huygens and Marin Mersenne).[4] This way crucial information about Stevin's unpublished chapters was preserved. However, in the processes of preserving the content of these valuable manuscripts of Simon Stevin parts were left out, adapted, replaced and represented in new contexts. Although the handling of these unpublished fragments often resulted in completely different formats of Stevin's writings, some authors in their reconstructions tried to do justice to the original content or to say it in Hendrik Stevin's words to reassemble them: 'na de Autheurs gevoelen' [as the author intended].[5] Others made notes and excerpts for own use as part of their research for publications or other purposes. Here, the various intentions of the scholars who translated Stevin's manuscripts in various formats (and sometimes partly in other languages) for reuse will be explored. Particular focus will be on the implications of the transformation and translations of Stevin's manuscripts and drawings that circulated in correspondences, notes, technical designs and scholarly publications for the representation and visualization of knowledge in the seventeenth century. Four aspects of translation will be discussed: 'the untranslatable', 'translation and transformation', 'translation and simulation', and finally 'translation and commodification'.[6] Before this discussion of translations of Stevin's work, we briefly zoom in on the various roles of the scientist, on the authorship of his most important publication and his views on language in the context of the rise of the vernacular in Early Modern Europe.

---

4 Beeckman Isaac, *Loci communes* etc [hereafter *Journal*], manuscript HS 6471 in Middelburg, Zeeuwse Bibliotheek, partially published in *Journal tenu par Isaac Beeckman de 1604 à 1634*, ed. C. de Waard, 4 vols. (The Hague: 1939–1953). Parts of the manuscripts of Simon Stevin were published posthumously by his son: Stevin Hendrik, *Materiae Politicae Burgherlicke Stoffen Vervanghende Ghedachtenissen der Oeffeninghen des Doorluchtichsten Prince Maurits van Orangie* (Leiden, Adryaen Rosenboom: 1649) and Stevin Hendrik, *Wisconstich Filosofisch Bedryf* (Leiden, Philips de Croy: 1667). For the correspondence between Huygens and Mersenne regarding the manuscripts of Stevin: Huygens Constantijn, *Adversaria*, The Hague, Koninklijke Bibliotheek, KA XLVII, fols. 146r–148r, and Mersenne Marin, *Correspondance du P. Marin Mersenne Religieux Minime*, ed. C. de Waard (Paris: 1932–1988). See on the crucial role of Mersenne in European scientific communication also the chapters by Jan van de Kamp, Meghan C. Doherty, Rodolfo Garau, and Fabien Simon in this volume.

5 Stevin H., *Materiae Politicae*, "Tytels en Cortbegrypen" [Titles and Tables of Content] 144.

6 Some of the examples for this book chapter are published in Heuvel, *De Huysbou* with translations by Sam Herman. These texts are reused here to illustrate various modes of translations of the works of Simon Stevin by himself and other authors.

Simon Stevin's Private and Public Roles as a Scientist

Simon Stevin [Fig. 5.1], born in Bruges in 1548, left the southern Low Countries for the province of Holland around 1580. In 1581 Stevin was registered in the city of Leiden and two years later, on 16 February 1583 he matriculated at its university. Here he probably met the future Stadtholder of Holland and Zeeland, Prince Maurice (1567–1625). It would be the starting point of a long-standing collaboration that would result in the well-protected career of Stevin at the court in The Hague until his death in 1620. Prince Maurice was seventeen years old when his father Prince William the Silent was assassinated by a bullet on 10 July 1584. Despite his success on the battlefield, some members of his family were concerned at the responsibilities Maurice had to take on as Stadtholder, Captain General and Admiral of the Fleet. But the young Maurice surrounded himself with good advisors such as Grand Pensionary Johan van Oldebarneveldt, who remained Maurice's counsellor in administrative and political affairs almost all his life. Around 1590 Simon Stevin became Maurice's private tutor and advisor on military affairs. Stevin accompanied Maurice for years in the numerous military campaigns and sieges in this turbulent period of the Eighty Years War between the Spanish Empire and the Dutch Republic. On request of Prince Maurice, Stevin wrote in 1600 the program for the *Duytsche Mathematique*, a training school for military engineers and land surveyors that provided pupils with elementary lessons in geometry and fortification.[7] Lessons were taught in Dutch to enable a rapid supply of engineers skilled in the practice of fortification in this period of conflict.

Despite his rather private and protected role as Maurice's tutor and advisor at the court in The Hague, Stevin sometimes went public. Stevin published several books on various topics such as mathematics, mechanics, astronomy, geography, navigation, fortification, military camps, civil engineering, book keeping and politics. Furthermore, he applied with success for at least fifteen different patents.[8] Almost all patents that were granted to Stevin by the States of Holland between 1584 and 1590 were for inventions for draining mills. Stevin analysed many existing mills of which he took notes, which he compared with calculations for his own inventions. However, he did not publish his analyses. Apparently Jacob Golius, orientalist and specialist in Arabic languages, found the hydrostatical notes and calculations of Stevin interesting enough to bring them together in a manuscript possibly with the aim of publishing them as a

7 Heuvel, "Le traité incomplet de l'Art Militaire".

8 Forbes R.J., "Contents and history of the Patents of Simon Stevin", in *The Principal Works of Simon Stevin* vol. V, 9–38.

FIGURE 5.1 *Anonymous, "Portrait of Simon Stevin" (no date). Oil on canvas, 63.5 × 49 cm. Leiden University Library, Icones 40.*

separate work.[9] Fragments with text and drawings with inventions of Stevin for cogs and wheels were published for the first time by Hendrik Stevin.[10] It is possible that Simon Stevin himself was hesitant to publish these notes. Competitors could easily run off with innovative ideas, before those could be protected by patents.[11] Apart from competition there might be other reasons to explain Stevin's decision not to publish his notes on mills. Although he had received several patents for his mill inventions, the applications hereof were, in practice, certainly not all successful and had even resulted in lawsuits against the engineer. In short, we can only guess why Stevin did not publish his inventions for these machines that were crucial for the safety and welfare of the Dutch Republic. However, it seems plausible that Stevin's various private and public roles had an impact on the selection and presentation of his publications.

Stevin, Prince Maurice and Authorship

The task that Hendrik Stevin set himself—of publishing the text of his father 'as the author intended'—was not an easy one, because the authorship of Stevin's writings was not always evident. In the preface of the *Wisconstighe Ghedachtenissen*, Stevin explains how Prince Maurice during military campaigns and sieges studied books and manuscripts and out of fear of losing the exercises and 'eygen vonden' [own inventions] of the Prince, 'die ick na mijn stijl beschreef' [that I described in my own style] decided to publish them.[12] This seems at first sight a rhetorical opening to praise the skills of Prince Maurice, but Beeckman's table of contents of the manuscripts of Simon Stevin that he compiled at the house of the widow, lists: 'Syn Excellenties teyckeningen ende

9 The collection of *Adversaria* of Constantijn Huygens, The Hague, Koninklijke Bibliotheek, KA XVLII, 163–179 contains a manuscript on mills with the title *Stevin, Vande Molens. Gereciviseert door den Professor Golius 1634*. This work was published in *The Principal Works of Simon Stevin*, vol. V. Other fragments were published by Bierens de Haan D., *'Vande Spiegeling der Singconst' et 'Vande molens'. Deux traités inédits* (Amsterdam: 1884).

10 Stevin H., *Wisconstich Filosofisch Bedryf*, Book VI "Van aldervolmaecste cammen en staven" [Of the most perfect cogs and staves], prop. 2 and X "Van den handel der Watermolens onses Vaders Simon Stevin" [On the Watermills of our father Simon Stevin].

11 In at least one occasion we know that Stevin in 1590 lodged a complaint of an infringement of his patent, probably for the cog and wheels of a wind mill by Cornelis Dircxz Muys, to whom several patents had been granted as well including for wind mills. *The Principal Works of Simon Stevin* vol. V, 14.

12 Stevin S., *Wisconstighe Ghedachtenissen* vol. IV. Voorreden [Preface].

schrift' [His Excellency's drawings and writings].[13] Some contemporary scholars also considered the *Wisconstighe Ghedachtenissen* as a work completely, or at least partly, written by Prince Maurice. The well-known scholar Hugo Grotius referred, in a poem of 1617, to *Mathematica Principis Mauritii* without mentioning the name of his friend Stevin, and wrote in the part on Language in his *Parallelon rerumpublicarum liber tertius* that:

> Ejus ipsius Principis nunc in lucem prodeunt Rerum Mathematicarum Commentarii, opus maximum et in unoquoque artium genere consummatissimum, cui nihil unquam aequale vetustas dedit. Hos libros cum ille Imperatorum sapientissimus, scriptos nostrate lingua, Latina Gallicaque veluti per interpretem loqui jusserit, dubitare non patitur, quae omnium exprimendae rerum proprietati sit aptissima [...].[14]

> From that Prince now appear in light commentaries on Mathematical Affairs, a most important work and the most complete of each art that one can wish for, and nothing similar has been delivered in Antiquity of a similar level. Because the wisest of all rulers has ordered that these books, written in our language, should speak as through an interpreter in Latin and French, he does not allow us to doubt which language of all languages is the best to express the nature of all things [...].[15]

The historian and theologian, later rector magnificus of the University of Groningen, Ubbo Emmius wrote in letter to the Calvinist theologian of the University of Franeker on 29 September 1608:

> Intelligo grande volumen prodiisse in lucem, habens inscriptum nomen principis nostri Mauritii, complexum eius mathematica, editum a Simone Stevino mathematico, qui principi est familiaris, cura huius politum et auctum. Editum accipio tribus linguis Latina, Gallica, Belgica. Stevinus

13 Heuvel, '*De Huysbou*' 503.

14 Grotius Hugo, *Batavi, Parallelon rerumpublicarum liber tertius* [etc.], ed. J. Meerman (Haarlem: 1801–1803) III, 69. However, Grotius praised Stevin for his work in this poem, Grotius Hugo, *Poemata Collecta & magnam partem nunc primum edita à fratre Guilielmo Grotio*, 12 books (Leiden, Andr. Clouquius: 1617) Book II, Sylvarem liber secundus II, 62–71. See further, *De dichtwerken van Hugo Grotius, oorspronkelijke poëzie deel I 2a/b, pars 3*, ed. A. Eyffinger (Assen: 1988) 118–148.

15 I am indebted to Henk Nellen for this translation of Grotius's text in the abovementioned Meerman edition.

> assuit et inseruit operi opiniones suas plus quam phanaticas et absurdas, tellurem moveri et esse luminare ac octavum planetam [..] Doleo principis nomen et labores his inquinamentis contaminari.[16]
>
> I gather that a large work has been published, which bears the name of our Prince Maurice; it contains his *Wisconstighe Ghedachtenissen*, [and] is published by Simon Stevin, mathematician and a friend of the Prince, and corrected and augmented by him. I have heard that it appeared in three languages, Latin, French and Dutch. Stevin has inserted his more than absurd and preposterous view that the earth is moving [...]. I regret that the good name and studies of the Prince are sullied by this filth.

It is hard to tell what the real impact of Maurice was on the content of the *Wisconstighe Ghedachtenissen*. But apart from these references to a potential co-authorship of Prince Maurice, it is evident that Stevin's function as a private tutor had an impact on the selection of specific content to be included in the *Wisconstighe Ghedachtenissen* and other works in order to make it accessible to a wider public. He had to translate both his own notes and the ones of Prince Maurice into a coherent text for publication.

Stevin on Didactical Clarity of Argument and Dutch as Scientific Language

Stevin's role as private tutor not only seems to have had an impact on the selection of the content, but also on the presentation thereof. A very explicit reference to the need of the didactical clarity of the argument we find in Stevin's work *De Sterctenbouwing* [Fortification]: 'op dat leering met gheen strijding verduystert en worde' [because the teaching should not be complicated by arguments].[17] Stevin's didactical note would not withhold translators trying

16 *Briefwechsel des Ubbo Emmius*, ed. H. Brugmans – F. Wachters, 2 vols. (The Hague: 1911–1923), vol. 2 (1608–1625) 51. The French translation of the *Wisconstighe Ghedachtenissen* by Jean Tuning was published as *Memoires Mathematiques, contenant ce en quoy s'est excercé* [...] *Maurice, Prince d'Orange* [..] (Leiden, Jan Paedts Jacobsz.: 1608) and the Latin translation by Willibrord Snellius was published as *Hypomnemata Mathematica, hoc est eruditus ille pulvis, in quo se exercuit* [..] *Mauritius Princeps Auraicus* [...] (Leiden, Ex Officina Ioannis Patii, Academiae Typographi: 1608).

17 Stevin Simon, *De Sterctenbouwing, beschreven door Simon Stevin van Brugge* (Leiden, Françoys van Ravelenghien: 1594) Figure 8. *The Principal Works of Simon Stevin* vol. IV, 67.

to impose their own interpretations. In an anonymous English translation of Stevin's work on fortification [Fig. 5.2a] an extra illustration was included with the explanation: 'This 9e figure added by the translator for the better expressing of the authors meaning' [Fig. 5.2b].[18] Stevin had indeed proposed, apart from the wooden model with two bastions that he described and included as figure 8 is his publication, to make two other wooden models, one of a complete regular fortress with its six bastions, the other only of half a bastion without the ditches. The latter model would enable the viewer to perceive better some details of the bastion, such as embrasures, merlons, vaults, stairs, doors in the flank, which are hard to read from the ground plan. This suggestion might have inspired the translator to add an extra three-dimensional representation. Strangely enough the translator did not add a figure of a half bastion but of a full bastion. The figure indeed provides far more details of the bastion than Stevin's figure 8 and is easier to read than his ground plan. However, if the translator had chosen for a half bastion, the reader not only would have been able to read the same details, but the section would have provided a sense of the dimensions of the bastion and allowed a view of its interior. For Stevin, a faithful translation not only implied doing justice to its didactical clarity or to its semantic meaning in another language. In his view Dutch has intrinsic qualities that cannot be found in other languages. This becomes clear from another work of Stevin, a treatise on music in which he not only shifts the content in various manuscript versions to find an optimal logical order to present his argument, but also reveals his views on the natural logic of the Dutch language.[19] Stevin's views on musical theory were perhaps not as innovative as

---

18 Cambridge, Trinity College, Ms. R 16.30. Symon Stephen of Bridge, *The Building of Fortes* (1604) Figure 9. Cited by Dijksterhuis, *Simon Stevin* 43. The illustration indeed shows flanking embrasures in the orillions and parapets that are missing in Stevin's descriptions and figures, and that are for the account of the translator. Quote from *The Principal Works of Simon Stevin* vol. IV, 32. Stephen Johnston mentions in *Making mathematical practice: gentlemen, practioners and artisans in Elizabethan England*, Ph.D. dissertation (Cambridge: 1994) http://www.mhs.ox.ac.uk/staff/saj/thesis/practitioners.htm#note73 that this manuscript is an anonymous copy of a translation by the English engineer Paul Ive presented to the Earl of Northumberland in 1600, nowadays in the Petworth House Archives (West Sussex Record Office) MS HMC 138. I was not able to check whether the latter manuscript contained the additional figure 9 as well.

19 The Hague, Koninklijke Bibliotheek, Ms. XLVII, fols. 624–705. While Bierens de Haan published the third part of the manuscript (fols. 687–705) in the hand of the clerk together with the excerpts on mills in "Van de Spiegheling der Singconst", Adriaan Fokker was responsible for the publication and editing of the second version of the manuscript in Stevin's own handwriting (fols. 655–672) that was included in *The Principal Works of*

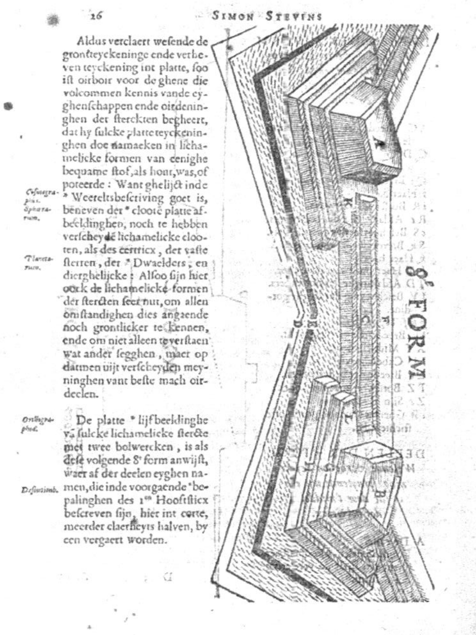

26 SIMON STEVINS

Aldus verclaert wesende de gronsteyckeninge ende verheven teyckening int platte, soo ist oirboir voor de ghene die volcommen kennis vande eyghenschappen ende oirdeninghen der sterckten begheert, dat hy sulcke platte teyckeninghen doe namaeken in lichamelicke formen van eenighe bequame stof, als hout, was, of poteerde: Want ghelijck inde * Weereltsbescriving goet is, beneven der * cloote platte afbeeldinghen, noch te hebben verscheyde lichamelicke clooten, als des eertrick, der vaste sterren, der * Dwaelders; en dierghelijcke: Alsoo sijn hier oock de lichamelicke formen der stercken seer nut, om allen omstandighen dies angaende noch grontlicker te kennen, ende om niet alleen te verstaen wat ander segghen, maer op datmen uijt verscheyden meyninghen vant beste mach oirdeelen.

De platte * lijfbeeldinghe vā sulcke lichamelicke stercte met twee bolwercken, is als dese volgende 8e form anwijst, waer af der deelen eyghen namen, die inde voorgaende * bepalinghen des 1en Hoofstlicx bescreven sijn, hier int corte, meerder claerheyts halven, by een vergaert worden.

FIGURE 5.2A *Simon Stevin, engraving, illustration of relief model of two bastions in Simon Stevin,* De Sterctenbouwing (*Leiden, François van Ravelenghien: 1594*) 4^{0}. *Print, illustration 8.*

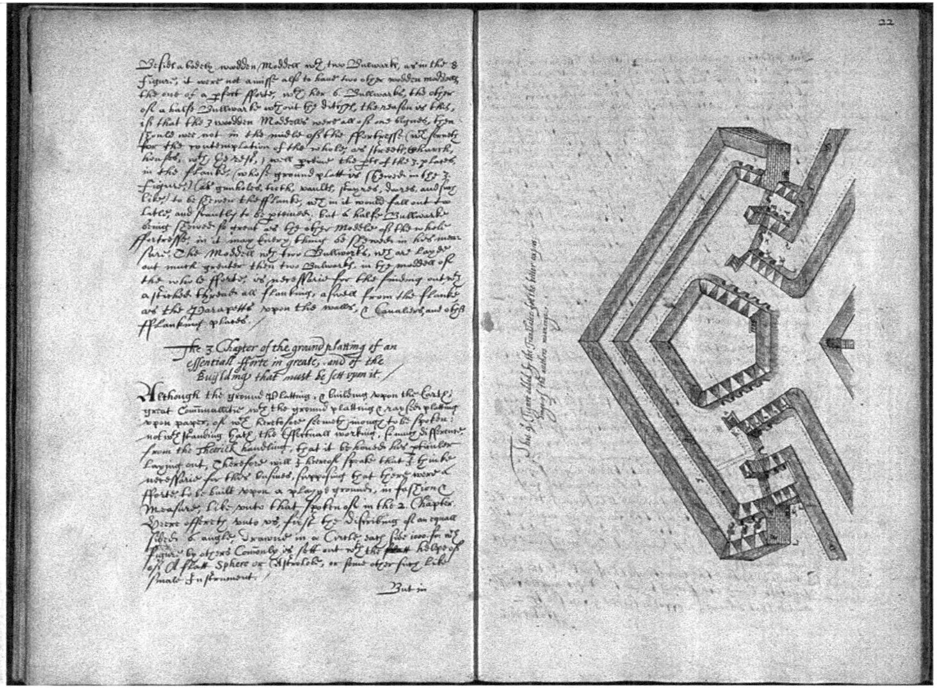

FIGURE 5.2B *Anonymous,* “Symon Stephen of Bridge, The Building of Fortes” [*ca. 1600*] *Manuscript, Cambridge, Trinity College, Ms. R 16–30, illustration 9.*

some authors have us believe, but they illustrate well how Stevin tried to translate and combine ideas worked out in other publications into new studies.[20] Stevin's ideas on music cannot be separated from his views on the role of mathematics and of the Dutch language that he propagated in most of his publications to reveal the true nature of the sciences:

> Want de redens inde stof des gheluijdts niet soo opentlick bekent en sijn als in ander stoffen daer sij ons ontmoeten, sullen om meerder claerheyt eerst segghen vande Redens en Evenredenheyt int ghemeen; daer nae vande ghedaente des redens der Singconst duer haer verlijcking met de bekende reden der meetconst. Ende ten laetsten van d'eijghen redens der singhelicke gheluyden. [...] Siet hier duytsche woorden licht om te verstaen ende slecht van ghelaet maer eyghenlick van onteindelick vermueghen. Want soomen ansiet het bepaelde te weten Everedenheijt tis als bepaling sijns grondts, wiens gheluijt alleen, int eerste anhooren ons vermaent ende anwijst dattet recht grontlick verstandt der Everedenheyt byde Griecken ende hun navolghers niet en gheweest en heeft. [...] Doirsaeck dier dwalinghen is dat hun spraeck dit word medtsgaders al d'ander Wisconstighe namen niet soo eyghentlick beteeckenen en conde als dese [...] Maer het DUYTSCH om de vrie consten daer in te leeren, om de natuerens verborghentheden daer in duergronden ende te bewysen dat wonder gheen wonder en is.[21]

> Because ratios in the field of sound are not as manifestly known as in other fields where we meet with them, for the sake of greater clarity we shall first speak about ratio's and equirationality [Stevin uses the Dutch neologism evenredenheijt] in general; subsequently about the aspect of ratio in singing by comparison with the familiar ratio in geometry. And finally of the ratios proper to musical sound. Ratio as defined generally is the relation between things of the same nature. [...] Look here: Dutch words, easy to understand and modest in appearance, but in reality of an

---

Simon Stevin vol. v, 413–464 with an English translation. Also the first part, fols. 624–654, is in Stevin's handwriting.

20 Fokker A., "An Introduction: Simon Stevin's views on music", in *The Principal Works of Simon Stevin* vol. v, 420. Several historians have pointed to Stevin's work as a very early plea for equal temperament. The meaning of Simon Stevin's musical theory has been brought back to right proportions by Cohen H.F., "Simon Stevin's equal division of the octave", *Annals of Science* 44, 4 (1987) 471–488.

21 *Principal Works of Simon Stevin* vol. v, 426–428.

> infinite power. For if one considers the thing defined, viz. equirationality, it is like a definition of substance, the mere ring of which, at first hearing, brings home to us and shows us that the very thorough understanding of equirationality was not found among the Greek and their ancestors. [...] The cause of these errors is that the Greek could not interpret this term together with all other mathematical terms as properly as this one [the Dutch language]. [...] But DUTCH serves to teach the liberal arts to fathom the hidden secrets of nature, and to prove that miracle is no miracle.

In short Stevin claims that Greek language, unlike Dutch, is not adequate to explain the order of nature expressed in mathematical—in this case geometrical—terms. This explains Stevin's view that all attempts to find the correct solution for the problem raised by Pythagoras—how to divide the string to suit the true musical scale—had failed. Stevin maintains that both the five major steps (whole tones), and the two minor steps (semitones) that are each half of a major step, must be equal. An octave in that view consists of six equal major steps. Contemporaries did not agree. A befriended organist, Abraham Verheyen in a letter to Stevin demonstrated in an experiment with the harpsichord that six whole tones do not make an octave, and produced an example of a song in two parts showing the difference between major and minor semitones.[22] Moreover Isaac Beeckman, who first supported Stevin's proportional division of the octave, would later reject it. He was followed by another scholar with access to the original manuscripts, Christiaan Huygens, and indirectly by Marin Mersenne.[23] It is indeed possible that in fear for the loss of his intellectual reputation Stevin decided in the end not to publish his theories on music.[24] However, despite this decision, Stevin remained convinced all his life that Dutch was the superior language to express scientific concepts and ideas.

22 This letter sits between the manuscripts of Stevin's text on music in the *Adversaria* of Constantijn Huygens, The Hague, Koninklijke Bibliotheek, KA XLVII, fols. 673r–677v.

23 See for this criticism by contemporaries, Cohen, "Simon Stevin's equal division of the octave" 486–487.

24 *Principal Works of Simon Stevin* vol. V, 420; Cohen, "Simon Stevin's equal division of the octave" 486.

Stevin on Dutch and the Rise of the Vernacular in Early Modern Europe

As Peter Burke has pointed out, printed praise of the vernacular became more and more common throughout Europe in the sixteenth and seventeenth centuries. Authors like Dante, Joachim du Bellay, Martin Opitz, Richard Carew, Bolhuslav Balbin, Otto Sperling and Vasily Trediakovsky defended respectively the Italian, French, German, English, Czech, Danish and Russian languages as being superior to Latin.[25] Such laudations often refer to the aesthetic or ancient character of their language. This was no different for the Dutch language. Stevin's views on the importance of the Dutch language were perhaps less extreme than those of Johannes Goropius Becanus, who in his *Origines Antwerpianae* [The Origin of Antwerp] had claimed that Dutch was the only language that directly stemmed from the first language that Adam and Eve spoke in Paradise. However, Stevin refers in his *Discourse on the Worthiness of the Dutch Language* to a period in the remote past when civilization and science was at its highest level, the *Wijsentijt* [The Age of Sages]. In this period all scholars collaborated in order to improve the level of the sciences. In order to restore this high level of the Age of Sages scientific ideas should be expressed in the vernacular. We already noticed that Stevin was adamant on the superiority of Dutch as a scientific language. It is therefore no coincidence that his essay *Uytspraeck van de Weerdicheyt der Duytsche Tael* [Discourse on the Worthiness of the Dutch Language] served as an introduction to his publication on mechanics, *De Beghinselen der Weeghconst* [The Elements of the Art of Weighing]: 'wiens diepsinnighe gedaenten duer slechter spraken ten eersten niet wel bedietlick en sijn' [the profound nature of which cannot be well expounded at once in inferior languages].[26] However, in the context of the rise of the vernacular in Early Modern Europe, Stevin's praise of the Dutch as a scientific language is rather exceptional. Ann Blair has noticed that Latin in the natural sciences (including philosophy) was more persistent in its battle against the vernacular than in many other disciplines.[27] Different from the supporters

25 Burke P., *Languages and Communities in Early Modern Europe* (Cambridge: 2004). In particular chapter 3: "Vernaculars in competition" 61–71.

26 "Discourse on the worthiness of the Dutch language", in *The Principal Works of Simon Stevin* vol. I, 58.

27 Blair A., "La persistance du latin comme langue de science 'a la fin de la Renaissance'", in Chartier R. – Corsi P. (eds.), *Sciences et langues en Europe* (Luxembourg: 1996) 33–39; and Deneire T., "Neo-Latin and the Vernacular: Methodological Issues", in Ford P. – Bloemendal J. – Fantazzi C. (eds.), *Brill's encyclopaedia of the Neo-Latin world* (Leiden – Boston: 2014) 275–285, in particular 283.

of other vernaculars that used ancientness or euphony as qualities to praise the superiority of their language, Stevin in the footsteps of Becanus, seems to emphasize an additional argument, i.e. the structural, grammatical qualities of the Dutch language.[28] Stevin was convinced that Dutch was superior to other languages, such as Greek, Latin or French to explain scientific concepts because it contains far more monosyllabic words which could be combined to create clear compound words. To support his view, Stevin included in this introductory discourse a list with hundreds of monosyllabic words in Dutch of which their Latin and French translations needed more syllables to express the same concept [Fig. 5.3]. Moreover, Stevin is famous for introducing Dutch neologisms where, in his view, scientific concepts were expressed in a limited way by existing terms.[29] In the first case of translations we discuss an untranslatable Dutch neologism of Stevin in a Latin didactic poem by Constantijn Huygens. It focusses on Stevin's use of the term 'leijckseijdicheydt' [mirror symmetry] within the context of his unpublished work *De Huysbou* [Architecture] of which excerpts in manuscripts circulated between scholars in the Dutch Republic. Three cases will follow in which scholars adapted Stevin's manuscripts and drawings for their own purposes and that focus on transformation, simulation and commodification as aspects of translation.

The Untranslatable

In May 1639 Constantijn Huygens (1596–1687) poet, musician and secretary to the successor of Prince Maurice, Frederic Hendrik, Prince of Orange, wrote a poem with the title *Domus* [House] to his three sons (respectively 11, 10 and 8 years old) in which he asked them to put their toys aside and to pay attention to his explanation of the house they lived in.[30] The house was built on the property donated to Huygens by Frederic Hendrik on the Plein, the prominent central square of The Hague, next to the residence of Count Johan Maurice of Nassau-Siegen, who built his house, the 'Mauritshuis' around the same time. Similar to Stevin, Constantijn did not publish this work on architecture. The manuscript of *Domus* in the National Library in The Hague did not appear in print until 1999. It consists of twenty-three folios that describe the house

28 Frederickx E. – Hal T. van, *Johannes Goropius Becanus (1519–1573). Brabants arts en taalfanaat* (Hilversum: 2015) 126–130 and 187–188.

29 Kool M., "De rekenkundige termen van Simon Stevin", *Scientiarum Historia* (1992) 91–107.

30 Blom F.R.E. – Bruin H.G. – Ottenheym K., *Domus: het huis van Constantijn Huygens in Den Haag* (Zutphen: 1999).

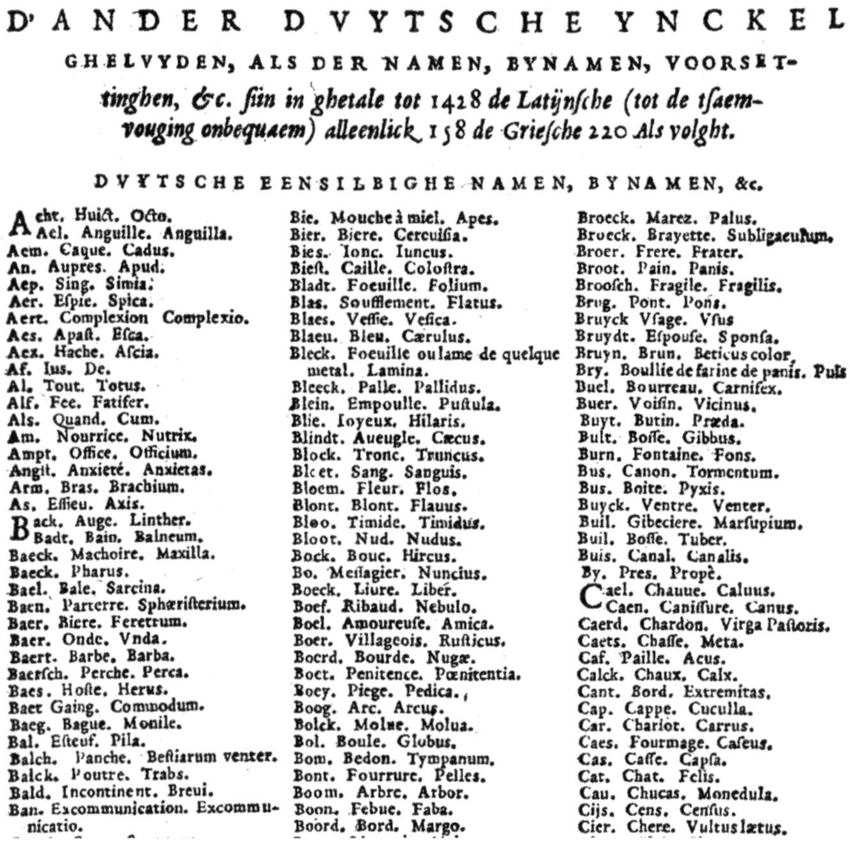

D'ANDER DVYTSCHE YNCKEL GHELVYDEN, ALS DER NAMEN, BYNAMEN, VOORSETtinghen, &c. ſiin in ghetale tot 1428 de Latijnſche (tot de tſaemvouging onbequaem) alleenlick 158 de Grieſche 220 Als volght.

DVYTSCHE EENSILBIGHE NAMEN, BYNAMEN, &c.

Acht. Huict. Octo.
Ael. Anguille. Anguilla.
Aem. Caque. Cadus.
An. Aupres. Apud.
Aep. Sing. Simia.
Aer. Eſpie. Spica.
Aert. Complexion Complexio.
Aes. Apaſt. Eſca.
Aex. Hache. Aſcia.
Af. Ius. De.
Al. Tout. Totus.
Alf. Fee. Fatifer.
Als. Quand. Cum.
Am. Nourrice. Nutrix.
Ampt. Office. Officium.
Angit. Anxieté. Anxietas.
Arm. Bras. Brachium.
As. Eſſieu. Axis.
Back. Auge. Linther.
Badt. Bain. Balneum.
Baeck. Machoire. Maxilla.
Baeck. Pharus.
Bael. Bale. Sarcina.
Baen. Parterre. Sphæriſterium.
Baer. Biere. Feretrum.
Baer. Onde. Vnda.
Baert. Barbe. Barba.
Baerſch. Perche. Perca.
Baes. Hoſte. Herus.
Baet Gaing. Commodum.
Baeg. Bague. Monile.
Bal. Eſteuf. Pila.
Balch. Panche. Beſtiarum venter.
Balck. Poutre. Trabs.
Bald. Incontinent. Breui.
Ban. Excommunication. Excommunicatio.
Bie. Mouche à miel. Apes.
Bier. Biere. Cercuiſia.
Bies. Ionc. Iuncus.
Bieſt. Caille. Coloſtra.
Bladt. Foeuille. Folium.
Blas. Soufflement. Flatus.
Blaes. Veſſie. Veſica.
Blaeu. Bleu. Cærulus.
Bleck. Foeuille ou lame de quelque metal. Lamina.
Bleeck. Palle. Pallidus.
Blein. Empoulle. Puſtula.
Blie. Ioyeux. Hilaris.
Blindt. Aueugle. Cæcus.
Block. Tronc. Truncus.
Bloet. Sang. Sanguis.
Bloem. Fleur. Flos.
Blont. Blont. Flauus.
Bloo. Timide. Timidus.
Bloot. Nud. Nudus.
Bock. Bouc. Hircus.
Bo. Meſſagier. Nuncius.
Boeck. Liure. Liber.
Boef. Ribaud. Nebulo.
Boel. Amoureuſe. Amica.
Boer. Villageois. Ruſticus.
Boerd. Bourde. Nugæ.
Boet. Penitence. Pœnitentia.
Boey. Piege. Pedica.
Boog. Arc. Arcus.
Bolck. Molue. Molua.
Bol. Boule. Globus.
Bom. Bedon. Tympanum.
Bont. Fourrure. Pelles.
Boom. Arbre. Arbor.
Boon. Febue. Faba.
Boord. Bord. Margo.
Broeck. Marez. Palus.
Broeck. Brayette. Subligaculum.
Broer. Frere. Frater.
Broot. Pain. Panis.
Brooſch. Fragile. Fragilis.
Brug. Pont. Pons.
Bruyck Vſage. Vſus
Bruydt. Eſpouſe. Sponſa.
Bruyn. Brun. Beticus color.
Bry. Boullie de farine de panis. Puls
Buel. Bourreau. Carnifex.
Buer. Voiſin. Vicinus.
Buyt. Butin. Præda.
Bult. Boſſe. Gibbus.
Burn. Fontaine. Fons.
Bus. Canon. Tormentum.
Bus. Boite. Pyxis.
Buyck. Ventre. Venter.
Buil. Gibeciere. Marſupium.
Buil. Boſſe. Tuber.
Buis. Canal. Canalis.
By. Pres. Propè.
Cael. Chauue. Caluus.
Caen. Caniſſure. Canus.
Caerd. Chardon. Virga Paſtoris.
Caets. Chaſſe. Meta.
Caf. Paille. Acus.
Calck. Chaux. Calx.
Cant. Bord. Extremitas.
Cap. Cappe. Cuculla.
Car. Chariot. Carrus.
Caes. Fourmage. Caſeus.
Cas. Caſſe. Capſa.
Cat. Chat. Felis.
Cau. Chucas. Monedula.
Cijs. Cens. Cenſus.
Cier. Chere. Vultus lætus.

FIGURE 5.3 *Simon Stevin, detail of a page with monosyllables in "Uytspraeck van de Weerdicheyt der Duytsche Tael", in Simon Stevin,* De Beghinselen der Weeghconst (*Leiden, Christoffel Plantijn by François van Raphelingen: 1586*), 4^{0} *Print* [*Not numbered; 17*]. *Stevin counts 1728 monosyllabic nouns and adjectives in Dutch compared to 158 in Latin and 220 in Greek.*

followed by notes that reveal that Constantijn had a work in mind that was approximately double its length.[31] From the notes with content that Constantijn intended to work out in *Domus*, it becomes clear that he had access to manuscripts of Stevin's unfinished work on architecture, *De Huysbou*:

> Locus. Deliberatum. Qua versum domus dirigenda esset. Ex Vitru<vio> De regionib<us> venti, et Principis autoritate et nostro Arbitrio orientem placuisse obliqua ad viam correcta.

31 The Hague, Koninklijke Bibliotheek, KA XLVIII, fols. 733r–752v.

> Cataracta replenda. Ratio fundamenti, nec sine vitio tamen. quod mirum. *Stevin van 't Wellsand.*] Eodem anno equile extructum, *gelycksydigh*, sed non accuratis membris, ut postea indicabitur.[32]
>
> The location. The question to which wind direction the house should be orientated. Vitruvius about the wind directions. The choice for the east with the consent of the Prince and following our own judgment. Correction of the bend in the road.
>
> The well that had to be filled up. The manner of making fundaments. That this was not without errors. What surprised us. Stevin on *Wellsand.* [my cursive] The building of the coach house in the same year, *gelycksydigh*, [my cursive] but with its part not exactly measured, as will be demonstrated.

The two quotations put in cursive by me in the Latin text, are the only two in the whole manuscript that are in Dutch. The first deals with a typical Dutch problem, the lay out of the foundations of the house on marshy, soggy grounds, where water fills up the spaces between the sand, 'well sand'; the second one, '*gelyckseydigh*', deals with the symmetrical composition of its ground plan.

It is possible that Constantijn could not immediately find a good translation in Latin for the rather specific phenomenon of 'well sand'. However the inclusion of the Dutch word '*gelyckseydigh*' probably was deliberate. It would have been logical to use the term '*symmetria*' but Stevin had explained in his work on architecture that he did not want to use this commonly used term in the Vitruvian tradition since in his view it did not correspond to the order in nature. Whereas Vitruvius interpreted *symmetria* as the harmonious ratio of the parts to the whole, Stevin preferred the use of neologism '*lycksijdcheyt*', literally like-sideness, which corresponds with our modern concept of symmetry: mirror symmetry.

Stevin argued that if the term *lycksijdicheyt* had been used in antiquity instead of explaining *symmetria* in terms of '*saemmaticheyt*' [proportionality], such a misunderstanding in architecture would never have arisen, noting that long before Vitruvius in the Age of the Sages the architects of antiquity had declared that:

32 The Hague, Koninklijke Bibliotheek, KA XLVIII, fol. 748r. An etching of Pieter Post (1639) shows the mirror symmetry in the ground plan of the Huygens House, The Hague, Gemeentearchief, Collectie Prenten, gr. B 1329. Italics mine.

> [...] 't Gesticht moet wesen als een dier, en wanneer ment wil maken na 't behooren, men moet daer in de natuer volghen. Waer by te verstaen is, dat gelyck de Natuer of Schepper der gedierten, de selve Lycksydicheyt geeft, also moet oock de Boumeester dat navolgen, en de gestichten met Lycksydicheyt veroirdenen.[33]

> [...] a building should be like an animal, and if one wishes to make it correctly, one should follow nature. By which it should be understood that just as nature or the Creator of animals produces like-sidedness, so the architect should emulate this and design buildings with like-sidedness.

Mirror symmetry in Stevin's view had its origins in nature and should, therefore, serve as the basis for logical architecture, unlike the proportional symmetry of Vitruvius and his followers. Since there is no proportional relationship between the parts of human and animal bodies there could be no question, in this instance, of a consistent natural order of the parts to the whole serving as a basis for architecture. The latter also explains the words directly following '*gelyckseydigh*' in *Domus*: 'sed non accuratis membris, ut postea indicabitur'.

In short, Constantijn Huygens, similar to Stevin, seems to have preferred to avoid the term *symmetria* in the Vitruvian proportional sense, since he had designed the ground plan of his house (similar to coach house mentioned in the quote here above) according to the rules of mirror symmetry [Fig. 5.4]. As such the untranslated neologism in Dutch was intentional because it expressed a specific meaning that could not be covered by the contemporary commonly used classical term.

Translation and Transformation

Whereas in the previous case a Dutch word intended for treatise on architecture seemed untranslatable after insertion into a Latin poem about a house without losing its semantic meaning, in the following case we will focus on the interaction between the translation of text in specific documents and the transformation of content into new documents. Translation must be read here in a very strict sense as the communication of the specific semantic meaning of a text in one language into another, the so-called target language. Transformation here implies the translation of the representation of content in one specific format

33 Heuvel, '*De Huysbou*' I, 1, 209. Stevin H., "I Onderscheyt van de Oirdeningh der Steden", in *Materiae Politicae* 13.

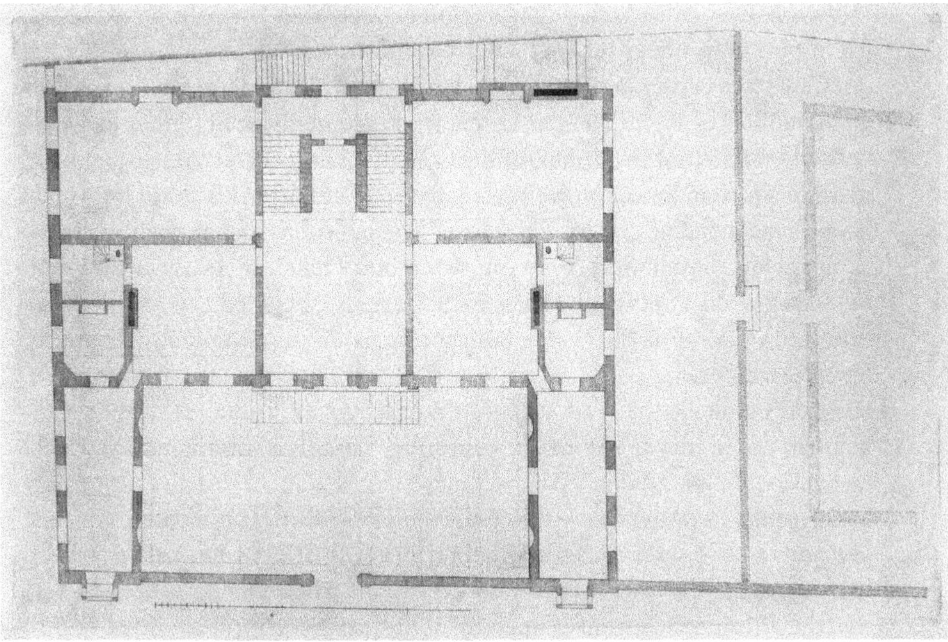

FIGURE 5.4 *Theodor Matham after Pieter Post, "House of Constantijn Huygens in year 1637" (1639). Etching, 51.3 × 57 cm. The Hague, Gemeentearchief, Collectie Prenten, gr. B 1329.*

or type of document (in the widest sense of the word) into another format or type of document. During this translation into another format or type of document, the semantic meaning does not necessarily change. The claim is put forward that texts and images with more or less the same content that are re-used in various contexts can have different authorative meanings.

In 1639, the same year in which Constantijn Huygens scribbled his notes for *Domus*, he received a letter of the renowned French Minorite scholar Marin Mersenne. In this letter Mersenne asked for information about the use of wells and pumps in Holland. Huygens replied on 1 April 1640 with some pride that a drilling method was already in use in his country that could draw up good water from deep under the ground, even in cities such as Amsterdam and Leiden where it was necessary to drill down to enormous depths. In this letter Huygens referred to some of Stevin's unpublished texts that he possessed in which the latter referred to these drills as the finest invention since antiquity.[34]

34 Heuvel, *'De Huysbou'* II, 1, 231. Huygens Constantijn, *Adversaria*, The Hague, Koninklijke Bibliotheek, Ms. KA XVLII fols. 146r–148r; and London, British Library, Add, Ms. 16912, fols.

Mersenne was so interested that he immediately sent Huygens an extensive list of questions about the method. Huygens dispatched an envoy to put the questions to a certain Job Adams, 'well driller' of Amsterdam. Job Adams was great-nephew of Pieter Pietersz Enten, who together with his father had successfully carried out the drilling at the Oudemannenhuis (an old-age home for men) in Amsterdam thirty-five years previously in 1605, with a drill for which he received a patent in 1602 [Fig. 5.5a].[35] The questions and answers, with accompanying commentary, were translated into French by Huygens and sent to Mersenne on 3 June 1640. Huygens's answer also included new and detailed drawings of the drill [Fig. 5.5b]. Both the questions and the drawings in this letter correspond in large measure to the descriptions and illustrations in Huygens's manuscripts. The questions themselves are definitely in Huygens's handwriting from which it may be concluded that this is the original list which he had sent to Job Adams.

Huygens was rather shocked by Job Adams's somewhat unscientific answers and pointed to discrepancies with Simon Stevin's work.[36] Adams, for example, had referred to the trouble caused by 'quicksand' when drilling, which showed that the artisan was obviously unaware of the use of water pressure as a means of overcoming this problem. On the other hand, the description of the drilling method in the *Adversaria* of Huygens reveals that the exceptional aspect of this invention lay precisely in the fact that the drilled hole always remained full of water. The outward pressure of this cylinder of water against the walls of the well not only prevented it from collapsing, it also stopped the drill and the well from clogging up with sand. Job Adams, as a well driller, was undoubtedly

181 ff. A copy of the document also sits in Bibliothèque National in Paris, Ms. fr. 13051. The document is published in Worp J.A., *Briefwisseling van Constantijn Huygens*, ed. J.A. Worp, 6 vols. (The Hague: 1911–1917) vol. III, 17–18 and in *Correspondance du P. Marin Mersenne*, ed. C. de Waard, 17 vols. (Paris: 1932–1988) vol. VIII (letter dated 26 August 1639) 490–495, esp. 493, and vol. IX (letter dated 1 April 1640) 257–261 and (3 June 1640) 378–384.

35 Doorman G., *Octrooien voor uitvindingen in de Nederlanden uit de 16de–18de eeuw* (The Hague: 1940) 107 (G. 68). Design for patent, National Archives of the Netherlands, 4VTHR 4795 A.

36 *Briefwisseling van Constantijn Huygens* vol. III, 38: Huygens to Mersenne: 'Tant y a que nous plaindre avec la Samaritaine, [followed by a reference to *New Testament, John* 4:12: 'Are you a greater man than our father Jacob, who gave us this well and drank from it himself with his sons and his cattle?', in Greek]. Cependant vous trouveres un peu de discrepance entre ces relations, et ce qu'en escrit Stevin, et nommement me choque, que cet artisan ne recognoist pas tant l'effect de l'eau tenue haute dans le puits comme faicts, veu les inconveniens qu'il raccompte du sable mouvant. Mon deputé—personne de bon sens—en raisonne encore autrement etc.'

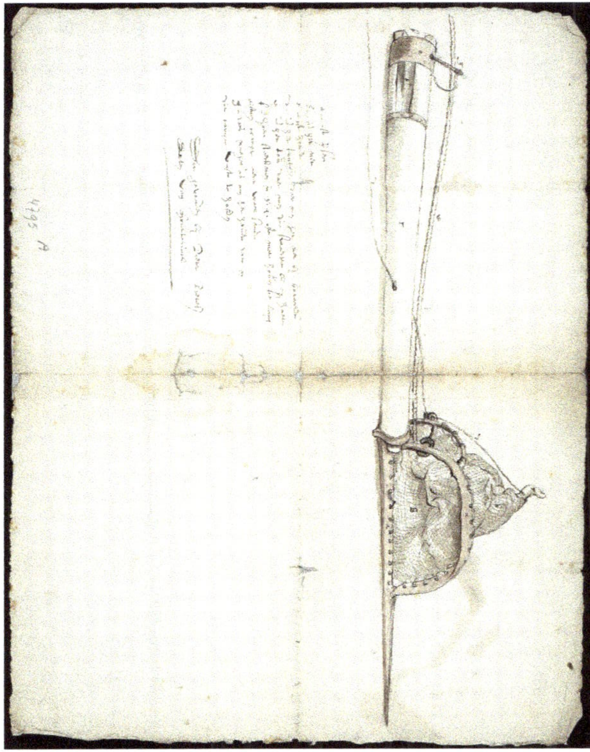

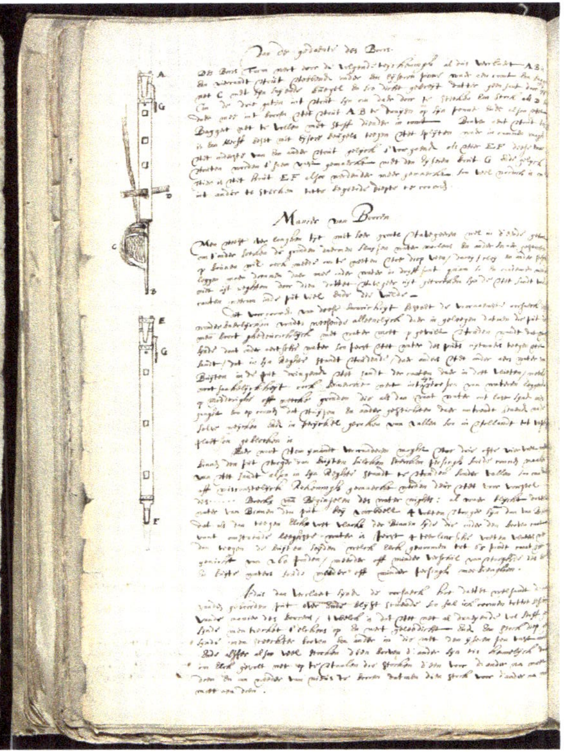

FIGURE 5.5A (*Left*) *Pieter Pietersz Enten, drawing of drill as part of in 1602 granted patent for drilling deep in cities in Holland used in Amsterdam in 1605 [1602]. Ink, 393 × 518mm, The Hague, National Archives of the Netherlands, 4VTHR – 4795A.*

FIGURE 5.5B (*Right*) *Constantijn Huygens, description and illustration of drilling method used to get fresh water in Amsterdam in 1605 based on the manuscripts of Stevin's Huysbou described in Huygens'* Adversaria *(1 April 1640). ink, 19 × 31 cm. The Hague, National Library of the Netherlands,* KA XVLII, *fol. 147v.*

acquainted with this particular drilling method and probably used the same type of drill used by his father and great-uncle thirty-five years before in Amsterdam. For that reason, Job Adams had no reason to conceal the origin of this efficient drilling method. It seems more probable that Stevin wished to propose drilling as practical proof of his previously expounded theories and therefore presented the method as favourably as possible.[37] Stevin may have wished to elaborate on the method in the discourses on the form of drills and method of drilling in Huygens's manuscripts and that for that reason the descriptions of the method of drilling used in Amsterdam in 1605 have become entangled with his own ideas. It is also possible that Stevin considered submitting an application for a patent for an improved method. Indeed his son referred to '*Van de schrijvers vont diep onder water ter booren*' [On the writer's invention for drilling deep under the water].[38] The advantages of the drilling method used to enter deep in the marshy soil were generally recognized, but apparently still required some adjustment in practice. Enten must have felt threatened by competition because in 1613, the year in which his patent was due to lapse, he produced a witness to reaffirm under oath that in 1602 he had reached a depth of 100 feet with his drill.[39] And although there is no evidence in the registry of applications of an extension of this patent being granted (nor rejected) Enten was still drilling wells in Amsterdam in the year 1622.[40] If Stevin did apply for a patent on the basis of the drawing of this instrument, it was probably not granted because it was too similar to the patented design. Nonetheless, Stevin's description became well known both nationally and internationally. Mersenne included it in his *Cogitata, physio mathematica* (Paris 1644), and on 10 June 1663, some sixty years after the drilling in Amsterdam, Constantijn and Christiaan Huygens presented the subject to the Royal Society in London and the text popped up in the English language in the history of this institution by Thomas Birch.[41]

---

37 Stevin had previously expounded this principle in *De Beghinselen des Waterwichts* (Leiden, François van Ravelinghen: 1586). A more detailed explanation appeared in "Van de weeghconst", in *Wisconstighe Ghedachtenissen.*

38 "Byvough der Stedenoirdening", in *Materiae Politicae* 127–128.

39 *Bronnen tot de geschiedenis van het bedrijfsleven en het gildewezen van Amsterdam* ed. J.G. van Dillen, 3 vols., *Rijks Geschiedkundige Publicatiën* (The Hague: 1929–1974), vol. II, 33, no. 64.

40 Dillen, *Bronnen* vol. II, 463, no. 805.

41 Mersenne Marin, *Cogitata, physio mathematica* (Paris, Antonius Bertier: 1644), "Hydraulica Pneumatica", prop. 53, 217- 221; *Oeuvres complètes de Christiaan Huygens*, ed. D. Bierens de Haan – J. Bosscha e.a., 23 vols. (The Hague: 1888–1950), vol. XXII, 270–272, note 35. Oldenburg wrote to Boyle on 10 June 1663 that Constantijn and Christiaan that afternoon

This case illustrates first that Stevin might have edited, even manipulated, the description of the drilling method in such a way that it created it him some space to bring in his own interests. From his letter to Mersenne it becomes clear that Huygens had more confidence in Stevin's more scientific interpretation at a distance, than in the direct answers of the artisan who was professionally related, and a personal relative, of the artisan directly responsible for the execution of the drilling method. The description of the patent had been translated in various formats over a period of sixty years. It was the authoritive translations rather than the factual descriptions in text and image of the experiment with the patented drill in Amsterdam that got a wider international recognition.

Translation and Simulation

Whereas in the previous case some deviation took place in the translation (in the sense of transformation) of the description of a historical event into scientific narration—possibly in order to create some space for the translator's own interests—in the next case, involving a patent, theoretical descriptions were translated into a simulation model to provide scientific proof of the claim of the patent application of someone else.

In 1624 Isaac Beeckman, who lived and worked in Middelburg in the province of Zeeland, travelled three times to the house of the widow of Stevin, in the village of Hazerswoude in Holland, to compile tables of content and to make extracts of Stevin's manuscripts. He included these in his *Journal* and added his own commentary. Not surprisingly, Beeckman was already acquainted with the work of Simon Stevin before he set out on his long journey to see whether there was more. Evidence hereof we find in his *Journal* in a description of an experiment that he witnessed in the house of a certain Daniel Nota in his home town Middelburg 1620.[42] The direct cause of this experiment was a

had visited the Royal Society and had asked for Robert Boyle. They were entertained with experiments. According to Birch Huygens in that year presented the description of the drilling method that was included in the English Language in: Birch Thomas, *The History of the Royal Society of London for improving of natural knowledge etc.* (London, A. Millar: 1756), vol. 1, 265.

42 For a fuller description of this paragraph: Heuvel C. van den, "'Tot meerder bewijs'. Een kaart en een model van Daniel Note van 1620 ter demonstratie van een nieuwe uitvinding om de haven van Middelburg zandvrij te schuren", in Hoftijzer P. – Ommen K. van – Warnar G. – Witkam J.J. (eds.), *Bronnen van kennis. Wetenschap, kunst en cultuur in de collecties van de Leidse Universiteitsbibliotheek* (Leiden: 2005) 100–108.

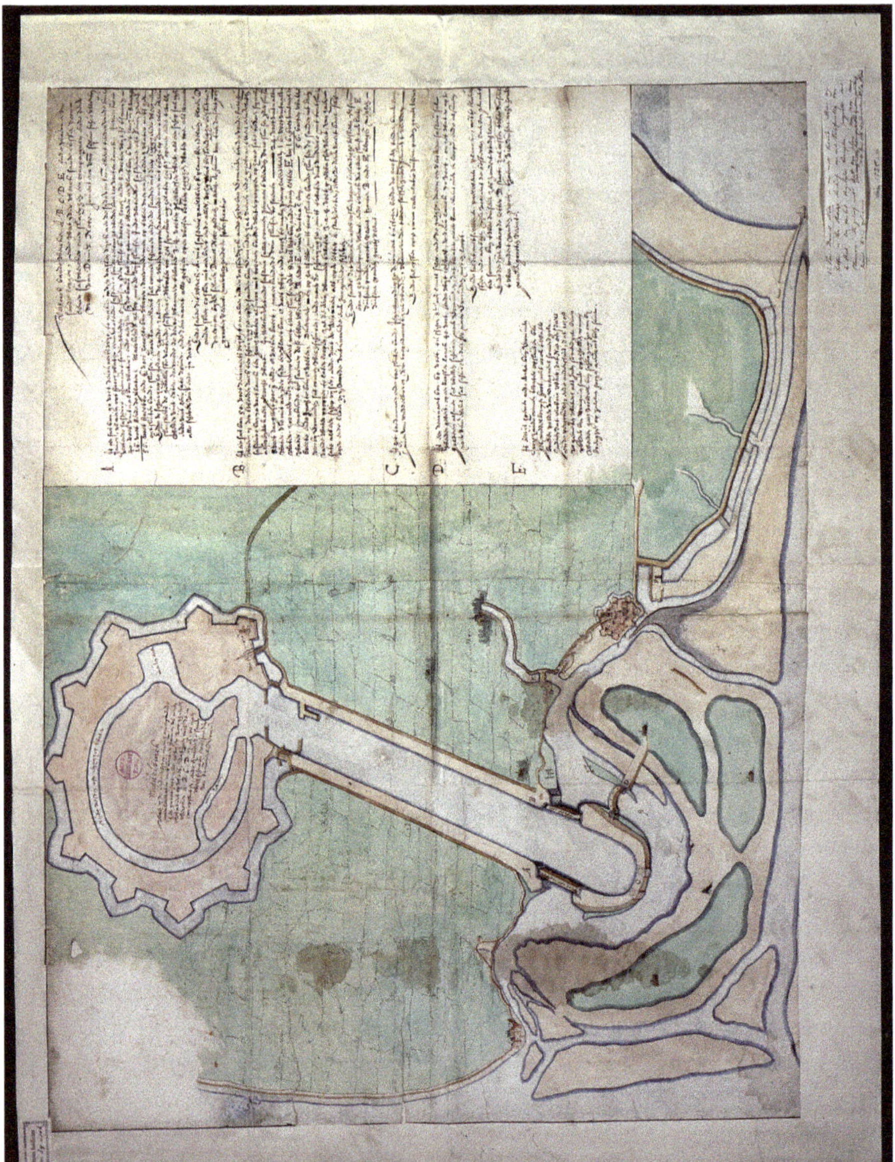

FIGURE 5.6 *Daniel Note, design for floatable sluices to scour the harbour of Middelburg (1619–1620). ink drawing, 65 × 86 cm, mounted on paper 74 × 96 cm. Leiden University Library, Collection Bodel Nijenhuis 051-28-001.*

dispute about the latter's invention for scouring the harbour of Middelburg. Nota had been asked to demonstrate to the town council the effectiveness of an invention that he had submitted a year before in 1619 with a request for the enormous sum of thirty-six thousand guilders if it were to be applied.[43] On 1 May 1620 Nota demonstrated the plan using a map [Fig. 5.6] and a model.[44] The burgomasters consulted the engineer Domenicus Melckebeke, who reported negatively on the invention. Playing safe, the burgomasters turned Nota's plan down. Nota clearly felt this to be a stain on his reputation and so he asked Isaac Beeckman if he and a number of others would care to see the invention. On 10 May 1620 Nota demonstrated it to 17 or 18 'leading enthusiasts in Middelburg'. Beeckman does not mention the map again, but describes in his journal how Nota for further proof used a '*modelleken*' [a small model] to demonstrate:

> [...] hoe dat een weynich water so sterck teghen eenen staenden bodem drangh als een groot water, dat even hooghe was; twelck hy toonde met een scheefwicht alsoock met staende bodemen van leder, twelck alles generaelick ende vastelick van Simon Stevyn in syn Waterwicht bewesen ende betoont is.

> [...] how the force of a small amount of water on a standing tank is as great as a large amount of water of equal height; which he showed with an oblique weight and standing tanks of leather, all of which is proven and shown generally and accurately by Simon Stevin in his *Waterwicht*.[45] [Fig. 5.7]

With this exposition Daniel Nota aimed to show that a floating sluice as large as 100 foot long and 20 feet high could be filled with water and sunk to the bottom, where it would remain full until ebb tide and than opened so that it would scour the whole length of the harbour for a mile in the direction of Rammekens and Arnemuiden [Fig. 5.8]. Although Beeckman believed that Melckebeeke 'die dynghen niet verstaen en heeft, gelyck ick selve ook wel weet, dat hy in de rechte nature van het waterwicht niet seer geoeffent en is' [had not understood these things, as I know personally that he is not well acquainted with

43 Nota (also known as Note) had founded a society for his invention a year before: Rijksarchief Zeeland, Register ten Rade Middelburg (1614–1622), 100–106 kz (11 March 1619); and ibidem 131 (5 May 1620).

44 Leiden University Library, Collectie Bodel Nijenhuis, 051-28-001.

45 Beeckman vol. II, 39.

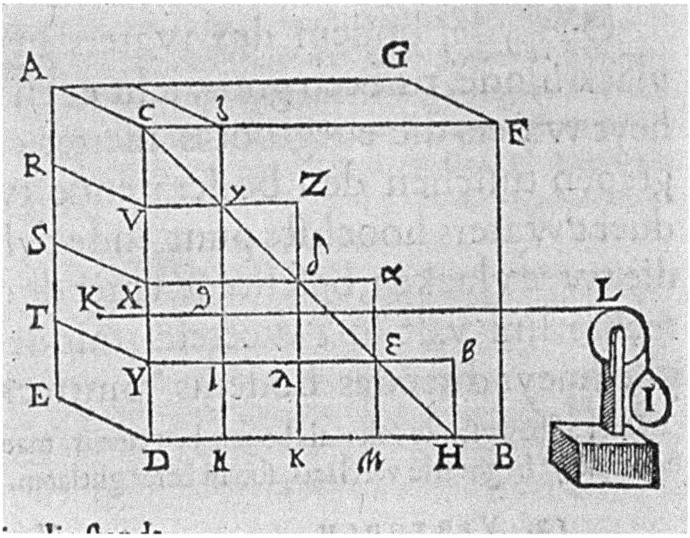

FIGURE 5.7 *Simon Stevin, model to test the force of water on an oblique face in Simon Stevin,* De beghinselen des Waterwichts *(Leiden, Christoffel Plantijn: 1586). print 4°, illustration at x Voorstel [10th Proposition]. Leiden University Library.*

the true nature of hydraulics],[46] unlike Daniel Nota, he feared that it would be difficult to move the enormous floatable sluice unless it would be on wheels or on a complete flat harbour floor. Whether he was right or not, what makes this significant is that the translation of Stevin's theories on hydraulics described and visualized in his *Beghinselen des Waterwichts* were translated into an experimental simulation for demonstration in support of Nota's claim to have found a practical solution for solving the problem of the silting up of the harbour of Middelburg.

The silting up of harbours was a common problem in the Low Countries and a direct menace to the Dutch economy that heavily depended on overseas trade. The silting up of the access to Ghent and Stevin's native town Bruges had spelled the end of prosperity of these most wealthy cities in the Middle Ages in favour of Antwerp and later Amsterdam and other seaports in Holland and Zeeland. We do not know whether Stevin for that reason had a specific interest in the topic, but he addressed the problem in a study with the title *Waterschuyring* [Waterscouring] that was published posthumously by his son Hendrik in his *Wisconstigh Filosofisch Bedryf* [Mathematical Philosophical Deed] of 1667. On the basis of characteristics of the harbours of Bruges,

46 Beeckman vol. II, p. 39.

FIGURE 5.8 *Daniel Note, design for floating sluices to scour the harbour of Middelburg (1619–1620). Manuscript map, 65 × 86 cm, mounted on paper 74 × 96 cm. Leiden University Library, Collection Bodel Nijenhuis 051-28-001. Detail, left below.*

Middelburg, Calais and Ostend, Stevin developed a typology of scouring solutions with sluices that could be used against the silting up of harbour cities with similar features. This typology was translated into schematic diagrams, of which the second solution (form 18[sic], 13), clearly shows the features of the design of Daniel Nota that was never executed [Fig. 5.9].

Translation and Commodification

Whereas in the case of Beeckman and Huygens, the manuscripts of Simon Stevin played in an instrumental role in their own studies and were not intended for publication, Hendrik Stevin really set out to publish the intellectual legacy of his father. Despite his claim to do so 'as the author intended', he did not follow his father's manuscripts to the letter. When Hendrik Stevin decided to publish some of his father's unfinished manuscripts originally

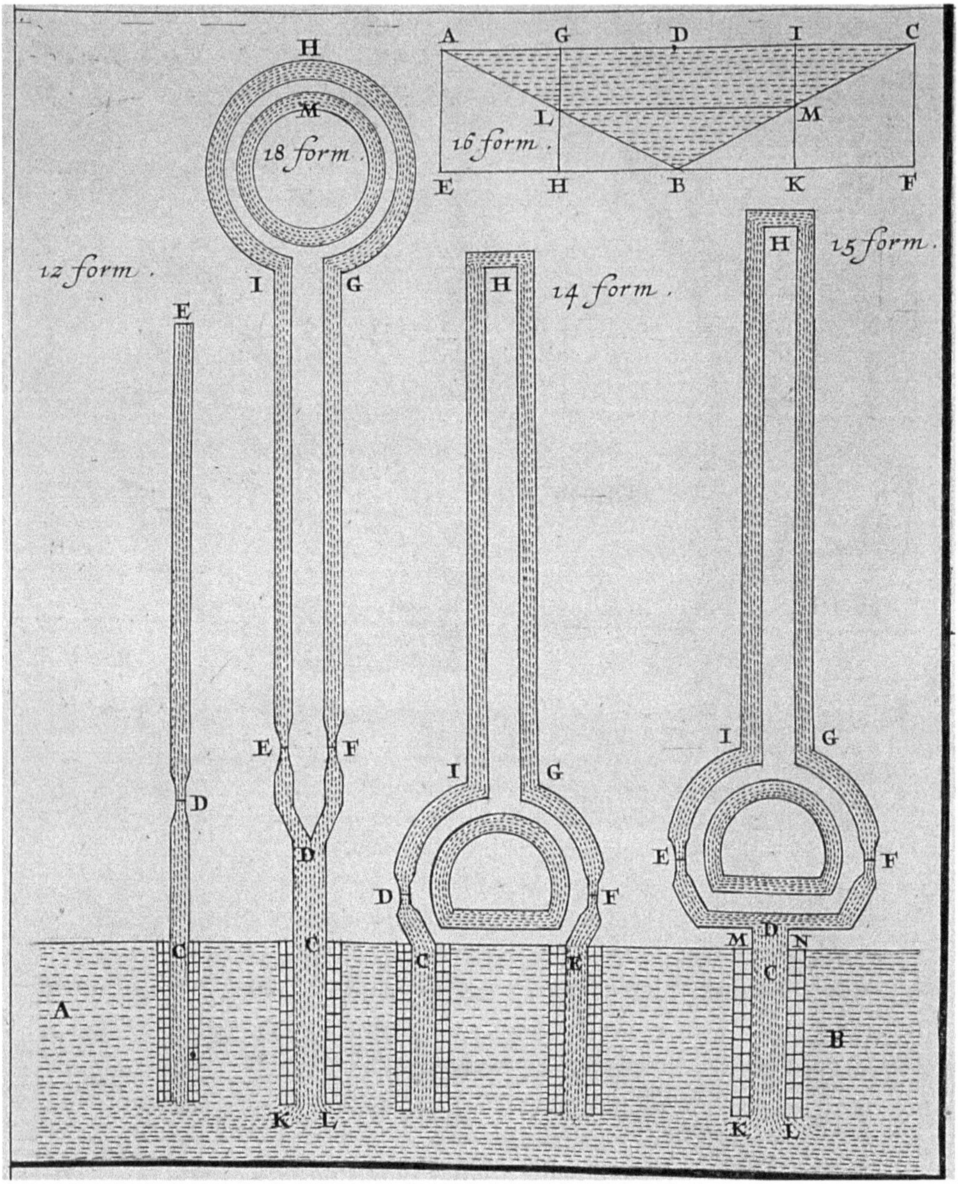

FIGURE 5.9 *Simon Stevin, designs for scouring of harbours illustration in Hendrik Stevin,* Plaetboec *(Leiden, Philip de Croy: 1688) print 2°. plate 22. Leiden University Library. The type used at Middelburg shown second from the left.*

meant for *Wisconstighe Ghedachtenissen*, he did not envisage an enlarged reissue of that work but rather a selection in his *Materiae Politicae: Burgherlicke Stoffen* [Political Matters: Civic Matters].[47] In a preface he set out his reasons for this choice:

> Ghemerckt het voornemen niet en was de eerstgedruckte Wisconstighe ghedachtenissen, welcke na des Schrijvers eyntlicke meyning gheschickt sijn, te herdrucken [...] soo heb ick my an d'oirden die sommighe der stucken of deelen van die in't Cortbegrijp des vijfden Stucks derselve eerstgedruckten ghegeven was, niet ghenoodsaeckt gevonden te verbinden, maer die na 't ghene my die de omstandighen en eyghen beweghinghen schenen te vereysschen, geschickt.[48]

> Given that it was not the intention to reprint the original edition of *Wisconstighe Ghedachtenissen* which is arranged according to the writer's own wishes, [...] I have not considered myself to be bound to the order of some of the pieces or parts as described in the summary of the fifth section of the aforementioned first edition, but have arranged these as I deemed necessary in the circumstances and according to their internal logic.

It goes beyond the scope of this paper to indicate all places in his publication where Hendrik deviated from his father's manuscripts, but focus on one case that might explain Hendrik Stevin's intentions best.

Before the publication of the *Materiae Politicae* in 1649, Hendrik made an excerpt of his father's text on town planning that was intended as an integral part of the original treatise of Simon Stevin on architecture, *De Huysbou*. However, from the introduction to the hand-written copies of his father's manuscripts that he had prepared for publication, it becomes clear that Hendrik intended to publish these texts on town planning in a completely different context, as part of Stevin's unfinished works on the military arts.[49] Therefore the order of the content table and of the text are quite different from the printed version of this part in the *Materiae Politicae*. Hendrik explains these differences in various

47 Stevin H., *Materiae Politicae Burgherlicke Stoffen Vervanghende Ghedachtenissen der Oeffeninghen des Doorluchtichsten Prince Maurits van Orangie* (Leiden, Adryaen Rosenboom: 1649).

48 Stevin H., "An den leser" [To the reader], in *Materiae Politicae* [not numbered].

49 The Hague, Koninklijke Bibliotheek, Ms. 128 A 9 – II: *Eenighe Stucken der Crychconst. Beschreven deur Simon Stevin*, fol. 1v.

places for the sake of clarity of the argument. However, these arguments do not explain differences in formal aspects in text and image as well. The latter difference becomes especially clear in Hendrik's translation of his father's ideal town plan from the manuscript to the printed form. At first sight the illustrations in *Materiae Politicae* appear to have been the same as the drawings copied from the original manuscript; yet there are some differences. For example, the town plan in the manuscript copy in the National Library in The Hague does not show the bridges depicted in the printed illustration in the *Materiae Politicae*.[50] Moreover, Hendrik Stevin gave the names of the public buildings in his father's town plan in full, whereas these only appeared as letters in the transcript [Figs 5.10a and 5.10b]. Using letters instead of full names was typical of the style of illustrations in Stevin's other publications. In his discourse on logic of 1585, *Dialeckticke Ofte Bewysconst*, he had explicitly suggested that letters could be used to represent terms.[51] Hendrik's additions might seem trivial, but by inserting full names they became independent designs, providing complete reproductions of the entire town, instead of diagrams to be understood in combination with the text and other illustrations. His father, however, never aimed to produce beautiful, artistic impressions of architecture. He preferred depictions that were reliable and provided information:

> Waer benevens noch gebeurt dat, dat onervaren eygenaers des gebous, meer siende na 't cieraet door de constige hant gewrocht, dan na de beste oirden het quatste verkiesen; Indervougen dat slechten gronttreck en stantreck in sulcken ansien bequamer is[52]
>
> Indeed, inexperienced owners of buildings may be more interested in artistically-fashioned decoration than in the best layout, making the worst choices. In that sense, simple elevations and ground plans may be regarded as more effective.

Stevin's original drawings should not be seen as independent objects with which to reproduce the imaginary ideal of a house or town. Instead Stevin's representation of architecture consisted of a series of diagrams to accompany texts, designed to illustrate the ideas and logical order for educational and didactic purposes. However, it seems that Hendrik had a different motivation which becomes clear in his note to the reader in his *Materiae Politicae*:

50 *Eenighe Stucken der Crychconst. Beschreve deur Simon Stevin*, fol. 6r.

51 Stevin S., *Dialecticke ofte Bewijsconst* (Leiden, Christoffel Plantijn: 1585), Definition XXXVI.

52 Heuvel *'De Huysbou'* 270; Stevin, "Byvough der Stedenoirdening", in *Materiae Politicae* 103–104.

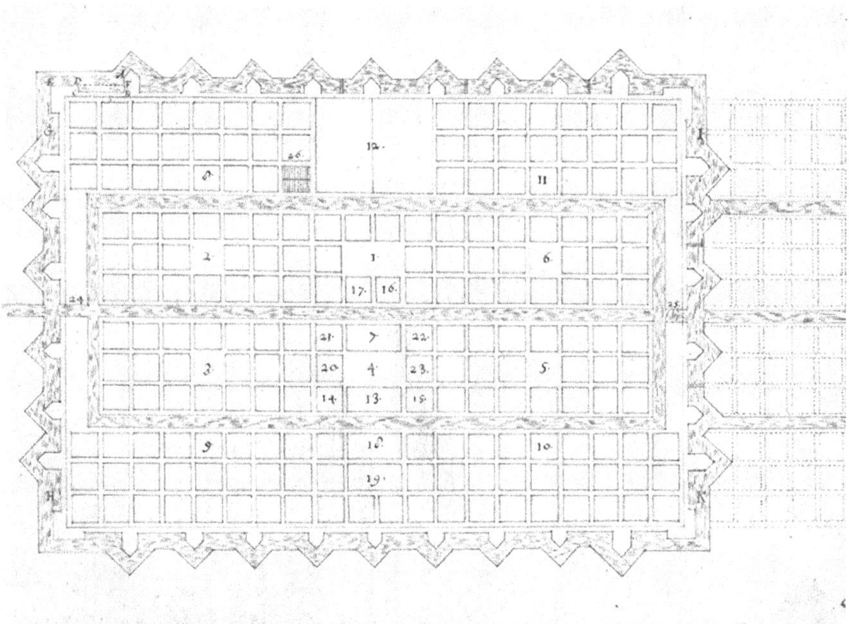

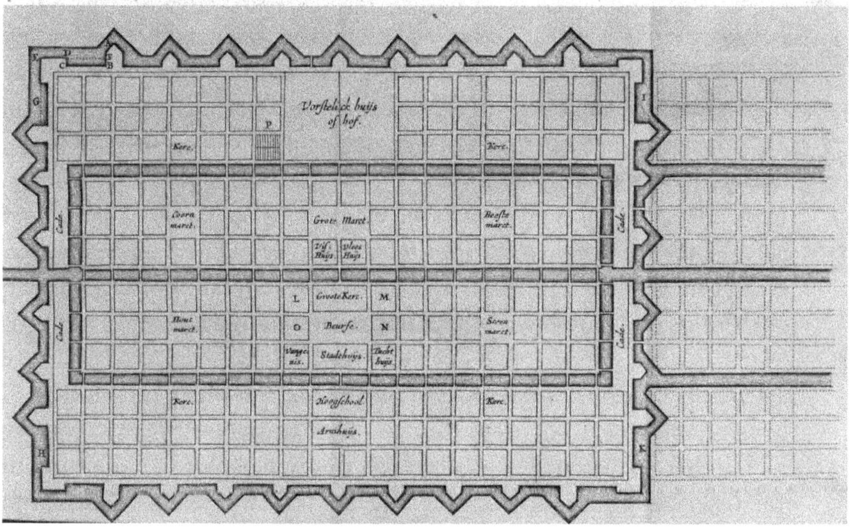

FIGURE 5.10A (*Top*) *Hendrik Stevin, ideal town plan according to an original design by Simon Stevin (no date). Drawing in ink, 21 × 33cm, in Hendrik Stevin,* Eenighe Stucken der Crychconst. Beschreve deur Simon Stevin, *Manuscript, The Hague, K.B. National Library of the Netherlands, 128 A 9 – 11, fol. 6r.*

FIGURE 5.10B (*Bottom*) *Hendrik Stevin, ideal town plan according to an original design by Simon Stevin, illustration in Hendrik Stevin,* Materiae Politicae (*Leiden, Adryaen Rosenboom: 1649*) *print 8º, "Byvough der Stedenoirdening". Leiden University Library.*

> Ghemerckt Geleerden haer tijt liever en dickwils nutter besteden met haer eyghen vonden en leeringhen te verclaren, dan 't hooft te breken in 't optoyen der Schetsen van anderen, en dat onder de ongheleerde ick eene was, soo scheent billich my dien last te aenveerden.[53]
>
> Given the fact that Scholars prefer to spend their time often more useful by explaining their own inventions and what they have learned, than with racking their brains over adorning the drafts of others and because I was one amongst the unlettered, it seemed fair to me that I would accept that burden.

This different motivation of Hendrik also explains changes to the original visualizations of his father. Whereas Simon Stevin had intended to include the text 'Vande ordeningh der steden' [On the order of cities] with his illustrations in his treatise on architecture as part of the fifth part of the *Wisconstighe Ghedachtenissen*, Hendrik brought the part on town planning to the foreground and published only a small part of the *De Huysbou* as an appendix:

> Int voorgaende is meest oirdening beschreven, waer van het beleyt den Overheden toebehoort. Maer anghemerckt het groot gebruyck, dat de huysoirdening onder de Burgerijen by elck een heeft, so sal ick daer af, als anhangsel der Steden, desen Byvoug formen; Doch, gelijck ick segge, alleen int ansien van oirdening, en volgens het alghemeen voornemen, alleenlick soo veel vant selve, als yder gesont oirdeel (al en isser geen kennis noch genegentheydt totte Wisconsten by) dat lesende, verstaen can: Waer af het dadelick maecksel en ander omstandigen inden eyghentlicken * Huysbou (*Architectura) beschreven sijn.[54]
>
> The above contains a description of most of the planning relevant to government policy. But given the extensive use made of house ordering by burghers individually I shall attach this appendix as an addition to 'Steden', although as I say, only as much of it as any healthy mind (even with no knowledge or love of mathematics) is able to read and understand, while the actual work of construction and other circumstances are fully described in De Huysbou.

---

53 Stevin H., "An den leser", in *Materiae Politicae* [not numbered].

54 Heuvel, *'De Huysbou'* 493; Stevin H., "Byvough" [Appendix], in *Materiae Politicae* 40.

This selection and change in the order of chapters of *De Huysbou* seems to have been motivated by the dedication of the part on town planning to the burgomasters, and aldermen and magistrates of Amsterdam. In this dedication we read Hendrik Stevin was inspired in this decision by the resemblance between the text of his father and the order of the city of Amsterdam that it seemed that is was based on its principles:

> Ja dat soo ymandt den last had, de grontreghelen van U Ed: Grootbaarheden Stadtoirdeningh by gheschrift te verclaren, hy voldoen soude connen, mits dese Stedenoirdening haer anbiedende.[55]
>
> Yes, if someone had the task to explain the order of the city [i.e. of Amsterdam] of you, your noble honoured, he could suffice with the presentation of this order of the town [i.e. his father's text on town planning]

In the same dedication Hendrik Stevin makes a direct link between "Vande ordeningh der steden" and the prosperity of the city of Amsterdam.[56]

> Vanden Hausbau Vervangende mede 't geen noodich is tot bevordering der welvaer, behoudenis en geduerige verbetering van Steden en Landen na des Authuers gevoelen.[57]
>
> Vanden Huysbou, replacing which is needed for the promotion of the wealth, defence and improvement of Cities and Countries, as the author intended.

These references to the wealth of Amsterdam and Dutch cities in general might also explain Hendrik's changes to his father's manuscript drawing in the printed version. By adding the names on the print they could stand on their own. Not only the city but also representations thereof became symbols of wealth. Whereas in the *Materiae Politicae* the illustrations still were inserted in the body of the text, Hendrik in his later publication of his father's manuscripts would add a separate: *Plaetbouc, vervangende de figuren of formen*

55 Heuvel, *'De Huysbou'* 491; Stevin H., "1 Onderscheyt vande Oirdeningh der Steden", in *Materiae Politicae* 4.

56 Stevin H., "1 Onderscheyt vande Oirdeningh der Steden", in *Materia Politicae* 4: 'The famous, very large and rich city of Amsterdam can serve us an example, where through skillful government the people is maintained in such a praiseworthy, laudable state.'

57 Stevin H., "Tytels en Cortbegrypen", in *Materiae Politicae* 144.

gehorig tottet Wisconstich Filosofisch Bedryf [Book of plates, replacing the figures or illustrations of Mathematical Philosophical Activity]. In short, Hendrik increasingly translated his father's intellectual property in such a way that it became a commodity.

Epilogue

Only a few works of Stevin, especially his earlier ones, appeared immediately in Latin and French. Some of his publications were translated later into other languages by others, but the greater part of Stevin's work appeared in Dutch. He was praised for it by seventeenth century Dutch historians of language, and even mentioned in one breath with famous poets, such as Joost van de Vondel, Hugo Grotius and P.C. Hooft. In fact, Stevin could be better associated with the Flemish advocate of the Dutch vernacular, Becanus, who used the grammatical argument of the high number of monosyllables before Stevin to stress the superiority of the Dutch language. Stevin's claim that Dutch was the most suitable language to express scientific ideas was rather unique in that sense that Latin resisted the raise of the vernacular, especially in works on the natural sciences. To some extent Stevin's texts were untranslatable, not only for their many neologisms, as we noticed in the example of the word *lycksijdcheyt* [like-sideness], but also for his views on the intrinsic qualities of the Dutch. Even in the case that a translation would cover the semantic meaning of his texts, it would be in most cases less concise. However, Stevin's fixation on the Dutch language also implied that he was less known abroad than many other contemporary scholars of the Low Countries. It was the transformations of his manuscripts and some translations of his printed work that gave Stevin some recognition at an international level, even if they were not always as the author had intended them.

Selective Bibliography

Archival Sources

Beeckman Isaac, *Loci communes*[*etc.*], Middelburg, Zeeuwse Bibliotheek, manuscript HS 6471.

Enten Pieter Pietersz., Drawing of drill as part of granted patent for drilling deep in cities in Holland used in Amsterdam in 1605, The Hague, National Archives of the Netherlands 4VTHR – 4795A.

Huygens Constantijn, *Adversaria*, The Hague, National Library, KA XVLII and XVLIII.

Nota Daniel, Design to scour the harbor of Middelburg, Leiden University Library, Collectie Bodel Nijenhuis 051-28-001.

Post Pieter, The Hague, Gemeentearchief, Collectie Prenten, gr. B 1329.

Stevin Hendrik, *Eenighe Stucken der Crychconst. Beschreve deur Simon Stevin*, The Hague, National Library in the Netherlands 128 A 9 –11.

Stevin Simon, Symon Stephen of Bridge, *The Building of Fortes* (1604) Cambridge, Trinity College, Ms. R 16.30.

Literature Before 1800 and Later Editions/ Reprints

Beeckman Isaac, *Journal tenu par Isaac Beeckman de 1604 à 1634 publié avec une introduction et des notes par Cornelis de Waard*, 4 vols. (The Hague: 1939–1953).

Birch Thomas, *The History of the Royal Society of London for improving of natural knowledge*, 4 vols. (London, A. Millar: 1756–1757).

Emmius Ubbo, *Briefwechsel des Ubbo Emmius*, eds. H. Brugmans – F. Wachters, 2 vols. (The Hague: 1911–1923).

Grotius Hugo, *Poemata Collecta et magnam partem nunc primum edita a fratre Guilielmo Grotio*, 12 books, Book 11, Sylvarum liber secundus: Mathematica Principis Mauritii (Leiden, Andreas Clouqius: 1617).

Grotius Hugo, *Batavi, Parallelon rerumpublicarum liber tertius*, ed. J. Meerman (Haarlem: 1801–1803).

Grotius Hugo, "Mathematica Principis Mauritii", in *De dichtwerken van Hugo Grotius, oorspronkelijke poëzie*, deel I 2a/b, pars 3, ed. A. Eyffinger (Assen: 1988) 118–148.

Huygens Christiaan, *Oeuvres complètes de Christiaan Huygens*, ed. D. Bierens de Haan – J. Bosscha e.a., 23 vols. (The Hague: 1888–1950).

Huygens Constantijn, "Briefwisseling van Constantijn Huygens", *Rijksgeschiedkundige Publicatiën*, ed. J. Worp, 6 vols. (The Hague: 1911–1917).

Huygens Constantijn, *Domus: het huis van Constantijn Huygens in Den Haag*, ed. F.R.E. Blom – H.G. Bruin – K. Ottenheym (Zutphen: 1999).

Mersenne Marin, *Cogitata physio mathematica* (Paris, Antonius Bertier: 1644).

Mersenne Marin, *Correspondance du P. Marin Mersenne Religieux Minime*, ed. C. de Waard, 17 vols. (Paris: 1932–1988).

Stevin Hendrik, *Materiae Politicae Burgherlicke Stoffen Vervanghende Ghedachtenissen der Oeffeninghen des Doorluchtichsten Prince Maurits van Orangie, &c. Gouverneur, Opperste Velt-Heer, Admirael-Generael vande Verenichde Nederlanden, &c. Beschreven deur Simon Stevin van Brugge, Superintendent van zijne Finance, Quartiermeester Generael van 't Leger,&c. En uyt sijn naegelate Hantschriften by een gestelt door Sijn Soon Hendrick Stevin, Heere van Alphen, van Schrevelsrecht,&c. Gedruckt tot Leyden* (Leiden, Adryanen Rosenboom: 1649).

Stevin Hendrik, *Wisconstich Filosofisch Bedryf* (Leiden, Philip de Croy: 1667).

Stevin Hendrik, *Plaetbouc, vervangende de figuren of formen gehorig tottet Wisconstich Filosofisch Bedryf* (Leiden, Philip de Croy: 1668).
Stevin Simon, *Dialektike ofte Bewysconst* (Leiden, Christoffel Plantijn: 1585).
Stevin Simon, *De Beghinselen der Weeghconst* (Leiden, Christoffel Plantijn by François van Raphelingen: 1586).
Stevin Simon, *De Beghinselen des Waterwichts beschreven duer Simon Stevin van Brugghe* (Leiden, Christoffel Plantijn by François van Raphelingen: 1586).
Stevin Simon, *De Sterctenbouwing, beschreven door Simon Stevin van Brugge* (Leiden, Françoys van Ravelenghien: 1594).
Stevin Simon, *Wisconstighe Ghedachtenissen, inhoudende t'ghene daer hem in gheoeffent heeft den Doorluchtichsten Hooghgheboren Vorst ende Heeren, Maurits, Prince van Oraengien* (Leiden, Jan Bouwensz.: 1605–1608).
Stevin Simon, *Hypomnemata Mathematica, hoc est eruditus ille pulvis, in quo se exercuit* [...] *Mauritius Princeps Auraicus* (Leiden, Ioannes Patius [Jan Paedts Jacobsz.]: 1608).
Stevin Simon, *Memoires Mathematiques, contenant ce en quoy s'est excercé*[..]*Maurice, Prince d' Orange* (Leiden, Jan Paedts Jacobsz.: 1608).
Stevin Simon, *The Principal Works of Simon Stevin*, ed. E. Crone – E.J. Dijksterhuis – R.J. Forbes – M.G.J. Minnaert – A. Pannekoek, 5 vols. (Amsterdam: 1955–1966).

Scholarly Literature

Blair A., "La persistance du latin comme langue de science `a la fin de la Renaissance", in Chartier R. – Corsi P. (eds.), *Sciences et langues en Europe* (Luxembourg: 1996) 33–39.
Blom F.R.E. – Bruin H.G. – Ottenheym K., *Domus: het huis van Constantijn Huygens in Den Haag* (Zutphen: 1999).
Burke P., *Languages and Communities in Early Modern Europe* (Cambridge: 2004).
Cohen H.F., "Simon Stevin's equal division of the octave", *Annals of Science* 44, 4 (1987) 471–488.
Deneire T., "Neo-Latin and the Vernacular: Methodological Issues", in Ford P. – Bloemendal J. – Fantazzi C. (eds.), *Brill's encyclopaedia of the Neo-Latin world* (Leiden – Boston: 2014) 275–285.
Devreese J.T. – Vandenberghe G., *The Wonderful World of Simon Stevin: Magic is No Magic* (Southampton: 2007).
Dijksterhuis E.J., *Simon Stevin* (The Hague: 1943).
Dillen J.G. van (ed.), "Bronnen tot de geschiedenis van het bedrijfsleven en het gildewezen van Amsterdam", in *Rijks Geschiedkundige Publicatiën*, 3 vols. ('s-Gravenhage: 1929–1974).

Doorman G., *Octrooien voor uitvindingen in de Nederlanden uit de 16de–18de eeuw* (The Hague: 1940).

Frederickx E. – Hal T. van, *Johannes Goropius Becanus (1519–1573). Brabants arts en taalfanaat* (Hilversum: 2015).

Heuvel C. van den, "Le traité incomplet de l'Art Militaire et l'ínstruction pour une école des ingénieurs de Simon Stevin", in *Simon Stevin (1548–1620) L'emergence de la nouvelle science*, Exhibition catalogue, Catalogue Bibliothèque Royale Albert I, Bruxelles, 17-09-2004/30-10-2004 (Turnhout: 2004) 101–111.

Heuvel C. van den, *'De Huysbou': A reconstruction of an unfinished treatise on architecture, town planning and civil engineering by Simon Stevin* (Amsterdam: 2005).

Heuvel C. van den, "'Tot meerder bewijs'. Een kaart en een model van Daniel Note van 1620 ter demonstratie van een nieuwe uitvinding om de haven van Middelburg zandvrij te schuren", in Hoftijzer P. – Ommen K. van – Warnar G. – Witkam J.J. (eds.), *Bronnen van kennis. Wetenschap, kunst en cultuur in de collecties van de Leidse Universiteitsbibliotheek* (Leiden: 2005) 100–108.

Kool M., "De rekenkundige termen van Simon Stevin", *Scientiarum Historia* (1992) 91–107.

CHAPTER 6

Bringing Euclid into the Mines: Classical Sources and Vernacular Knowledge in the Development of Subterranean Geometry

Thomas Morel

Introduction

During the early modern period, subterranean geometry (*Markscheidekunst*) developed into a consistent body of knowledge in the mining regions of central Europe. Although geometry and geometers were depicted very early in legal and technical mining texts [Fig. 6.1], little is known about the sources of this peculiar discipline.[1] The hypothesis presented here is that vernacular and practical knowledge progressively came to use classical mathematical sources over the course of the seventeenth century. In the following chapter, I study how this process took place and to what extent these interactions helped reshape the discipline.

More broadly, this case study should be seen as an example of how mathematical knowledge could circulate between very different social contexts in the early modern period. The Ore Mountains were ruled by the Electors of Saxony, who showed great interest in encouraging the sciences and the arts.[2] The Saxon State was an intellectual and technical centre in which various people dealt with mathematical sciences: instrument makers and *Hofarithmetici* [court arithmeticians] worked at the court of Dresden (most notably for the art chamber, the *Kunstkammer*) while professors and astronomers were active in the universities of Leipzig and Wittenberg.

1 The first mining laws (*Bergordnung*) in central Europe were written in the thirteenth century. Officials responsible for underground surveying were mentioned as early as c. 1360 and considerably expanded their influence during the sixteenth century. See Nehm W., "Die rechtliche Stellung des Markscheiders auf dem Oberharz während des 16. Jahrhunderts", *Mitteilungen aus dem Markscheidewesen* 45 (1934) 65–74.

2 The most famous example was Augustus of Saxony (1526–1586), founder of the *Kunstkammer*. He had a passion for surveying and ordered numerous instruments, some of which he personally used in surveying operations. See Dolz W., "The Waywisers of Elector Augustus of Saxony", in Strano G. et al. (eds.), *European Collections of Scientific Instruments 1550–1750*, Scientific Instruments and Collections 10 (Leiden – Boston: 2009) 43–60.

 | DOI 10.1163/9789004349261_008

Subterranean geometry was a body of knowledge that included not only surveying skills, but also the understanding of the legal system of the mines and of the structure of the earth. The underground surveyors [*Markscheider*] were often mining officials with great responsibilities, and could be described as 'superior craftsmen' or 'technical experts'.[3] While they obviously needed practical geometry for their work, they used it not as a science but as a technology-related knowledge, in order to plan and direct mining operations, draw maps or write reports for the administration. The relationship between the *Markscheider* and scholars dealing with mathematics is indeed a 'complex interaction' that is still to be fully understood.[4]

Little is known about the way this mathematical knowledge was acquired, produced and circulated in the Ore Mountains of Saxony.[5] A common criticism against subterranean geometers was their unwillingness to share their knowledge. In the sixteenth century, one could hear that 'der mehrertheil / iha fast alle Marscheider ihrer Kunst so neidisch und mißgünstig seind / das sie niemandes wollen lassen zusehen' [the majority / indeed almost all the subterranean geometers are so jealous and begrudging of their art / that they don't want to let anyone see it].[6] Indeed, not a single book written by a practitioner would be published before the very end of the seventeenth century.

This paper is therefore a contribution to the early history of subterranean geometry. It investigates the genesis of a coherent body of practical mathematical knowledge, in which translation, broadly conceived, played a crucial

3 E. Zilsel used the terms 'superior craftsmen' and 'artist-engineers' to describe those 'who needed more knowledge for their work than their colleagues did [...] Many of the artist-engineers wrote—in the vernacular and for their colleagues—diaries and papers on their achievements. For the most part these papers circulated as manuscripts only'. Zilsel E., *The Social Origins of Modern Science*, Boston Studies in the Philosophy of Science 200 (Dordrecht – London: 2000) 11. Thomas Kuhn also used the term 'artist-engineers' to describe the 'new intellectual milieu' of the early modern period (Kuhn T., "Mathematical vs. Experimental Traditions in the Development of Physical Science", *The Journal of Interdisciplinary History* 7, 1 (1976) 1–31. For a recent discussion on the knowledge of artisans and practitioners, see Smith P.H., *The Body of the Artisan* (Chicago: 2004) esp. 6–8, 17–20.

4 Smith, *The Body of the Artisan* 149, speaks of the 'complex interaction [...] between [...] the active knowledge of artisans and other handworkers and [...] the textual knowledge of the scholars'.

5 About the teaching of subterranean geometry in the early eighteenth century, see Morel T., "Le microcosme de la géométrie souterraine: échanges et transmissions en mathématiques pratiques", *Philosophia Scientiæ* 19, 2 (2015) 17–36.

6 Reinhold Erasmus, *Vom Marscheiden kurtzer und gründlicher Unterricht* (Erfurt, Bawman: 1574) "Dedication". All translations are by the author of the present chapter, unless otherwise stated.

role. I focus on the *De geometria subterranea*, a group of manuscripts that were widely circulated in the mines of Saxony during the seventeenth century. These texts were written in German with an abundant use of Latin and *Bergmannsprache*—the language of the miners. Why did classical excerpts, most notably taken from Euclid's *Elements*, find their way into these manuscripts? How did this knowledge interact with the vernacular, essentially artisan and sometimes esoteric methods accumulated by generations of miners? How can this inform us about the way mathematics was conceived and used by these practitioners?

In the first part of this paper, I question the influence of scholarly knowledge about mining, for example Agricola's *De re metallica* (1556), on the manuscripts written and used by the geometers. Interestingly, this presumably obvious relation proves to be deceptive. *Per contra*, I then show that several gateways existed between the geometry of Euclid and the miners' manuscripts. One of them is a short book written in 1563 by Christoph Puehler, teaching surveying methods based on Euclid's propositions.

Puehler's work, which turns out to be itself a pseudo-translation of a *Geometria practicæ*, a manuscript written by Hugh of Saint Victor (c. 1090–

FIGURE 6.1 *Drawing of surveying operation in a mine, using a wax disk and a compass. Manuscript in the* Schwazer Bergbuch *(c. 1554). Bochum, Bergbau-Museum, Entwurfexemplar 872, fol. 31v. Detail.*

1141), exerted an unexpected influence on the underground geometry in the Ore Mountains of Saxony. This is shown by studying an exemplar of Puehler's work, annotated during the early seventeenth century, as well as transcriptions of this book in several manuscripts. I finally discuss the role of translations and reformulations of the Euclidean geometry in the construction of a corpus of practical knowledge, to show that it was a central component of the appropriation process.

Agricola and Reinhold, Scholarly Knowledge about Underground Surveying

During the seventeenth century subterranean geometers often kept their know-how secret.[7] This is not to say that no instruction about this art was available in print. Indeed, some knowledge about subterranean geometry could be found in at least two books in the sixteenth century: in Agricola's *De re metallica* as well as in a surveying work by Erasmus Reinhold.[8] Both men were respected scholars living in Saxony and therefore sensible to the use of mathematical sciences in mining. Georgius Agricola (1494–1555) was a physician and scholar working in the Ore Mountains, the leading mining centre of the time. Although his *De re metallica* is mainly remembered for its contribution to the mining and smelting sciences, books 4 and 5 dealt with the use of mathematics in the mines, and were the first printed works about subterranean geometry. Agricola used the Latin methods of the ancient *agrimensores*, to which he added his own methods and reflections about specific instruments.[9]

Erasmus Reinhold the Younger (1538–1592), was a mathematician and physician.[10] Reinhold published in 1574 two books on surveying, above and

7 See for example Briggs H., "The Development of Mine Surveying Methods from Early Times to 1850", *Transactions of the Institute of Mine Surveyors* VI (1925) 120.

8 Agricola Georg, *De re metallica* (Basel, H. Froben: 1556).

9 In Ancient Rome, the *agrimensores* were the official land-surveyors. The mathematics of the *De re metallica* has to my knowledge not been studied for itself, probably because it belongs to the history of mathematics in a very specific context. I hope to produce such an analysis as soon as possible. In the fourth book, Agricola mentioned the *De limitibus agrorum*, see for example Agricola Georg, *De re metallica*, trans. H.C. Hoover – L.H. Hoover (New York: 1950) 77–90.

10 He was the son of Erasmus Reinhold the Elder (1511–1553), an important astronomer of the sixteenth century, professor of mathematics at the university of Wittenberg and mainly remembered for his *Tabulæ prutenicæ*, astronomical tables that helped promote Copernicus calculation methods.

underground, claiming he was only finishing a work his famous father had undertaken.[11] His intended aim was to produce a geometry treatise for practitioners. To this end, he purposely avoided using Euclid, although it was the standard work of the time, to focus on the technical surveying operations.[12]

Previous historians have therefore assumed that these books had actually been used by the geometers: 'Agricola's works as well as Reinhold's books "On Subterranean Geometry" were of great help for the many-sided mining enterprises of the Electors of Saxony'.[13] The assertion sounds plausible: after all, these books had been written by important Saxon scholars and printed in Saxony. But this does not seem to have been the case. Reinhold's book seems to have been quickly forgotten and rediscovered only late in the eighteenth century.[14] Agricola's *De re metallica*, despite its widespread diffusion and its cultural importance, was not used for this purpose.

This can be shown using several kinds of evidence. Firstly, the subterranean geometers of the eighteenth century abundantly wrote on the history of their discipline. Agricola and Erasmus Reinhold were either missing from their report or explicitly described as works having been rediscovered only recently.[15] Secondly, a single mention of Agricola and none of Reinhold are to be found in the manuscripts that were used by the subterranean geometers in Saxony during the seventeenth century, at least in the handful of copies that survived

11 See Reinhold, *Vom Marscheiden*, "Vorrhede" [introduction]. There is no evidence to support or reject this claim, although presenting an important astronomer and scholar as the author of the book might have been a mere rhetorical statement.

12 Reinhold, *Vom Marscheiden*. The section about subterranean geometry is very detailed and forms an independent part of the book, the mathematical content of which has been partly studied in Wilkening W., "Erasmus Reinholdt, der Verfasser der ersten deutschen Markscheidekunde", *Mitteilungen aus dem Markscheidewesen* 67 (1960) 13–15; 58–74.

13 Wunderlich H., *Kursächsische Feldmeßkunst, artilerische Richtverfahren und Ballistik im 16. und 17. Jahrhundert* (Berlin: 1977) 38. Wunderlich does not give any source to support his statement. He may have thought that the mere availability of the books implied they had been used.

14 See for example Calvör Henning, *Acta Historico-Chronologico-Mechanica*, 2 vols. (Braunschweig, Waysenhaus-Buchhandlung: 1763), vol. 2, 4, according to whom Reinhold's books 'was reprinted in 1615, but fell into oblivion over the years' so that the instruction written by N. Voigtel in 1686 was considered as the first manual ever on that topic. See Voigtel Nicolaus, *Geometria Subterranea, oder Marckscheide-Kunst* (Eisleben, Dietzel: 1686).

15 The only other explanation would be a major break between in the transmission of knowledge between the sixteenth and the eighteenth centuries, but the art of *Markscheiden* rather shows a clear continuity.

until today and have until now not be studied by historians.[16] This would of course not be conclusive if these texts contained no mentions of scholarly books whatsoever, but they actually mention several contemporary mathematicians. However, the single most quoted work is, surprisingly enough, Euclid's *Elements*.[17]

The absence of Agricola and Reinhold, although they both wrote precisely on subterranean geometry, shows that nothing should be taken for granted when talking about circulation of knowledge. Pamela Long has convincingly argued that many mining books of the sixteenth century were written in a patronage relationship.[18] Our analysis further shows that, at least in the case of the subterranean geometry, the learned discipline that scholars created should not be mistaken for the methods that practitioners actually used. Their well-polished rhetoric of usefulness has to be carefully questioned using other sources written by the technicians themselves. In the case of subterranean geometry, the content of their manuscripts shows little relation to Agricola's and Reinhold's works, as will be shown later in this paper. The development of printing techniques did not end the debates about secrecy, and both printed and handwritten books coexisted with various audiences and different purposes.[19]

Surprisingly enough, the practitioners manuscripts contain numerous mentions of Euclid's *Elements*, noticeably absent in both Agricola's and Reinhold's works. This provides puzzling questions about the sources used by the subterranean geometers. They seem to have ignored two important scholars of the time, both of them active in Saxony, whose works had allegedly been written for them and available in vernacular. And yet they used propositions taken from Euclid's *Elements* to determine in which direction mine shafts should be dug, although this milestone of geometry was hardly practical. Why were classical sources so important to geometers who enjoyed little theoretical education? And how did they find their way into these manuscripts in the first place?

16 Interestingly, this single occurrence of Agricola's name is to be found in a context of translation. Describing the miner's compass (*Hang-Compass*), a *Markscheider* indicates that 'Agricola names it in latin *Instrumentum cui index* est' (TU BAF – UB XVII 11, fol. 4v).

17 See on the translation of Euclidian mathematics also the chapter by Richard J. Oosterhoff in this volume.

18 Long P., *Openness, Secrecy, Authorship, Technical Arts and the Culture of Knowledge from the Antiquity to the Renaissance* (Baltimore – London: 2001) 175–191.

19 Long, *Openness* 181: after the invention of the printing press 'manuscript books, including those pertaining to mining and metallurgy, continued to be written throughout the [sixteenth] century'. The example of subterranean geometry shows that this was still the case in the following century, and indeed manuscript knowledge stayed a relevant means of circulation until the very end of the eighteenth century.

To answer this last question, I will now analyse in detail one of the sources used by the *Markscheider*, a practical geometry written by the mathematician Christoph Puehler.

A Practical Geometry Treatise, Christoph Puehler's Anlaytung (1563)

Not much is known about Christoph Puehler (c. 1500–?), who was active in the middle of the sixteenth century, and was therefore a contemporary of both Agricola and Reinhold. His only known work is a short book entitled *Ein kurtze und grundliche Anlaytung zu dem rechten Verstand Geometriæ* [A short and systematic introduction to the right understanding of geometry] published in 1563, here referred to as *Anlaytung*.[20] This work fortunately contains key biographical information about him. Puehler was born in Syclas and later lived and worked in Passau. He claimed to have learned mathematics from Petrus Apian in Vienna in the early 1520s, before Apian became printer, and professor of mathematics in Ingolstadt.[21]

In 1561, Puehler finished the manuscript of his *Anlaytung* and subsequently sent it to Apian's son Philip, himself a mathematician and cartographer, who had it published two years later.[22] Highly relevant for us is the introduction of his book, in which the author described its genesis. Puehler explained having contracted a serious disease four years before:

> Als mir nun seidhero vergebenlich ein alt geschriben büchlein inn die händt geraten / darinn ein klein un ordenlichs *Compendium de Practica Geometriae*, jedoch ohn inscription und namen des Authors / inn lateinischer sprach begriffen. Hab ich / umd die verdrossen unnd langweilige zeit zuvertreiben / dasselbig verteutschen / auch andern damit dienen wöllen / Gleich aber zu anfang / hat mich mein kleiner unnd schlechter verstandt / vil ein andern weg / auch zu mehrer erweyterung und gantz newer *Tradition* gefürt / inn massen dann gegenwertiges büchlein nach lenges außweiset.

---

20 Puehler Christoph, *Ein kurtze und grundliche Anlaytung zu dem rechten Verstand Geometriæ* (Dilingen, Mayer: 1563).

21 Petrus Apian (1495–1552) was one of the leading mathematician and astronomer of his time, and was the personal mathematician (*Hofmathematiker*) of the Holy Roman Emperor Charles v.

22 For more information about Puehler's biography, see Launert D., *Wer erfand die Transversalteilung? Brahe, Ursus, Hommel, Pühler. Pühlers Practica Geometriae*, Algorismus 79 (Augsburg: 2014).

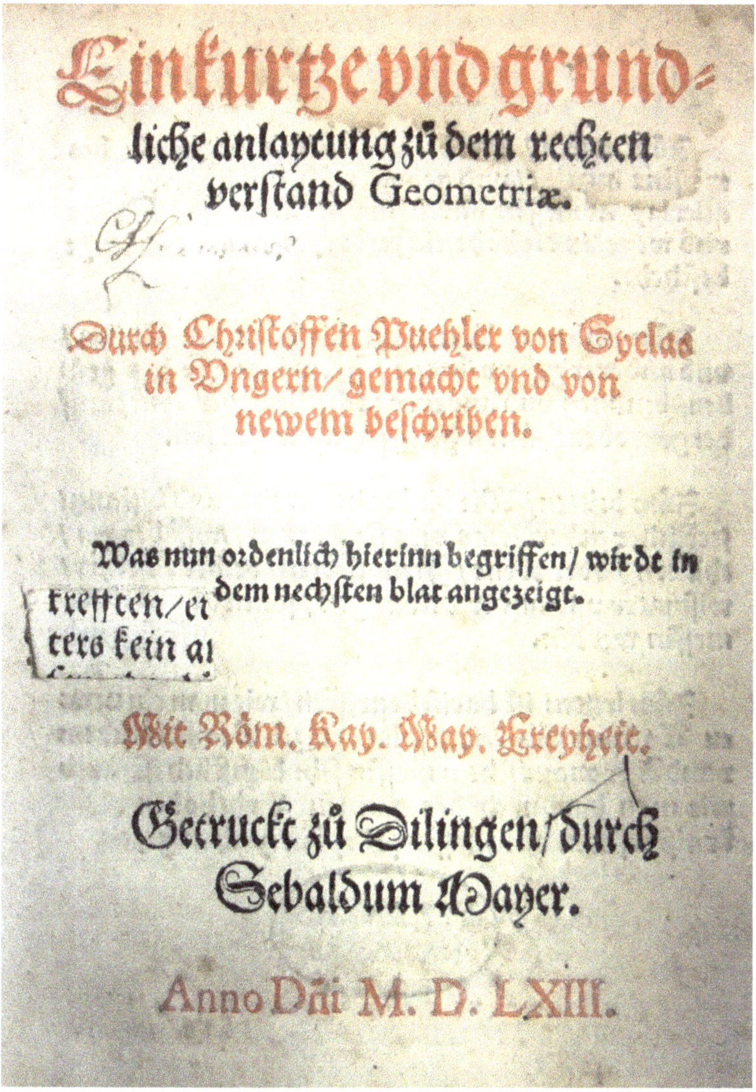

Ein kurtze vnd grundliche anlaytung zů dem rechten verstand Geometriæ.

Durch Christoffen Puehler von Syclas in Vngern/ gemacht vnd von newem beschriben.

Was nun ordenlich hierinn begriffen/ wirdt in dem nechsten blat angezeigt.

Mit Röm. Kay. May. Freyheit.

Getruckt zů Dilingen/ durch Sebaldum Mayer.

Anno Dñi M. D. LXIII.

FIGURE 6.2 *Title page to Christoph Puehler,* Ein kurtze und grundliche Anlaytung zu dem rechten Verstand Geometriæ *(Dillingen, Mayer: 1563). Freiberg, TU BAF – UB II 186.*

This is when an old written booklet fell into my hands for no particular reason / containing a short and decent *Compendium de Practica Geometriae*, but without any inscription or name of the author / composed in Latin. I translated it into German [*verteutschen*] / to pass the annoying and boring time / and to be useful to others. But from the very beginning / my short and poor understanding / made me take another way / and led me to many ameliorations and to a whole new *tradition* / that expended to form the present book.[23]

This introduction should of course not be taken literally: Puehler was here rhetorically justifying his decision of writing another compendium on practical geometry.[24] The illness serves as a pretext, in both senses of the word, to justify both the alleged goal (translation) and the final result (a new book). The readers were not fooled, as a scholar of the late eighteenth century ironically noted: 'among the countless medical observations, I don't know of a single one in which a sick person recovered with the help of mathematics. Mathematics would usually be something that physicians most certainly forbid to a convalescent'.[25] Two important questions should be answered: Was there an original source? In that case, which improvements were added by Puehler?

Studying Puehler's *Anlaytung*, D. Launert recently suggested that his book was a translation of the *Practica geometriae* written in the twelfth century by Hugh of Saint Victor.[26] There are doubtless many similarities between the two works, since both belonged to the tradition of practical geometry.[27] I nevertheless think that the *Anlaytung* better qualifies, and should be studied, as a pseudo-translation; meaning that Puehler should be considered as the legitimate author—an author who had good reasons for presenting his own work as being the translation of another (anonymous) person. The two texts differ in many respects, the *Anlaytung* being much more detailed. Puehler sometimes chose to use more recent authors, most of them German, thus showing both his comprehensive knowledge of the mathematics of his time and the originality of his enterprise.[28] It should be noted that he named the authors he was referring to, which was not the case in Hugh's *Geometria*. Puehler used

23 Puehler, *Anlaytung*, introduction, unpaginated.

24 Many such books had been published in the seventeenth century by mathematicians such as Peter Apian, Jacob Köbel, Johannes Stöfler, Wolffgang Schmid and Augustin Hirschvogel.

25 Kästner, A.G., *Geschichte der Mathematik*, 4 vols. (Göttingen: 1796–1800), vol. 1, 671. Kästner was explicitly referring to Puehler's *Anlaytung*.

26 See Launert, *Wer erfand die Transversalteilung?* 28–44. For a recent edition of his geometry, see Hugh of Saint Victor, *Opera propaedeutica*, ed. R. Baron (Notre Dame University: 1966).

27 On practical geometry seen as a literary tradition, seen Raynaud D., *Géométrie pratique. Géomètres, ingénieurs et architectes XVIe–XVIIIe siècle* (Besançon: 2015).

28 Among them Georg von Peuerbach (1423–1461), Regiomontanus (1436–1476), as well as Georg Joachim Rheticus (1514–1574) and Nicolaus Copernicus (1473–1543), as recognized by D. Launert.

the text to present several ideas and methods that would be considered innovative, most notably about the use of trigonometry in surveying—what would eventually become 'triangulation'.[29] As studies on 'pseudo-translation' have shown, presenting one's text as a translation is a common way to make novelties acceptable.[30] At the same time, the author also wanted to underline his originality: being both in the tradition of the *Geometriae practicœ* of the Middle Ages and outside of it.

Let us analyse for example the sixth chapter of Puehler's *Anlaytung*, entitled 'Wie der rechtwincklig Triangel / der das fürnembst mittel ist / dardurch alles messen geschicht / gemacht / gebraucht / unnd verstanden soll werden' [How should the right-angled triangle, which is the finest instrument to measure everything, be made, used and understood].[31] It is closely based on the third section of Hugh's *Geometria practicae.*[32] It begins with a description of the right-angled triangle and its sides, before explaining some of its properties that could be used as formulas. In both cases (Hugh and Puehler) the text was made more intelligible by adding a figure representing the triangle [Fig. 6.3].

The differences between the Latin text and Puehler's 'translation' are nevertheless meaningful. Although Puehler himself was probably not a practitioner, he clearly set his text in a context of application, as for example the title he gave to his sixth chapter suggests. Moreover, as was common at the time, this is only a partial translation: one can see in Fig. 6.3 that many terms were still written in Latin, using different fonts. Most notable are the names of the sides of the triangle (*basis, cathetus* and *hypothenusa*), and more generally all the mathematical terms (*superficies* for *area,* even *Triangel,* a mix between the Latin *triangulus* et the German *Dreieck*). Puehler himself justified this decision, explaining:

---

29 See Launert, *Wer erfand die Transversalteilung?* 102–105 as well as Bialas V., "Entwicklung und erste Anwendungen des Triangulationsverfahrens in der Geodäsie des frühen 17. Jahrhunderts", in Seck F. (ed.), *Wissenschaftsgeschichte um Wilhelm Schickhard* (Tübingen: 1981) 116.

30 Toury G., "Enhancing Cultural Changes by Means of Fictitious Translations", in Hung, E. (ed.), *Translation and Cultural Change: Studies in History, Norms and Image-Projection* (Amsterdam – Philadelphia, 2005) 3–17: one of the roles of a pseudo-translation is precisely to 'present a text as if it were translated, thus lowering the threshold of resistance to the novelties it may hold in store and enhancing their acceptability' (page 4).

31 Puehler, *Anlaytung,* fol. 11r.

32 Hugh, *Opera propaedeutica, Practica geometriae* §3; Launert, *Wer erfand die Transversalteilung?* 33–36.

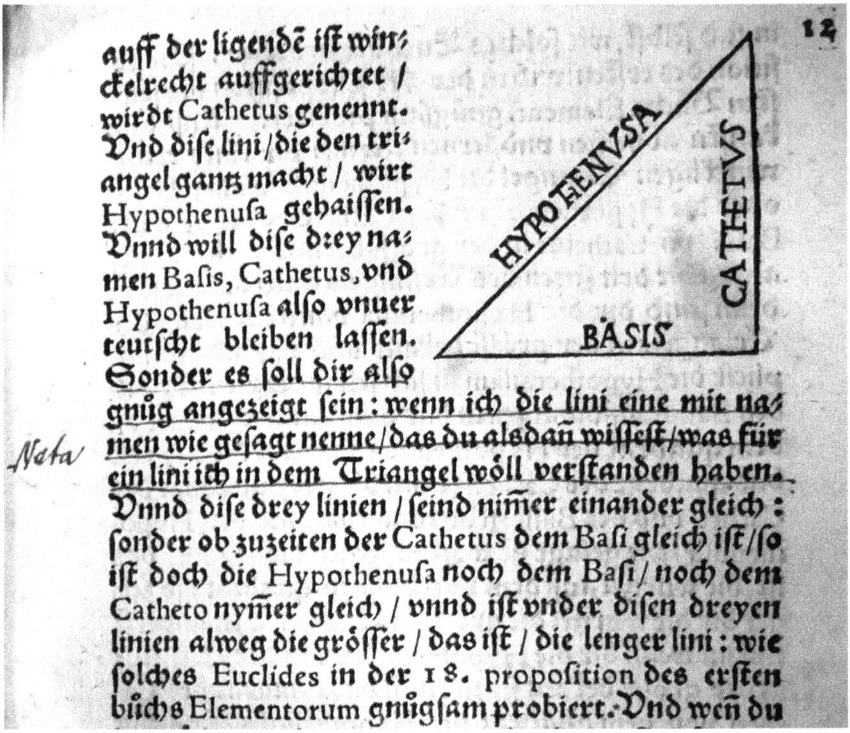
12
auff der ligendē ist win-
ckelrecht auffgerichtet /
wirdt Cathetus genennt.
Vnd dise lini/die den tri-
angel gantz macht / wirt
Hypothenusa gehaissen.
Vnnd will dise drey na-
men Basis, Cathetus, vnd
Hypothenusa also vnuer
teutscht bleiben lassen.
Sonder es soll dir also
gnůg angezeigt sein: wenn ich die lini eine mit na-
men wie gesagt nenne/das du alsdañ wissest/was für
ein lini ich in dem Triangel wöll verstanden haben.
Vnnd dise drey linien / seind niñer einander gleich:
sonder ob zuzeiten der Cathetus dem Basi gleich ist/so
ist doch die Hypothenusa noch dem Basi / noch dem
Catheto nyñer gleich / vnnd ist vnder disen dreyen
linien alweg die grösser / das ist / die lenger lini: wie
solches Euclides in der 18. proposition des ersten
bůchs Elementorum gnůgsam probiert. Vnd weñ du

FIGURE 6.3 *Christoph Puehler, woodcut illustration to Christoph Puehler,* Ein kurtze und grundliche Anlaytung zu dem rechten Verstand Geometriae *(Dillingen, Mayer: 1563), fol. 12r. Freiberg, TU BAF – UB II 186.*

Nachdem ich aber derselben namen in Teutscher sprach gar wenig gelesen / noch nennen gehört hab: wil ich die bey jren Lateinischen namen bleiben lassen: aber dermassen / jrem jnnhalt nach / beschreibe und außlegen / daß du eines jeden worts guten und rechten verstand wirst haben können / was die alten mit solchem wort gemainet und verstanden wollen haben.[33]

Since I have barely read / or heard using these very names written in German: I will therefore keep using their latin name, but describe and explain their content / in such a way that you will get a good and clear understanding / of what the Ancients meant and wanted to be understood using each of these words.

33 Puehler, *Anlaytung*, fol. 3r.

Most relevant for our topic are the additions concerning Euclid's *Elements*. Puehler chose to ascribe propositions to most of the procedures or formulas given by Hugh. In the sixth chapter, for example, Puehler described the relative length of the various sides and added 'wie solches Euclides in der 18. *proposition* des ersten buchs *Elementorum* gnugsam probiert' [as such has been sufficiently shown by Euclid in the eighteenth *proposition* of the first book *Elementorum*].[34] In the *Anlaytung*, about sixty references to Euclid can be found, most of them taken from the books I, V and VI (i.e. the books containing the theory of proportion and similar figures).

Most of these occurrences were similar to the example just given: the actual Euclidean proposition was not quoted, but reformulated to serve the author's purposes. The reformulation could sometimes be read as a list of instructions. This free rewriting can indeed be seen a practical translation of Euclid, in order to make it both easier to understand and immediately suited for instrumental practices. Puehler's presentation of the Pythagorean theorem illustrates this. After describing the relation between the squares of the sides, he stated that it had been 'sufficiently shown' by proposition I.47 and VI.4 of the *Elements*, adding:

> Auß dem kanstu abnemen und lernen / wenn dir eines rechtwinckligen Triangel die *Hypothenusa* un der *Basis*: oder die *Hypothenusa* un der *Cathetus*: oder bede der *Basis* un *Cathetus* in der grösse bekant seind / so wirt alweg die drit seiten des Triangels auch bekant sein.
>
> From which [propositions] you can derive and learn / when in an equiangular triangle you know the length of the *hypothenusa* and the *basis*: or the *hypothenusa* and the *cathethus*: or both the *basis* and the *cathetus* / then the third side of the triangle will also always be known.[35]

No other author than Euclid was mentioned more than a handful of times, which shows that one major goal of Puehler's 'pseudo-translation' was to show the correspondence between practical operations and propositions of the *Elements*. Other attempts to produce practical versions of Euclid were made in the following decades, most notably by the Saxon scholar Lucas Brunn (1572–1628) in his *Elementa practica, oder Ausszug aller Problematum*

34 The text of the eighteenth proposition is 'In any triangle the greater side subtends the greater angle'; translation by Thomas L. Heath, *The thirteen books of Euclid's Elements*, 3 vols. (Dover: 1956) vol. 1, § 18.

35 Puehler, *Anlaytung*, fol. 12v.

und Handarbeiten auss den 15. Büchern Euclidis (Halbmayer, Nürnberg: 1625). Director of the collections (*Kunstkammer*) of the Elector of Saxony, his edition presented the *Elements* as a list of geometrical problems to be solved using mathematical instruments.

An Unexpected Journey and a Careful Reader

The library of the *Bergakademie Freiberg* owns a good number of rare printed books from the sixteenth and seventeenth centuries, many of which belonged to mining officials before the creation of the academy in 1765. Most of these books are related to mining sciences, but a few deal with geometry. Among them, is a copy of Puehler's *Anlaytung* [Fig. 6.2], bound together with another *Geometria* published in 1568.[36] The leather binding bears the somewhat obscure inscription:

§ A§G § A§S §
§ 16 § 25 §

We can therefore suppose that the two volumes were bound together in 1625. The exemplar has been abundantly annotated, almost half of the pages containing comments, underscored sentences or calculations in the margins. We can make the assumption that someone living in the Ore Mountains of Saxony possibly bought, read and worked on this volume at some point between its publication in 1563 and its binding in 1625.[37] This fact is important because, as we will see, a chapter of Puehler's *Anlaytung* found its way into Saxon manuscripts of the early seventeenth-century dealing with subterranean geometry.

Several pieces of evidence indicate that the book was used, possibly by several persons, as a working or learning copy. One finds numerous 'NB' (for *nota bene*) and 'Nota' in the margins, while corresponding sentences are underscored, as can be seen in Figs 6.3 and 6.4. Moreover, many of the problems presented by Puehler were 'solved' anew by the reader. The data, be it names of the triangle sides, numbers or details of the operations, was reported on the

36 TU BAF – UB II 186. Rensberger Nicolaus, *Geometria, Das ist, wie man recht vn behend, eines jeden Dings höhe, lenge vnd breyte* [...] *abmessen soll* (Augburg, Willer: 1568).

37 A distinct possibility is of course that the comments were made after the binding. Nevertheless, evidence indicates that these comments were written in the early seventeenth century: the handwriting, the spelling that was used as well as the fact that the two books were annotated by the same hand.

figures for a clearer understanding. Comments were written using red ink and two types of black ink.[38]

The annotations, or the lack of them, help us understand why this book was considered interesting by its reader. Chapters one to four, which deal with the measuring units used by the Roman *agrimensores* as well as in various European countries, do not bear a single annotation.[39] The reader was probably looking for knowledge he could directly use in the mines, and not for a general lecture on the length system of other countries. In the margins of chapter 54, the reader wrote 'der Beschluß des Meß nach der Höhe, lange und Teiffe' [end of the measurement in height, length and depth].[40] It seems to have been the end of the part he wanted to work on, since the chapters 55–72 bear only two short comments.

The annotations are therefore heavily focussed on the chapters describing measuring operations. The reader seems to have been interested in the use of instruments and how applying Euclid's propositions in such cases could help determine the wanted lengths. The fourth proposition of the sixth book of the *Elements* (about similar triangles) was especially relevant for that purpose. Its usual formulation states: In equiangular triangles the sides about the equal angles are proportional, and those are corresponding sides which subtend the equal angles.[41] This version was nevertheless barely understandable for a practitioner with little, if any, mathematical education. This is why Puehler reformulated it [Fig. 6.4], both to make it easier to grasp and to make its practical use clearer:

> Denn alle weil der klein triangel in dem Instrument gemacht dem grossen änlich ist: so folget nach anzeigen *Euclidis* in der vierdten *Proposition* des 6. buchs *Ele*: das sich auch der *Basis* des grossen triangels zu seinem *Cathetum* oder *Hypothenusam* wirdt halten: wie sich der *Basis* des kleinen triangels zu seinem *Cathetum* od *Hypothenusam* ist halten.

> Because the smaller triangle in the instrument has been made similar to the bigger [one]: thus it follows according to Euclid in the fourth proposition of the 6th book of the Ele[ments] that the *basis* of the big triangle

38 The red comments were in all likelihood the first to be made, since the black comments occasionally cross out the red text, or correct calculations (see for example, fol. 47r).

39 Puehler, *Anlaytung*, exemplar conserved in Freiberg, TU BAF – UB II 186, fols. 3v–10r.

40 Ibidem, 87r.

41 Euclid, *Elements*, book 6, proposition 4.

will be to its *cathetum* or its *hypothenusam*, as the *basis* of the small triangle to its *cathetum* or *hypothenusam*.[42]

Puehler's account of Euclid was therefore a translation in three different ways. Firstly, he gave a German version of the original proposition, making it available to the unskilled reader who did not master Greek or Latin. It should be remembered that Puehler wrote this book in 1561, before the publication of Xylander's translation of the *Elements*, the first of its kind in German.[43] Secondly, instead of a literal translation, the author reformulated the proposition in a practical setting. The measuring instrument represented the small triangle, the measured object the bigger one, so that Euclid proposition was given as a procedure. This reformulation of Euclid made it suitable for practitioners. And the anonymous reader of Puehler's *Anlaytung* underscored this passage [Fig. 6.4], marking it with a NB (*nota bene*). This proposition was used time and again by Puehler with varying formulations to show its practical interest, and occasionally underscored by the reader.[44]

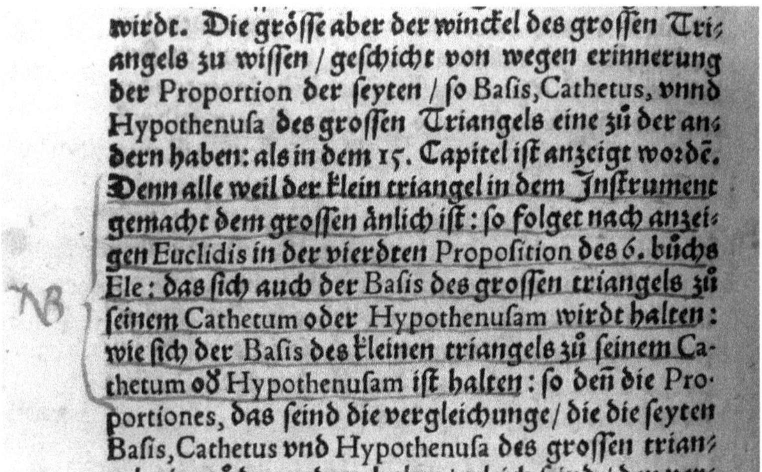
wirdt. Die gröſſe aber der winckel des groſſen Tri-
angels zu wiſſen / geſchicht von wegen erinnerung
der Proportion der ſeyten / ſo Baſis, Cathetus, vnnd
Hypothenuſa des groſſen Triangels eine zů der an-
dern haben: als in dem 15. Capitel iſt anzeigt wordē.
Denn alle weil der klein triangel in dem Inſtrument
gemacht dem groſſen ånlich iſt: ſo folget nach anzei-
gen Euclidis in der vierdten Propoſition des 6. bůchs
Ele: das ſich auch der Baſis des groſſen triangels zů
ſeinem Cathetum oder Hypothenuſam wirdt halten:
wie ſich der Baſis des kleinen triangels zů ſeinem Ca-
thetum od̄ Hypothenuſam iſt halten: ſo dēn die Pro-
portiones, das ſeind die vergleichunge/ die die ſeyten
Baſis, Cathetus vnd Hypothenuſa des groſſen trian-
gels eine zů der andern haben/ geleich ſeind / den ver-

FIGURE 6.4 *Christoph Puehler, detail of a page with underlinings in Christoph Puehler,* Ein kurtze und grundliche Anlaytung zu dem rechten Verstand Geometriae *(Dillingen, Mayer: 1563), fol. 31v. Freiberg, TU BAF – UB II 186.*

42 Puehler, *Anlayting*, fol. 31v.

43 It should be noted that Puehler's 'translation' differs greatly from Xylander's, which closely follows the original. Xylander Guilielmus, *Die Sechs Erste Bücher Euclidis, vom anfang oder grund der Geometrj* (Basel, Kündig: 1562).

44 For another example see Puehler, *Anlytung*, fol. 44v.

Thirdly, figures were occasionally used to exemplify the propositions, or make their practical use clearer. Chapter 34 deals with the classical problem of measuring the height of a tower using its shadow and a stick. It is based on the corollary of proposition VI.2 of the *Elements*: If a straight line is drawn parallel to one of the sides of a triangle, it will cut the sides of the triangle proportionally; and, if the sides of the triangle are cut proportionally, the line joining the points of intersection will be parallel to the remaining side of the triangle. Puehler introduces the chapter by describing the concrete situation, with a tower, a plane ground and a stick. The proportionality of the sides is described as practical relation between the stick and its shadow: 'Denn wie sich die rutten zu jrem schatten / also haltet sich auch der Thurn zu seinem schatten' [for the tower is to its shadow in the same relation the stick is to its shadow].[45] He only alludes to the proposition afterwards, writing 'wie solches *Euclides* in der andern Proposition des 6. Buchs zuverstehen gibt / unnd in nachfolgender figur einen augenschein kanst haben' [as Euclid suggests in the second proposition of the sixth book / and as you can get a glimpse of in the following picture].[46] In this case, the picture was used together with the practical reformulation and put in place of the actual proposition, which was literally translated, meaning that its content was transferred into another idiom more appropriate for the reader.

These various translations aimed at making Euclid useful in a new context. There are many more cases, for example when Puehler refers to 'der 19. *Proposition* des 7. Buchs *Ele*: welche *Regula detri*: genennt' [the 19. proposition of the 7th book of the Elements, which is called *Regula detri*].[47] Euclid's proposition is here considered as equivalent to the rule of three, thus getting the status of a mathematical formula.[48]

The annotations of the reader, on the other hand, are very instructive for us. They can be seen as a testimony of what was considered important by a mathematical practitioner working in mining administration in the early seventeenth century. For example, chapter 36 is entitled 'Eines gebews/so auff eim

---

45 Puehler, *Anlaytung*, fol. 52v. As usual for the time, the author does not give any mathematical formula, but enumerates the three cases where the shadow is respectively of the same size, shorter or longer than the stick.

46 Puehler, *Anlaytung*, fol. 52v–53r.

47 Puehler, *Anlaytung*, fol. 32r.

48 The original proposition of Euclid states: 'If four numbers be proportional, the number produced from the first and fourth will be equal to the number produced from the second and third; and, if the number produced from the first and fourth be equal to that produced from the second and third, the four numbers will be proportional'. Euclid, *Elements*, book 7, proposition 19.

berg gelegen/höhe auβ dem thal abzumessen' [To measure from the valley the height of an edifice located on a mountain].[49] An inaccessible tower is located on a mountain, and in order to find its height, two sets of measurements are made from two different points in the valley using Peuerbach's geometrical quadrant. Various propositions from the fifth book of the *Elements* (17, 22 and 23) are then used to determine the height of the tower using the known proportions as well as the distance between the two points in the valley. The reader underscored many parts of this chapter and added comments: it is clear that a way to calculate the height of an inaccessible point was of interest to him.[50] Although the content of the fifth book of the *Elements* was actually an abstract theory of ratio, Puehler's mediation and reformulation produced procedures easily available to a German reader, along with handy indications for using the instruments, writing down the measurements and computing the results.

A last example shows the appropriation of the text in the context of mining sciences. Chapter 48 presents a conventional problem, i.e. the determination of the length of a field. Its resolution is then illustrated with the engraving shown in Fig. 6.5.[51] Puehler's method was once again based on the use of similar triangles, the base of the bigger one (ABC in the figure) being the wanted length, and the smaller triangle (bCc) being obtained by adequately placing the moving rule of the quadrant. Puehler mentioned in passing that the quadrant could be placed either vertical ('auffrecht') or horizontal ('liegend').[52]

This was far from anecdotal for the reader, who showed a specific interest in this problem and commented 'diß sind Zweierley meßung / Die eine mit ligenden / Die ander mit Auffrechten Instrument' [these are two types of measurement, the first one with the horizontal, the other with the vertical instrument].[53] He then solved the problem, naming new points (in red on the figure) and using values he chose arbitrarily.[54] The importance of the figure for surveying in mining operations becomes obvious if we reverse the picture: if the observer is not placed at point B, but at point A, this picture indicates how to find the position of a point, with a decomposition in two plane measurements, one horizontal and one vertical.

49 Puehler, *Anlaytung*, fols. 55v–57r.
50 See the annotated copy in Freiberg, TU BAF – UB II 186.
51 Puehler, *Anlaytung*, fols. 72r–73r.
52 Puehler, *Anlaytung*, fol. 72r.
53 See the annotated copy in Freiberg, TU BAF – UB II 186, fol. 73r.
54 Namely 300, 1200 and 4800. These values are not to be found in the text explaining the figure.

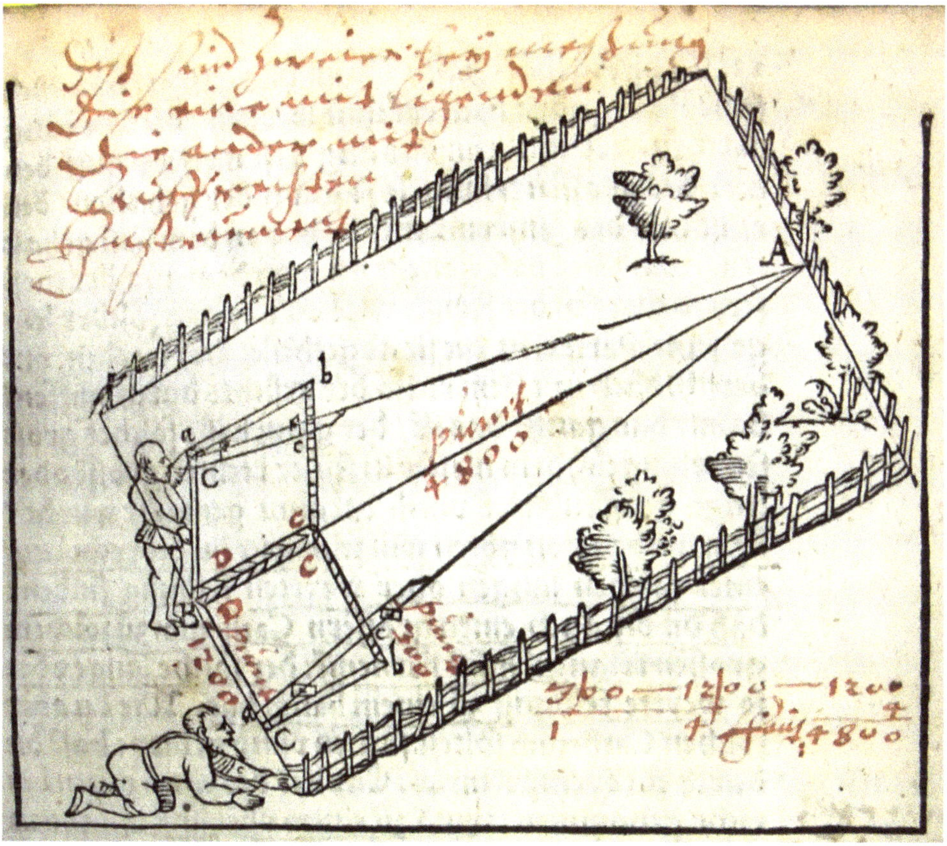

FIGURE 6.5 *Christoph Puehler, woodcut illustration with annotations in Christoph Puehler,* Ein kurtze und grundliche Anlaytung zu dem rechten Verstand Geometriae *(Dillingen, Mayer: 1563), fol. 73r. Freiberg, TU BAF – UB II 186.*

The *De Geometria Subterranea* and its References to Puehler

Puehler's *Anlaytung* was carefully read and commented upon by an anonymous reader, probably at the beginning of the seventeenth century. But the story does not stop here. One chapter was then copied into a manuscript written in the early seventeenth century by a subterranean geometer, of which several copies have survived to this day.[55] It would be tempting to imagine that the unknown annotator of the Freiberg exemplar of the *Anlaytung* was himself

55 TU BAF – UB II 186, fols. 11v–12v.

the surveyor who decided to include a chapter in his own manuscript; unfortunately no particular evidence supports this hypothesis.[56]

The existence of an annotated version of the *Anlaytung* and the contemporary inclusion of its sixth chapter in a surveying manuscript is nevertheless compelling. It shows that practitioners living in the Ore Mountains of Saxony at the time were eager to understand the mathematics needed to improve their practices. In that process, they gave a new importance to the *Elements*, despite the fact that this milestone of theoretical geometry seemed at first sight completely disconnected from their most immediate and practical concerns.

The anonymous manuscript *De geometria subterranea* was written in the seventeenth century, probably between 1617 and 1669.[57] The author might have been Balthasar Rößler, an important mining official and *Markscheider* in Saxony, or one of his students.[58] It was at the time common to use handwritten books for subterranean geometry; these kinds of manuscripts were used both for teaching and to note new methods.[59] These texts contained the knowledge needed by a subterranean geometer, including practical mathematics, and knowledge of the mining veins and mining laws.

Most of these manuscripts, including the copy that we are studying here, had a similar structure. They were structured around a body of 'propositions', i.e. practical problems that a geometer should be able to solve, such as the determination of the direction of a tunnel, or the digging of a shaft to meet a specific point.[60] The resolution of these problems was extensively

56 In this chapter, I use the *Geometria subterranea: Unterirdische Erdmäßung, oder so genannte Marck-Scheide-Kunst*, 17th century, author unknown, Freiberg, TU BAF – UB XVII 11. The chapter copied from Puehler is on fols. 38r–39v. Other copies are to be found in the TU BAF – UB, with the signatures XVII 677 and in the *Nachlass* (collected papers) of J.A. Scheidhauer, 300m. Other copies might have been destroyed, so it is not sure that the manuscript we will study in the following part of this chapter was indeed the first in which Puehler's thoughts were applied to the geometry of the mines.

57 It was likely written after 1617, since a passage on fol. 28v is taken from a book written by Bramer Benjamin, *Bericht und Gebrauch eines Proportional-Lineals* (Marburg, Egenolff: 1617) 11. In 1669, a modified version of this manuscript, entitled *Neu-Markscheide Buch* [New Book on Subterranean Geometry] was written by Adam Schneider, TU BAF – UB XVII 18.

58 On Balthasar Rößler (1605–1673) and his influence on the development of mining sciences, see Meixner H. et al. (eds.), *Balthasar Rösler: Persönlichkeit und Wirken für den Bergbau des 17. Jahrhunderts* (Leipzig: 1980).

59 Morel, "Le microcosme de la géométrie souterraine" 18–23.

60 This copy of the *De geometria subterranea* (TU BAF – UB II 186, fols. 2r–3r) contains 18 propositions, although later additions brought the total number to 25.

described, together with the instruments of the geometer: concrete instruments—suspended compass, water level and wax disks—as well as intellectual instruments—trigonometric tables and the geometry of the triangle.

This is precisely where the chapter of Puehler's *Anlaytung* was introduced [Fig. 6.6]. The author of the manuscript made the origin of this passage very clear, as he wrote 'Christoph Pühlers von Syclas in Ungern Geometriae Cap. Sextum', before giving the full title of this sixth chapter: 'Wie der rechchtwincklig Triangel / der das fürnembst mittel ist / dardurch alles messen geschicht / gemacht / gebraucht / unnd verstanden soll werden' [How the right-angled triangle, which is the finest instrument to measure everything, should be made, used and understood].[61]

The text runs on four handwritten pages and is almost identical to the *Anlaytung* of 1563. The few differences come from the evolution of the spelling, an indication that the text had been written in the seventeenth century. The slashes of Puehler's text that can be seen in Fig. 6.3 have been replaced by commas, following the evolution of the *Fraktur* font of the time; many words have a slightly different spelling.[62] Interestingly enough, some of the Latin terms that Puehler had Germanized are now written in Latin again, as can be seen comparing Figs 6.3 and 6.6: 'triangulus' (in Hugh's manuscript) was translated by Puehler ('Triangel'), while the author of the manuscript wrote 'Triangul' using a Latin font.

Concerning the description of the right-angled triangle ('recht wincklichter *Triangul*' or '*Triangulum rectangulum*'), the anonymous copyist followed Puehler's approach to use the Latin names 'ungermanyzed' ('unverteutscht', as he puts it): *basis, cathetus* and *hypothenusa*. Nevertheless, he must have felt the need to clarify their meaning since he added every time the equivalent in the language of the miners: 'Sohle' for 'basis', 'Seigerteiffe' for 'cathetus' and 'Dohnlege' for 'hypothenusa' (compare the triangle in Fig. 6.7 to the original in Fig. 6.3). Further evidence that this manuscript was used as a technical text: the numerous occurrences of 'quadrat', to speak of the square of a length, are abbreviated by the symbol of a square '□'.[63]

61 Puehler, *Anlaytung*, fol. 11r.

62 Most notably, the nouns are capitalized, which was not previously the case: 'lini' is now written 'Linie', 'aigenschafft' is written 'Eigenschafft', 'maß' is written 'Maaß', 'wincklen' is written 'Winkeln' and so on.

63 This cannot be seen on Figs 6.6 and 6.7, but appears several times on the following pages. See TU BAF – UB XVII 11, fols. 39r–39v.

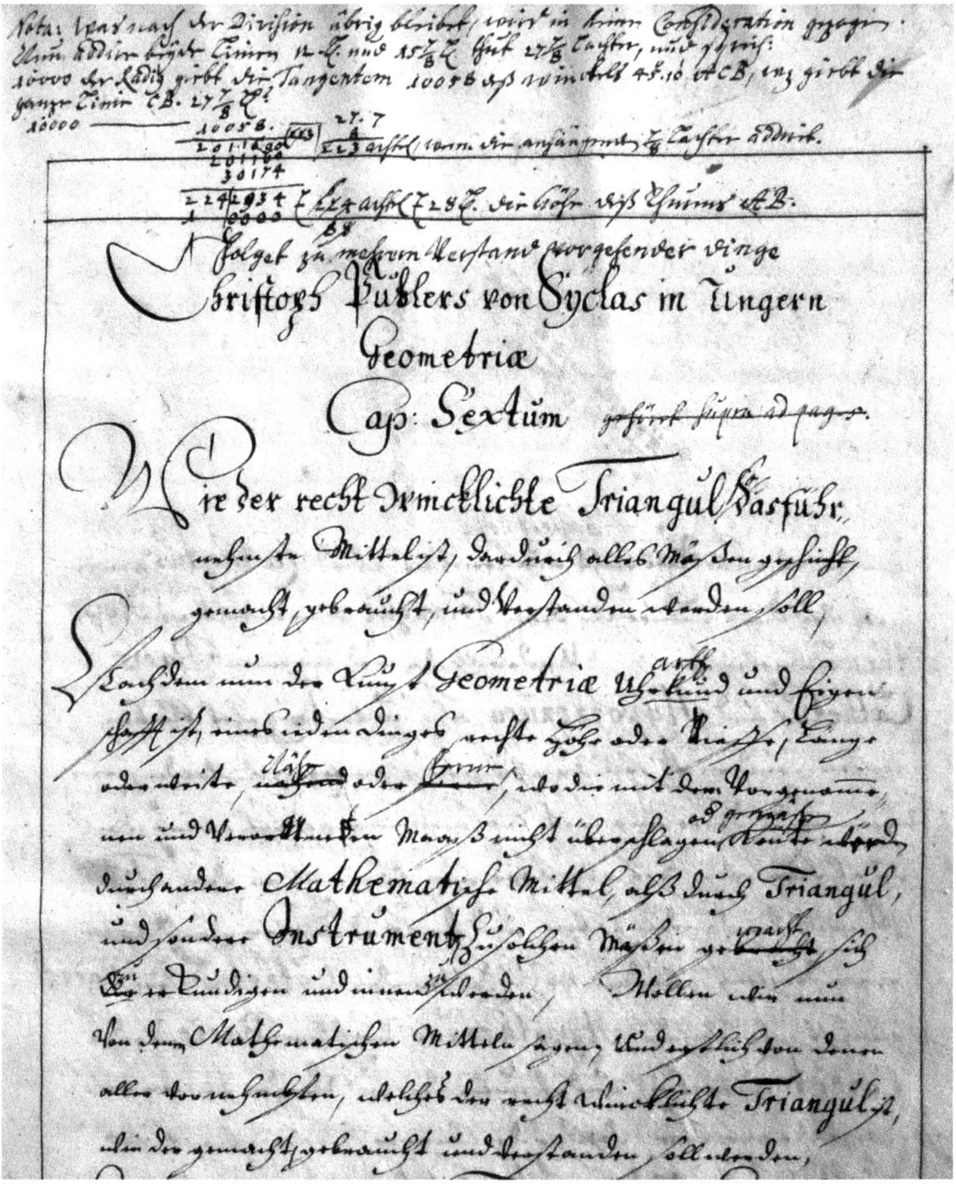
Christoffs Puhlers von Syclas im Ungern
Geometriæ
Cap: Sextum

FIGURE 6.6 *Anonymous, Introduction to Puehler's* Anlaytung, *in* Geometria subterranea: Unterirdische Erdmäßung, oder so genannte Marck-Scheide-Kunst (*17th century*). *Manuscript. Freiberg,* TU BAF – UB XVII 11, *fol. 38r.*

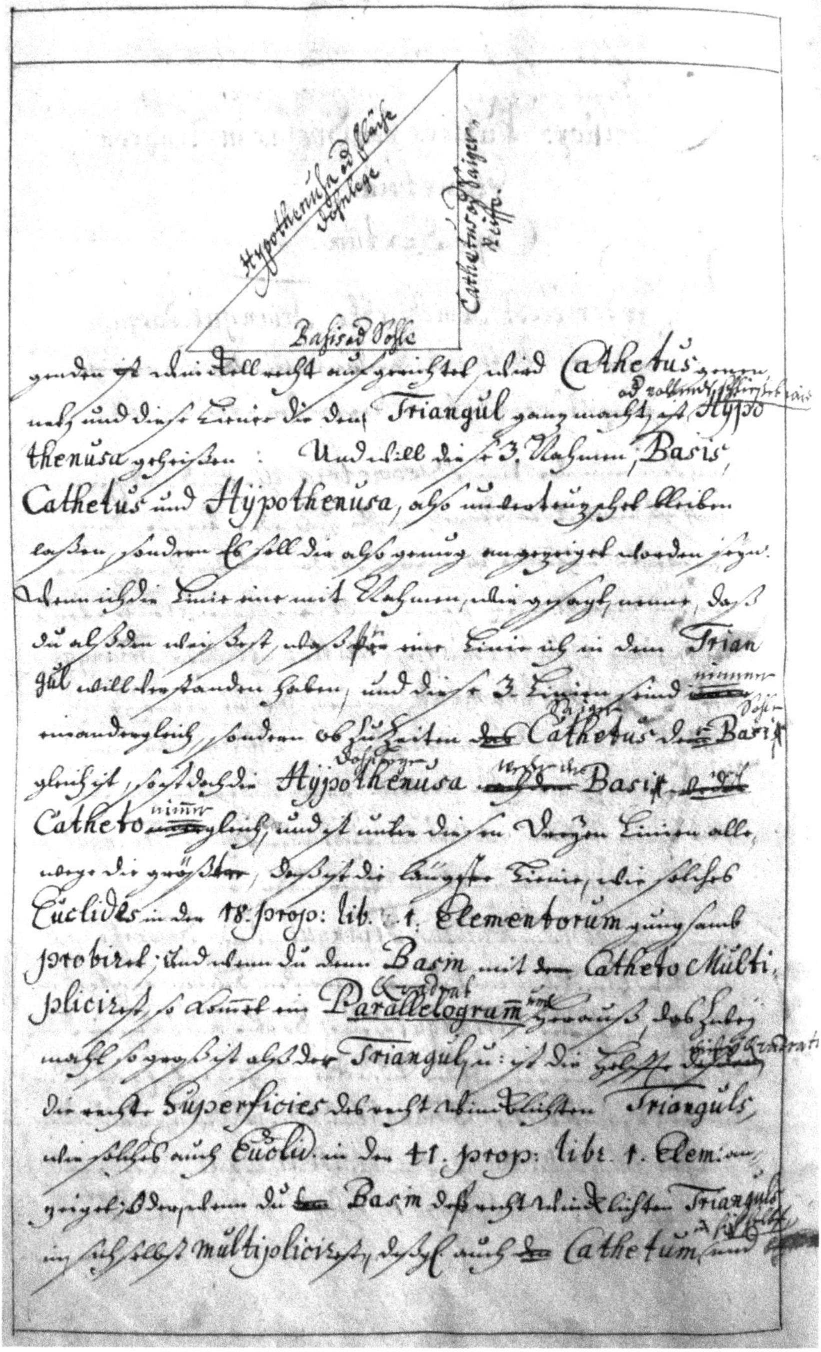

FIGURE 6.7 *Anonymous, A right-angled triangle in* Geometria Subterranea Unterirdische Erdmäßung, oder so genannte Marck-Scheide-Kunst (*17th century*). *Manuscript. Freiberg, TU BAF – UB XVII 11, fol. 38v.*

An important question is to understand why this chapter was integrated into a manuscript about practical geometry in the mines. It is pretty clear that you did not need to understand Euclid's *Elements*, even in the practical reformulation of Puehler, to apply geometric knowledge in the context of mining. This chapter was not directly used in the rest of the text, and contained only the archetypical 3–4–5 triangle example of Pythagoras's theorem. Our interpretation is that it nevertheless played an important role in legitimating the general method of subterranean geometry, i.e. the resolution of triangles.

Translating Mathematics and Using Euclid in the Mines

The *De geometria subterranea* contains other references to Euclid, which will help understand the use of 'classical' geometry in the mines of central Europe. The references always underline the importance of the right-angled triangle, and its close connection to actual surveying practices. The *Markscheider* used his instruments to obtain data: with the geometer's chain, one could measure the *hypothenusa*, while the suspended compass would measure horizontal angles, and the quadrant vertical angles [Fig. 6.8]. Puehler's *Anlaytung*, then, presented Euclid as the key to obtain the desired *basis* and the *cathetus* (in most cases). The propositions I.47 and VI.31 were reformulated as the Pythagorean theorem, thus allowing for a quick determination of the third side of a right-angled triangle whose *basis* and *cathetus* were known. Metaphorically, these theoretical instruments, used together with practical instruments, would transform the mine into a Euclidean space, whose frame was quite literally represented in Fig. 6.8. This paved the way for a calculative resolution of the problems.

The right-angled triangle had long had a special importance for geometry in general, and practical geometry in particular. In this manuscript it was presented as the 'true foundation' of subterranean geometry: 'Der wahre Grund dieser Kunst beruht in einen *Triangulo Rectangulo*, welche *Magister Matheseos* genannet wird' [the true foundation of this art relies on a right-angled triangle that is called master of mathematics].[64] The use of two Latin expressions is further evidence for the influence of scholarly geometers on the early development of the *Markscheidekunst*. The term 'magister matheseos' was commonly used in universities during the Middle Ages.[65]

64 TU BAF – UB XVII 11, fol. 3v.

65 See Cajori F., "Historical note", *The American Mathematical Monthly* 6, 3 (March 1899) 72–73.

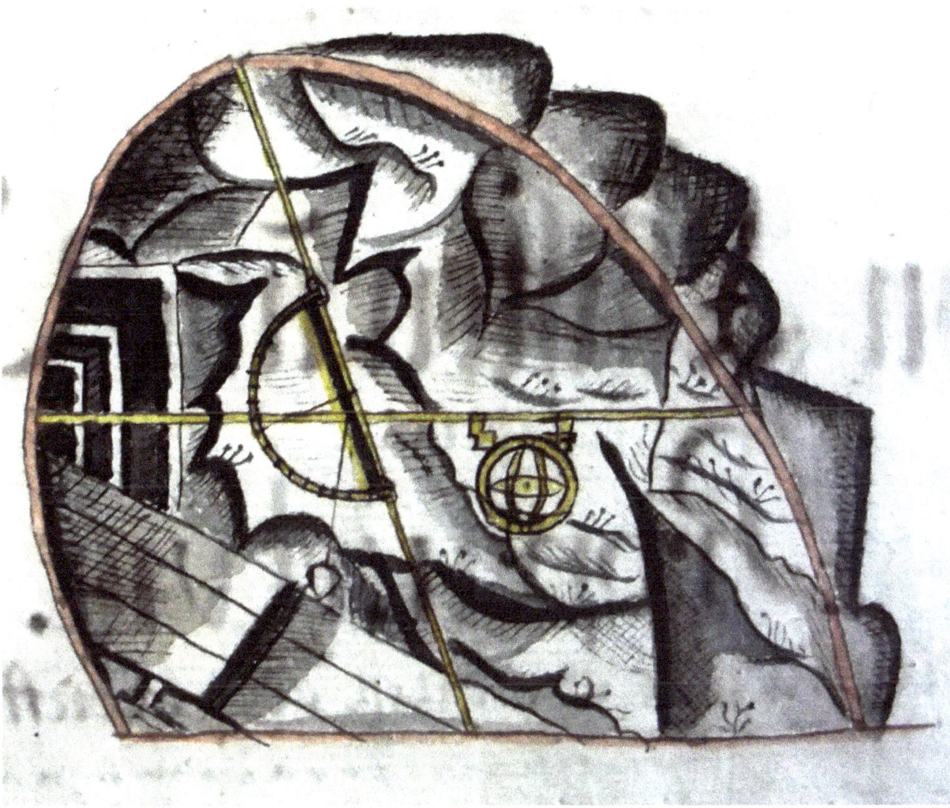

FIGURE 6.8 *Adam Schneider, drawing with instruments of the subterranean geometer, from the* Neu-Markscheide Buch *(c. 1669). Manuscript. Freiberg,* TU BAF – UB XVII *18, fol. 50v.*

The repartition of Euclid's propositions in the manuscript gives important information about their role. It closely matches the use of Latin terms, as could be expected. Both are almost absent from the first part of the book describing the 'propositions'. These technical operations were logically described using the *Bergmannsprache* (dialect of the miners), focussing on the concrete procedures rather than their mathematical bedrock. These methods could be used simply as algorithms by carrying out the successive tasks, in order to obtain the wanted length or angle. The reader was given instructions such as: take this instrument, use it in that way, read the result and look in this table.[66] No allusion was to be found to the underlying geometry of the triangle, and the term 'triangle' often did not even appear. The only Latin terms in these parts

66 TU BAF – UB XVII 11, fols. 40r–41v.

described the tangible tools the geometer was using: instruments (*compass*) and their scaling units (*gradus*), or the concrete mathematical tables (*tabula sinuum, columna*).

Latin is mostly used in the second part of the book, where more abstract considerations concerning the geometry of the triangle are given.[67] This is also where Euclid's *Elements* are quoted and commented upon. Besides the use of Puehler's *Anlaytung*, the most significant passage is a comment on Xylander's version of Euclid, presented in a section entitled 'Doctrina':

> Weil die Wage an der Schnur allezeit eines *Triangulum Rectangulum* giebet, darinnen zwene Winckel bekant sind; alß (1.) Rectg. So allezeit 90°. halt, und (2) der eine Acutg, welchen allezeit die angehangte Waßer-Wage giebt, durch deßen abziehung denn auch sein *Complementum* alß der dritte Winckel bekannt wird, Und dann (3) die thonlegte Schnur oder *Hypothenusa* woran die Wage gehangen wird, So werden auß diesen *Datis* oder bekanten Dingen, die zwo unbekanten Seiten alß *Cathetus*, oder Seiger Teuffe, und *Basis* oder Sohle erkundiget, nach 4. Prop: lib: 6. Euclid: welche also lautet: *Aequiangulorum triangulorum proportionalia sunt latera, quae circum aequales angulos, et Homologa sunt latera quae aequalibus angulis subtenduntur.*

> Because the compass on the string always gives a *triangulum rectangulum*, in which two angles are known, as 1) *a right angle* that always has 90° and 2) the *acute angle* that is always given by the water level, through the subtraction of which its *complementum* is determined as the third angle and then 3) the *thonledgte Schnur* or *Hÿpotenusa* to which the compass is attached. From this *datis* or known things it is then enquired about the two missing sides, i.e. *chatetus* or *Seiger-Teŭffe* and *basis* or *Sohle*, according to *Prop*: 4 *lib*: 6 of Euclid's [Elements], which says: *Aequiangulorum triangulorum proportionalia sunt latera, quae circum aequales angulos, et Homologa sunt latera quae aequalibus angulis subtenduntur.*[68]

The author then gave his own German translation of the proposition, adding comments made by two German scholars, Victorinus Strigel (1524–1569) and Guilielmus Xylander (1532–1576), author of the first printed German version of the *Elements*. This excerpt is interesting: the author of the manuscript shows how surveying operations produced data that was then mathematically pro-

---

67 TU BAF – UB XVII 11, fols. 26v–43v.

68 TU BAF – UB XVII 11, fol. 30r.

cessed. It directly leads to the next section, an '*operatio*' giving two concrete applications of this method.[69]

The author could have stopped here, having presented a method and then shown how to apply it. He nevertheless chose to add a section entitled 'Demonstratio Figurarum præcedentium' [Proof of the previous figure], which is the only 'proof' contained in the manuscript.[70] It is not, as one could have expected, a copy of Xylander's proof of the proposition VI.4, but is much closer to a description of the law of sines.[71] This is a compelling and meaningful difference: there is indeed a 'demonstration', but neither in a strict mathematical sense nor about the proposition in itself. What is demonstrated is the validity and legitimacy of using such a procedure for calculations in the mines. The author did not refer to Euclid in order to give a university course or to prove eternal truths. He wanted to show that mathematics could be so useful that it was worth learning his more elaborate method. I think this is precisely what he meant in the closing remarks of the demonstration: 'Die Übung hierinnen, thutt die Augen deß Verstandes weiter und beßer auff' [the exercise in that [matter] opens the eyes of the understanding further and better].[72]

Conclusion

Subterranean geometers seem to have used Euclid with a specific goal. Their manuscripts were made of several parts, which had very different scopes. This was reflected by the use of three different languages (German, Latin and the *Bergmannsprache*) in varying proportions. Classical sources such as the *Elements* were mainly used in the more complex parts, dealing not only with the actual practical way of solving problems but introducing new procedures based on the geometry of the triangle.

The analysis of translations, including the reformulation and the adaptation of scholarly works to the real conditions of the mines, turns out to be a very useful tool to understand why and how Euclid's *Elements* were brought into the Ore Mountains of Saxony. More broadly, it gives us an insight into the sources

69 Having used Euclid's proposition VI.4 to prove that the angles measured with the instruments were equal to the angles of the ideal triangle representing the gallery, the sine and cosine tables of Simon Stevin were used to calculate the *cathetus* and the *basis*.

70 TU BAF – UB XVII 11, fol. 33r–33v.

71 Xylander Guilielmus, *Die Sechs Erste Bücher Euclidis, vom anfang oder grund der Geometrj* (Basel, Kündig: 1562) 158–160.

72 TU BAF – UB XVII 11, fol. 33v.

and the constitution of the corpus of subterranean geometry manuscripts at the turn of the seventeenth century. Rather than using available books written in Saxony about mining technologies, a mathematical substructure was built by adapting works from contemporary mathematicians (Simon Stevin, Benjamin Bramer), using a pseudo-translation (Christoph Puehler), translations and commentaries (Guilielmus Xylander, Victorinus Strigel).

These sources help us propose a new hypothesis about the early history of subterranean geometry. Long-existing practical processes seem to have been progressively elaborated as juridical systems (*Bergordnungen*) were codified. This gave the body of propositions or duties that the geometer had to perform. No Latin was to be found since no theoretical mathematics or calculations were needed at the time. It was enough to use similar triangles and wax circles to transfer the angles.

These tasks were at once juridical and technical, and would at times lead to situations where artisan procedures would reach their limits. At the turn of the seventeenth century, more advanced bits of knowledge would then be incorporated. Instead of drawing on the mining literature of Agricola, the *Markscheider* turned to the mathematical literature of their time. Puehler's *Anlaytung* (1563) must have been an important source, since it was literally copied. To this first gateway to Euclid, one should add the German adaptation of Xylander (1562). These sources were used both for the methods they presented and to provide some general remarks on the geometry of the triangle. To cope with the increasingly difficult mathematical content, selected theoretical considerations were studied, adapted and thus found their way into the corpus of subterranean geometry.

Selective Bibliography

Agricola Georg, *De re metallica* (Basilae, H. Froben: 1556).

Anonymous, *Geometria Subterranea*, manuscript (17th century), Technische Universität Bergakademie Freiberg, wissenschaftlicher Altbestand, XVII 11 / XVII 677 / Nachlass Scheidhauer 300m.

Heath, Thomas l., *The thirteen books of Euclid's Elements*, 3 vols. (Dover: 1956).

Launert D., *Wer erfand die Transversalteilung? Brahe, Ursus, Hommel, Pühler. Pühlers Practica Geometriae*, Algorismus 79 (Augsburg: 2014).

Long P., *Openness, Secrecy, Authorship, Technical Arts and the Culture of Knowledge from the Antiquity to the Renaissance* (Baltimore – London: 2001).

Morel T., "Le microcosme de la géométrie souterraine: échanges et transmissions en mathématiques pratiques", *Philosophia Scientiae* 19, 2 (2015) 17–36.

Puehler Christoph, *Ein kurtze und grundliche anlaytung zu dem rechten verstand Geometriæ* (Dillingen, Mayer: 1563).
Schneider, A., *Neu-Markscheide Buch*, manuscript (c. 1669), Technische Universität Bergakademie Freiberg, wissenschaftlicher Altbestand, XVII 18.
Zilsel E., *The Social Origins of Modern Science*, Boston Studies in the Philosophy of Science 200 (Dordrecht: 2000) 3–22.

CHAPTER 7

Image, Word and Translation in Niccolò Leonico Tomeo's *Quaestiones Mechanicae*

Joyce van Leeuwen

Introduction

The humanist Niccolò Leonico Tomeo (1456–1531) is mostly renowned for his teaching of Aristotle directly from the Greek text. In the year 1497 he was officially appointed as lecturer in Aristotelian philosophy at the University of Padua.[1] While lecturing at the universities of Padua and Venice, Tomeo displayed a particular interest in Aristotle's scientific treatises. He published a translation and commentary of the *Parva Naturalia* in the year 1523, followed by his *Opuscula* (1525), which contain amongst others paraphrases of Aristotle's *Progression of Animals* and *Movement of Animals*, as well as a Latin translation of the Aristotelian *Mechanics*, the *Quaestiones Mechanicae*. The *Quaestiones Mechanicae* were accompanied by Tomeo's explanatory comments and diagrams to render the difficulties of the text more accessible.

Tomeo's *Quaestiones Mechanicae* were not the first Latin translation of the *Mechanics*, as the Venetian Vittore Fausto had already produced one in the year 1517. However, it was Tomeo's work that made the treatise available to the Latin world in its most popular version. His translation went through many re-printings and became the standard translation in the sixteenth century.[2] It was used by several commentators on the text, such as Alessandro Piccolomini (1547) and Bernardino Baldi (1621). These commentators praise Tomeo's translation and especially his philological expertise, but occasionally mock his comments for being of negligible meaning.[3] Indeed, rather than presenting a thorough mathematical analysis of each of the individual ques-

1 See Geanakoplos D.J., "The Career of the Little-known Renaissance Greek Scholar Nicholas Leonicus Tomaeus and the Ascendancy of Greco-Byzantine Aristotelianism at Padua University (1497)", *Byzantina* 13 (1985) 361.

2 See Rose P.L. – Drake S., "The Pseudo-Aristotelian Questions of Mechanics in Renaissance Culture", *Studies in the Renaissance* 18 (1971) 79–80.

3 See Baldi Bernardino, *In mechanica Aristotelis problemata exercitationes* (Mainz, Johannes Albinus: 1621), "Praefatio" 6.

 | DOI 10.1163/9789004349261_009

tions, Tomeo's commentary consists of an extended paraphrase of the treatise. Tomeo's aim is to establish a sound textual basis by paying attention to alternate readings and by offering textual remarks. In doing so, Tomeo paved the way for future commentators in many respects.

This chapter focuses on Tomeo's philological activities in his translation project of the Aristotelian *Mechanics*. Along with the final product, the printed text of the *Quaestiones Mechanicae* that was published in Venice in 1525, we also possess a manuscript in the Vatican Library containing two different versions of the Latin translation. Codex *Vat.Reg.Lat.* 1291 by the hand of Tomeo illustrates several stages of revisions in the Latin text in advance of the printed edition. Interestingly, Tomeo employed similar revision procedures for the Greek text when he aimed to establish the original text of the *Mechanics*. This undertaking is visible in the Greek codex *Bern.* 402 that was copied by Tomeo himself and emended at some point after completion of the manuscript. Moreover, Tomeo added diagrams in the margins of the text that underwent various alterations and transformations before they were finally adopted in the *Quaestiones Mechanicae*. These different sources give evidence for Tomeo's practices as editor and translator of ancient scientific texts. In this chapter I will explore the development in Tomeo's translation of Greek into Latin, the alterations from manuscript to print, and the role of the different diagrams and images in the spread of mechanical knowledge in the sixteenth century.[4] Scholars in the early modern period could use different modes of visualization, depending amongst others on the functions of their texts and their target audiences. I will argue that Tomeo in his translation project consciously employed various visual strategies, sometimes changing them from one source to another. He included mathematical diagrams supplying proofs of the problems described in the *Mechanics*, but also images that were only intended as supplements to the text.

Humanist Translation

In his endeavor to reconstruct original Greek texts, Tomeo's philological activities can occasionally be connected with the printer and publisher Aldus Manutius who had founded the Aldine Press at Venice. A manuscript copied by Tomeo for example provided the basis for Manutius' edition of Galen from

4 For the translation of images from manuscript to print see also the chapter by Charles van den Heuvel in this volume.

the year 1525;[5] Tomeo also collaborated on the Aldine editions of Philoponus (1504)[6] and Plutarch (1509).[7] The aforementioned codex *Bern.* 402 provided the source for the Aldine of Theophrastus,[8] and also of several Aristotelian treatises, amongst which the *Mechanics*.[9] The appearance of the complete edition of Aristotle that was printed by Manutius in the final years of the fifteenth century was certainly helpful for Tomeo's teaching of Aristotle's philosophy, even though it did not contain any of the mechanical diagrams from *Bern.* 402. Tomeo's own library, which has for the most part been reconstructed by Fabio Vendruscolo (1996) and Eleonora Gamba (2014), further displays his interest in Aristotle's natural philosophy, as well as in other ancient scientific works. He possessed, amongst other things, manuscript copies of Euclid, Ptolemy, Galen and other Greek medical writers.[10]

Before being able to describe Tomeo's practices in editing and translating the Aristotelian *Mechanics*, the precise relationships between the different sources need to be determined. The available literature is indecisive as to which Greek manuscript provided the exemplar of Tomeo's *Quaestiones Mechanicae*. Tomeo himself makes a remark in this context in his preface to the Latin translation in codex *Vat.Reg.Lat.* 1291. He claims to have used a single copy of the text that was partly corrupt and without diagrams.[11] From this, Gamba infers that the translation was not made from the manuscript *Bern.* 402, but from its source.[12] It might be that Tomeo here referred to his exemplar rather than to *Bern.* 402 itself; another possibility is that he points to the Bern manuscript

5 See Vendruscolo F., "Manoscritti greci copiati dall'umanista e filosofo Niccolò Leonico Tomeo", in Funghi M.S. (ed.), *ΟΔΟΙ ΔΙΖΗΣΙΟΣ: Le vie della ricerca. Studi in onori di Francesco Adorno* (Florence: 1996) 550 and Gamba E., "Un nuovo manoscritto copiato da Niccolò Leonico Tomeo (Par.gr. 1833): Appunti per la ricostruzione della sua biblioteca", *EIKASMOS* XXV (2014) 346.

6 See Vendruscolo, "Manoscritti greci copiati" 552 and Gamba, "Un nuovo manoscritto" 349.

7 See Vendruscolo, "Manoscritti greci copiati" 546, n. 13 and Gamba, "Un nuovo manoscritto" 343.

8 See Burnikel W., *Textgeschichtliche Untersuchungen zu neun Opuscula Theophrasts* (Wiesbaden: 1974) 22–25.

9 See Sicherl M., *Griechische Erstausgaben des Aldus Manutius: Druckvorlagen, Stellenwert, kultureller Hintergrund* (Paderborn: 1997) 96 and Van Leeuwen J., *The Aristotelian Mechanics: Text and Diagrams* (Dordrecht: 2016) 47.

10 See Gamba, "Un nuovo manoscritto" 333.

11 See De Bellis D., "Niccolò Leonico Tomeo interprete di Aristotele naturalista", *Physis* 17 (1975) 91.

12 See Gamba, "Un nuovo manoscritto" 341. See Van Leeuwen, *The Aristotelian Mechanics* 47 on the source of *Bern.* 402 for the Aristotelian *Mechanics*.

before it was emended and diagrams were added. In any event, Tomeo's remark does not rule out the possibility of *Bern.* 402 being the source for the Latin translation. In the course of the chapter, I will show that the Bern codex was designed by Tomeo to provide the basis for his *Quaestiones Mechanicae* on different levels. Because of the fact that Tomeo did not find any diagrams in his Greek source, he had to draw these himself. Diagrams were not included only then when he translated the Greek text into Latin, but he designed them already for his basis text in *Bern.* 402. These diagrams served as clear suggestions for the Latin translation, without being completely identical to them.[13] Moreover, we will see that there are several instances in the Greek manuscript where Tomeo states that he considers a certain passage to be a scholium or commentary. Accordingly, these passages have been omitted from the Latin translation.

Another relationship in need of clarification is that of codex *Vat.Reg. Lat.* 1291 and the 1525 printed edition of the *Quaestiones Mechanicae.* I already mentioned that the manuscript in the Vatican Library contains two different versions of the Latin translation. The first version contains Tomeo's translation in the right column of the page, while his comments are arranged in the left column. However, this version contains only Tomeo's commentary to the introduction of the Aristotelian *Mechanics*, and from the first problem of the text onwards we find nothing but his translation and diagrams. The second version presents both the text and the comments in the right column of the page, and the diagrams are then inserted to the left where they accompany the corresponding textual descriptions. The first version has been correctly identified by Heribert Maria Nobis as the first draft of the Latin translation, whereas the second version represents Tomeo's ready-for-press manuscript.[14] When comparing the two texts with each other, we notice revisions in the text and diagrams which show that the second version in *Vat.Reg.Lat.* 1291 was specifically prepared by Tomeo for the print. The fact that we have both a reworked Greek manuscript *Bern.* 402 by Tomeo and two handwritten versions of his Latin translation enables us to precisely analyze the consecutive editorial stages in the process from Greek manuscript to Latin printed edition.

---

13 See Van Leeuwen J., "Thinking and Learning from Diagrams in the Aristotelian *Mechanics*", *Nuncius* 29 (2014) 73–86 and Van Leeuwen, *The Aristotelian Mechanics* 166–187.

14 See Nobis H.M., "Über zwei Handschriften zur frühneuzeitlichen Mechanik in italienischen Bibliotheken", *Sudhoffs Archiv* 53, 3 (1969) 328.

Codex *Bern.* 402 was copied by Tomeo at some time before the year 1497.[15] At a later point after its completion, Tomeo intensively revised the Greek text of the *Mechanics* and added diagrams.[16] These revisions give evidence for Tomeo's philological expertise. Some of them even leave their traces in modern critical editions of the Aristotelian *Mechanics*, as they were adopted in the Aldine, and thence influenced subsequent editions of the treatise. Tomeo's emendations of the text are presented at different levels, involving variant readings that are added within the text or in the margins, additions of omitted lines, or marginal remarks on a specific passage. Part of the emendations were already adopted in the first version of the Latin translation in codex *Vat. Reg.Lat.* 1291, and then found their way into the 1525 printed edition of the *Quaestiones Mechanicae*. At line 849a23 of the text we find for example a marginal addition specifying the center of two circles: 'περὶ τὸ αὐτὸ κέντρον add. τὸ A' [about the same center A]; in both the Latin manuscript and print it is translated by Tomeo as follows: 'circa idem centrum A'.[17] On fol.99r of *Bern.* 402 Tomeo added a remark in the margin that he considered lines 847a 27–28 to be a scholium. This passage carries further an earlier statement of the text that the mechanical discipline is equally concerned with mathematics and physics by specifying this relationship: 'τὸ μὲν γὰρ ὥς διὰ τῶν μαθηματικῶν δῆλον, τὸ δὲ περὶ ὃ διὰ τῶν φυσικῶν' [For the *how* is demonstrated by mathematics, but the *about* by physics]. Consequently, we see that this passage was omitted from the Latin manuscript and print. In this example a reader of the printed edition does not notice that several lines are missing from the text, as Tomeo's instruction on the apparent scholium is only present in the Greek manuscript. Occasionally, however, similar remarks were included by Tomeo in the final print. An asterisk on fol.26v marks a statement that has already earlier been mentioned by the author of the *Mechanics*, namely that an object that is moving in no fixed ratio and in no fixed time will not move along a straight line. According to Tomeo, this repetition in the text has the status of an inserted gloss. Another example can be found in the notes to problem 20 of the text on the working of a steelyard. Here, Tomeo describes that he left out several lines from the Greek

15 See Vendruscolo, "Manoscritti greci copiati" 549–550 and Andrist P., *Les manuscrits grecs conservés à la Bibliothèque de la Bourgeoisie de Berne—Burgerbibliothek Bern: catalogue et histoire de la collection* (Zurich: 2007) 188–196.

16 See Van Leeuwen, *The Aristotelian Mechanics* 47–48. I have shown that some of Tomeo's emendations derive from a Byzantine paraphrase of the treatise by Georgius Pachymeres, while others involve his own revisions.

17 See *Bern.* 402, fol. 101r. All translations from Greek and Latin are mine, unless otherwise stated.

text in his translation as these seemed to be small glosses. The relevant lines 853b 38–854a 3 that were omitted from Tomeo's translation offer a reiteration of what has been said before on the measuring of a weight. In these remarks Tomeo presents himself as a careful reader of the text, who expected brevity on the part of the original author. Those passages that involved clear repetitions of the text were therefore considered by Tomeo to be inserted glosses.

Other emendations in *Bern.* 402 were adopted only in the second version of *Vat.Reg.Lat.* 1291. In a passage from the first problem of the *Mechanics*, the author gives an analysis of movement in a circle. He compares the motions of the radii in two concentric circles along the circumference, that of the radius AB in the larger circle and AX in the smaller one. He shows that in the same time the point B on the radius is moved along the circumference to H, the point X will be moved to Θ. The Greek text is corrupted at this point in all manuscripts and lines 849b11–12 read in Bern as follows: 'ὥστε τὸ B ἐνηνέχθαι ἂν τὴν BH ἐν τῷ [lac.] ἐφ' οὗ X σημεῖον' [that B is moved along BH in the [lac.] the point X]. In the margin an emendation was added by Tomeo: 'ὥστε τὸ B ἐνηνέχθαι ἂν τὴν BH ἐν τῷ τούτῳ χρόνῳ ἐφ' οὗ τὸ X σημεῖον τὴν XΘ' [that B is moved along BH in the same time as the point X is moved along XΘ]. We notice that this emendation was not adopted until the second version of the Latin translation in *Vat. Reg.Lat.* 1291. The first version in this codex is similar to the uncorrected text in Bern: 'ita quod B sit latum ipsam BF in quo X signum est'; in the second version the words 'in quo X signum est' were crossed out by Tomeo and corrected in: 'ita quod B sit latum per ipsam BF in tanto tempore in quo M punctum per ipsam ML' (whereby point M in Tomeo's text and diagram corresponds with X in the Greek text). This emendation of a corrupted passage demonstrates Tomeo's understanding of the text and gives evidence for his accomplishments as an editor of ancient texts. Another good emendation was made in the next line 849b 13, where we see that the Greek word for center, τὸ κέντρον, was corrected by Tomeo in the margin of *Bern.* 402 into KH. While the first version of the Latin manuscript translates *centrum*, in the second version the labels *XF* were added above a deleted *centrum*. These examples further show that Tomeo made the second version of his Latin translation not on the basis of his first translation only, but he referred to the Greek text at each step in the translation process. Tomeo's practices as editor and translator were thus aimed at optimizing the Greek and the Latin text, and several rounds of revisions are still visible in both texts.

An examination of the Latin text in the different versions of codex *Vat.Reg. Lat.* 1291 further displays some accurate revision work on the part of Tomeo. In the first version for instance he consistently translated the Greek term for weight, βάρος, with *onus*, while he changed it in the second version into *pondus*.

When we compare the first version with the second version and the printed text of the *Quaestiones Mechanicae*, other alternative translations from the beginning of the *Mechanics* include: 847a 26 *speculationum*: *contemplationum*; 847b 22 *accidentes*: *contingentes*; 848a 4 *secundum*: *alterum*; 848a 12 *reducuntur*: *referuntur*; 848a 33–34 *moto*: *comoto*; 848a 34 *deprehendentes*: *animadvertentes*; 848b 2 *accurationes*: *exactiores*; 848b16 *feratur a*: *latum sit a*; 848b16 *versus d*: *ad d*; 849a24–25 *cd et be*: *cd be*; 849a 36–37 *ad rectam diametrum*: *perpendiculares diametro*; 849b 36 *in medio*: *ad medium*; 849b 38 and 850a 1 *truncus*: *nodus*. While these changes primarily involve a different diction or style, others present Tomeo's interpretation of the text. For example in problem 20 of the text on the working of a steelyard, Tomeo translated lines 854a 13–14 in the first version of the Latin manuscript as follows: 'hic autem sacoma facit' [but here it makes counterpoise], thereby offering a literal translation of the Greek term for counterpoise, σήκωμα. In the second version and the printed text, Tomeo translated more freely and added meaning to this sentence: 'hic autem aequilibrium facit' [but here it makes equilibrium]. Tomeo's corrected translation makes more sense in this context, as it is the counterpoise in the balance that induces a situation of equilibrium. We notice that Tomeo every now and then abandoned a literal translation of the Greek in order to add clarity to the text. It is in such changes that his translation serves equally as a commentary. Tomeo's activities in his translation project may be best characterized as textual criticism. Tomeo tried to reconstruct the original text of the *Mechanics* by comparing variants in manuscripts and by proposing emendations of corrupted passages. He extended his philological activities to the Latin translation as well, of which the different versions show several rounds of revisions with the aim of achieving an accurate translation.

The observation that Tomeo stayed faithful to the Greek text is supported by the fact that he frequently preferred to use transliterations of Greek technical terms rather than looking for a Latin equivalent. Tomeo translated for instance the Greek word for fulcrum, ὑπομόχλιον, with *hypomochlion* (problem 3), κρόκη, pebble, with *croca* (problem 15), τροχιλεία, pulley, with *trochlea* (problems 8, 9 and 18), or κήλων, swing-beam, with *celonia* (problem 28). At times early modern translators and commentators had trouble to find appropriate expressions in Latin, for which reason they may have chosen to use transliterations of Greek terminology. The Italian humanist Piccolomini in his 1547 Latin paraphrase of the *Mechanics*, for example, complains at several points in his text that the Latin lacks a word for a particular term.[18] However, in the

18 See e.g. Piccolomini Alessandro, *In mechanicas quaestiones Aristotelis, paraphrasis paulo quidem plenior* (2nd ed., Venice, Traianus Curtius: 1565), fol. 37r.

specific examples given above, Latin equivalents were available to Tomeo. In the first Latin translation of the *Mechanics* from the year 1517, Fausto translated for example ὑπομόχλιον with *pressio*, κρόκη with *scrupulus* and *calculus*, or κήλων with *tolleno*. From Tomeo's comments in the *Quaestiones Mechanicae*, we notice that he knew the Latin words but decided against using them in his translation. In his notes to problem 3, Tomeo mentions that the Latin word for what the Greeks name *hypomochlion* is *pressio*, although he did not use it in his translation. Another example can be found in the comments to problem 8, where Tomeo states that *rechamus* would be the appropriate term for pulley, while he himself in his translation sticks with the transliterated Greek word *trochlea*. It is thus a conscious decision by Tomeo to remain close to the Greek text by not making use of existing Latin vocabulary or inventing new words for Greek technical terms. This practice conforms to the university circles in which Tomeo moved. His work was targeted at a humanistically-educated audience of people who could understand very well the Greek terminology. We could define Tomeo's *Quaestiones Mechanicae* as an erudite and humanist translation, while Fausto's work can rather be described as a working translation addressed to practitioners in the field, as is illustrated by his consistent use of Latin colloquial words for mechanical vocabulary.

Tomeo's philological activities are also reflected in the notes to his translation, in which he makes detailed remarks on the meaning and use of words. For instance, problem 20 of the *Mechanics* describes the working of a φάλαγξ, or steelyard. Tomeo starts his comments by saying that the Latin terms for the weighing device that is called a *phalanx* in Greek are *statera* or *trutina*. Before entering into the topic of the Greek text, Tomeo discusses other meanings of the words *phalanx*. Apart from its most common use as a line of battle, it also signifies a carrying-pole, by means of which carriers are able to divide heavy loads equally among them. Depending on the number of carriers, which are either four, six, or eight persons, they are called in Greek *tetraphori, hexaphori*, or *octophori*. Another problem of the text aims to explain the round shape of pebbles on the beach (problem 15). The Greek word used here for pebble by thc author of the *Mechanics* is κρόκη. Tomeo notes that he cannot remember to have read this word anywhere else in this context; the word for pebble that is usually found in Greek poets is κροκάλη. Occasionally, in the *Quaestiones Mechanicae* we also find digressions on words that are derived from a specific term. In his comments to problem 23 of the text, Tomeo gives a definition of the properties of a rhombus, followed by the statement that there is a certain type of fish that owes its name to this geometrical figure because of its similar shape. These and similar remarks show that Tomeo in his commentary set a clearly different focus apart from explaining the mechanical subject matter of the treatise. His concern for textual criticism is even extended to other classical texts, as the comments to problem 28 point out. Tomeo remarks that, although the common Latin term for the Greek word for swing-beam, κήλων, is *tolleno*, there were some older Latin authors who had used the Greek form *celonia*. Due to transmission errors and misreadings, these words, however, have not been correctly transmitted. In the remainder of his notes to this problem, Tomeo quotes passages from Plautus' *The Rope*, Columella's *On Agriculture* and Pliny's *Natural History*, for which he proposes an emendation and believes that the variant *celonia* would have provided the correct reading.

Detailed philological analysis is therefore at the heart of Tomeo's translation project of the Aristotelian *Mechanics*. For Aristotle's zoological works, it has been shown that Tomeo's method is similar to that of an ancient commentator, who is primarily concerned with the original text rather than with accumulating the positions of previous commentators.[19] An interpretation of a text must therefore be firmly grounded in philology, which also turns out to be Tomeo's guiding principle for the *Quaestiones Mechanicae*. With his Latin translation of the treatise, Tomeo thus paved the way for later sixteenth- and seventeenth-century commentators of the *Mechanics*, who could fully focus on the interpretation of the mechanical contents and thereby make an important contribution to the development of modern science. Before turning to some of the problems in more detail, I will first look at Tomeo's diagrams, as these occasionally present divergences from the text and consist of his commentaries on the mechanical content.

Visualizations of Mechanical Knowledge

In the prefatory letter to codex *Vat.Reg.Lat.* 1291, Tomeo mentions that there were no diagrams present in his source, for which reason he had to devise these himself. He included diagrams in his manuscript *Bern.* 402 that served as a model for the ones in the printed translation. In the Greek manuscript Tomeo did not only add the diagrams that were prescribed by the author, but he also inserted visualizations to different passages in the text. Some of the diagrams from Bern were adopted in the same form in *Vat.Reg.Lat.* 1291 and the printed text of the *Quaestiones Mechanicae*. The diagram of a steelyard (problem 20),

19 See Perfetti S., "Three Different Ways of Interpreting Aristotle's *De Partibus Animalium*: Pietro Pomponazzi, Niccolò Leonico Tomeo and Agostino Nifo", in Steel C. – Guldentops G. – Beullens P. (eds.), *Aristotle's Animals in the Middle Ages and Renaissance* (Louvain: 1999) 307.

a dental forceps (problem 21), or a nutcracker (problem 22) found their way unchanged into the 1525 print. There are also diagrams that are not present in Bern, but included only in the codex from the Vatican Library and the printed text, such as a diagram of two concentric circles to problem 8 of the text, which asks why circular objects are easily moved. Several diagrams in *Bern.* 402 turn out to be trials for the ones in the Latin translation. The second problem of the text involves a question on the different suspension points of balances. While two diagrams are required by the textual description, one of a balance suspended from above and the other of a balance supported from below, Tomeo included five diagrams of balances in his Greek manuscript [Fig. 7.1]. The two diagrams contained at the bottom of the page [Fig. 7.1], one representing each case, appear to be Tomeo's drafts for the diagrams of balances in his Latin translation. It also occurs that diagrams from the Bern manuscript were adopted in the first version of the Latin manuscript, and subsequently altered in the second version and printed edition. To problem 25 of the text for example, a question concerned with the cording of beds, *Bern.* 402 and the first version in *Vat.Reg.Lat.* 1291 both contain three diagrams of beds. These diagrams correspond with the defective description in the text and depict only partially corded beds. Two of them were copied into the second version, and, in addition, Tomeo added two more diagrams of fully corded beds. Apparently, he did not consider the diagrams based on the text sufficient to illustrate the principle and added two diagrams of actual beds from his own experience. Interesting in this context is the fact that Tomeo did not make a remark on it in his text, but only commented upon the bed problem by means of these additional diagrams. Together, these four diagrams were adopted in the printed edition. The diagrams in Tomeo's translation clearly demonstrate that the second version in *Vat.Reg.Lat.* 1291 represents the ready-for-press manuscript of the printed text. This observation is confirmed by the letter labels in the different versions: the labels in *Bern.* 402 and the first version of *Vat.Reg.Lat.* 1291 are all in lower-case letters, whereas the second version and the printed text contain upper-case letters.

The different stages of revisions, from the Greek text to the Latin translation and within the different versions of the Latin, underline Tomeo's deep engagement in philology. However minor the changes in the diagrams from the Bern manuscript to the printed translation may appear sometimes, they illustrate a different conception of diagrams in the early modern period. When we look again at the second problem of the text, we notice the different appearance of the balances in the Latin translation. Figure 7.2 from the 1525 printed edition of the *Quaestiones Mechanicae* depicts a balance that is suspended from above, and corresponds with the bottom left diagram from

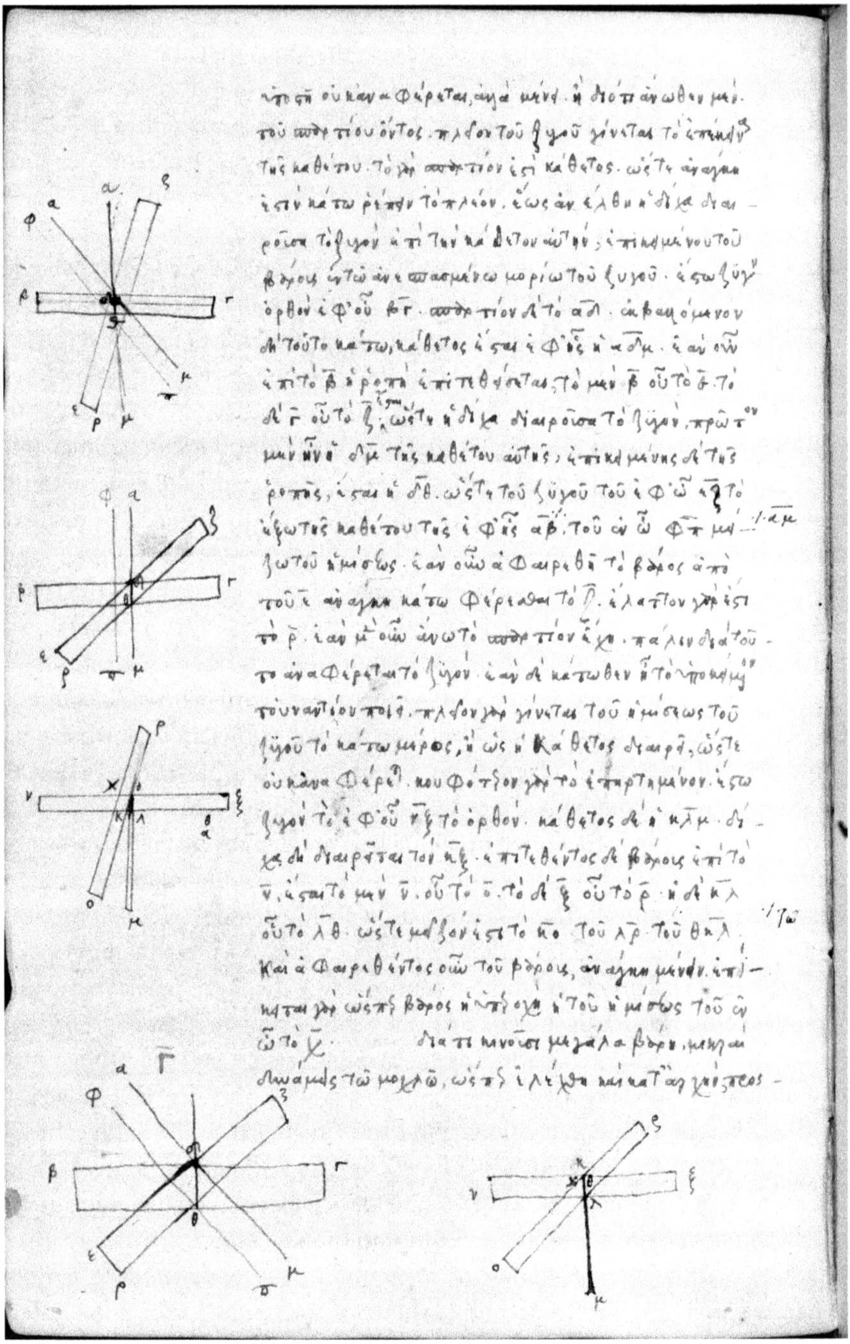

FIGURE 7.1 *Niccolò Leonico Tomeo, "Five diagrams of balances" (15th century, before 1497). Pen on paper, Burgerbibliothek Bern, Cod. 402, fol. 102v.*

Bern. 402 [Fig. 7.1]. In the Bern manuscript Tomeo depicted the balances by double lines and thereby paid attention to the physicality of the device. In response to the question why a balance suspended from above moves back to its original position after removal of the weight, the author of the *Mechanics* explains that in this case the raised part of the balance beam is heavier than the lowered part. The answer can be read off from the diagram in Tomeo's Greek manuscript as the thickness of the beam acts as a weight pushing down the balance. In codex *Vat.Reg.Lat.* 1291 and the printed translation, however, we do not find such a relation between the text and the diagrams. Now the balances are depicted by single in place of double lines and there is no longer an argument from the diagram. This example illustrates the different status of at least some of the diagrams in the *Quaestiones Mechanicae*. These diagrams play no part in the demonstration of the mechanical questions, but are turned into pictures merely illustrating the text.[20] Their altered status is further enforced by the addition of pictorial elements, such as the scales and the weights that are drawn in a three-dimensional representation. When tracing the developments of the diagrams in the different sources—from geometrical depictions of balances in the Greek manuscript to pictures in the Latin translation—we notice that Tomeo himself was responsible for making the changes in the print. The diagram of a wedge inserted at problem 17 of the text looks very similar in all sources apart from a stylistic difference. In the second version of *Vat.Reg.Lat.* 1291, Tomeo added three-dimensional effects to the diagram that were adopted in the same form in the printed text. Therewith, Tomeo purposely devised different looking diagrams for his Latin translation. When compared with the strictly geometrical diagrams from earlier traditions, in Tomeo we find, in addition, images and pictures illustrating certain mechanical devices. These visualizations foreshadow a different kind of diagrams which becomes more prominent in later commentators from the early modern period.[21] There we often find pictures of machines and other mechanical devices that bear only a very loose relation to the textual description. These pictures reflect the altered scope of mechanics in the sixteenth century. Whereas the primary focus of the ancient text is on the theoretical foundations of the mechanical discipline, the early modern interest in the treatise lies precisely at the intersection of theoretical and practical knowledge.[22]

20 See Van Leeuwen, "Thinking and Learning" 85.

21 See Van Leeuwen, *The Aristotelian Mechanics* 170–178.

22 See Valleriani M., "The Transformation of Aristotle's *Mechanical Questions*: A Bridge Between the Italian Renaissance Architects and Galileo's First New Science", *Annals of Science* 66, 2 (2009) 183–208 on the early modern interplay between theoretical and practical knowledge. Problem 16 of the *Mechanics* is used as a case study to illustrate the transformation of mechanical knowledge in the early modern period.

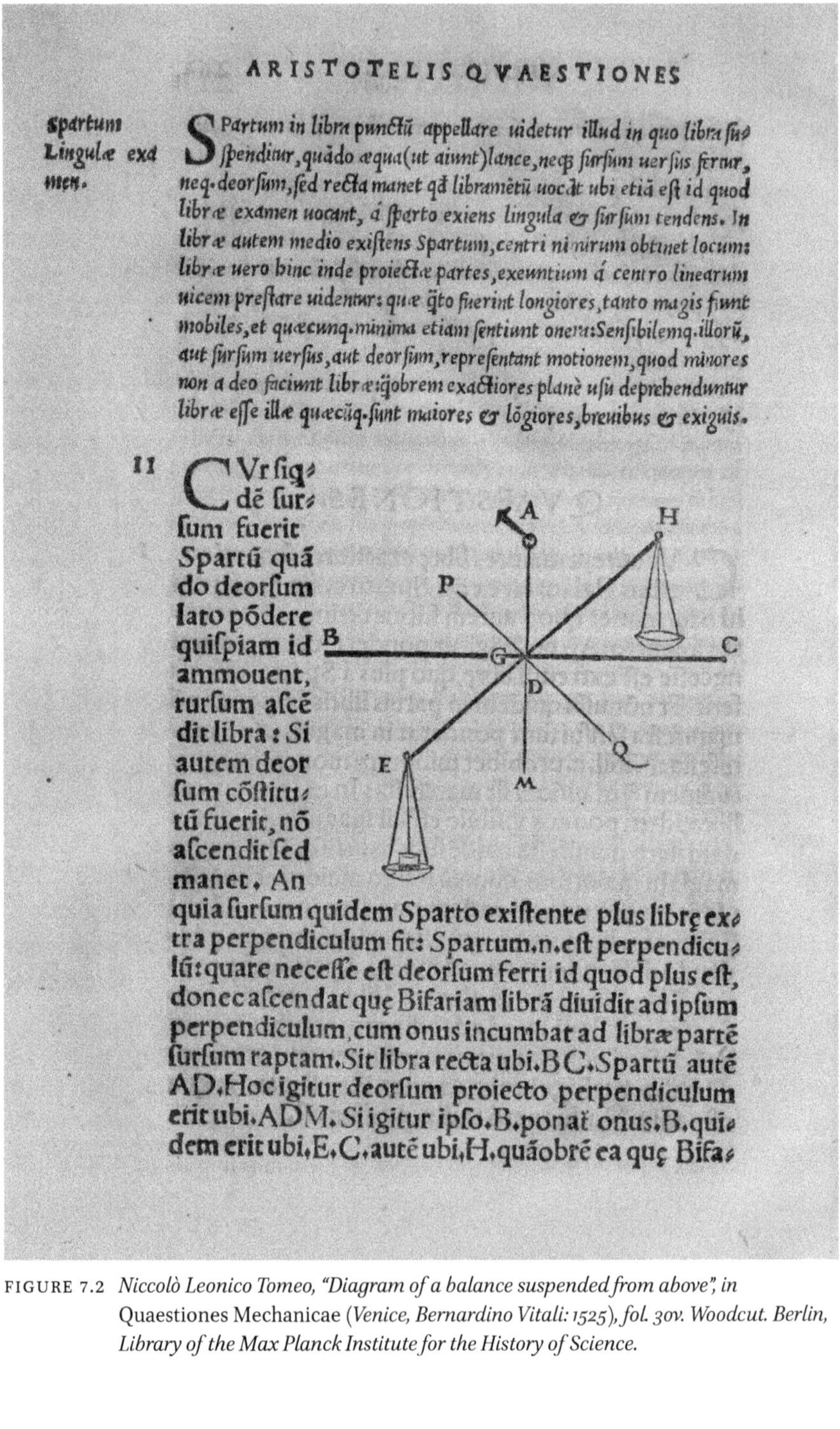

ARISTOTELIS QVAESTIONES

Spartum LinguIæ examen.

SPartum in libra punctũ appellare uidetur illud in quo libra ſuſpenditur, quãdo æqua (ut aiunt) lance, neq; ſurſum uerſus fertur, neq. deorſum, ſed recta manet qđ libramẽtũ uocãt ubi etiã eſt id quod libræ examen uocant, á ſparto exiens lingula & ſurſum tendens. In libræ autem medio exiſtens Spartum, centri nimirum obtinet locum: libræ uero hinc inde proiectæ partes, exeuntium á centro linearum uicem preſtare uidentur: quæ q̃to fuerint longiores, tanto magis fiunt mobiles, et quæcunq. minima etiam ſentiunt oneru: Senſibilemq. illorũ, aut ſurſum uerſus, aut deorſum, repreſentant motionem, quod minores non a deo faciunt libræ: q̃obrem exactiores planè uſu deprehenduntur libræ eſſe illæ quæcũq. ſunt maiores & lõgiores, breuibus & exiguis.

II CVr ſiquidẽ ſurſum fuerit Spartũ quãdo deorſum lato põdere quiſpiam id ammouent, rurſum aſcẽdit libra: Si autem deorſum cõſtitutũ fuerit, nõ aſcendit ſed manet. An quia ſurſum quidem Sparto exiſtente plus librę extra perpendiculum fit: Spartum.n.eſt perpendiculũ: quare neceſſe eſt deorſum ferri id quod plus eſt, donec aſcendat quę Bifariam librã diuidit ad ipſum perpendiculum, cum onus incumbat ad libræ partẽ ſurſum raptam. Sit libra recta ubi.BC. Spartũ autẽ AD. Hoc igitur deorſum proiecto perpendiculum erit ubi.ADM. Si igitur ipſo.B. ponat onus.B. quidem erit ubi.E.C. autẽ ubi.H. quãobrẽ ea quę Bifa-

FIGURE 7.2 *Niccolò Leonico Tomeo, "Diagram of a balance suspended from above", in* Quaestiones Mechanicae *(Venice, Bernardino Vitali: 1525), fol. 30v. Woodcut. Berlin, Library of the Max Planck Institute for the History of Science.*

While Tomeo in his translation included diagrams to each passage in the *Mechanics* in which the Greek author prescribed one by using letter labels in the text, we only rarely find visualizations in the commentary section of his work. Tomeo's notes referring to the meaning and use of Greek and Latin terminology do not require any diagrams, but this is different for those passages in which Tomeo describes specific mechanical devices. In his remarks to problem 18 of the text on the working of pulleys, Tomeo discusses a system that is in use in building construction for the lifting of heavy weights. In order to explain the details of this system of pulleys and levers, Tomeo quotes a passage from Book X of Vitruvius' *On Architecture*. Tomeo's readers would certainly have benefitted from a diagram of this mechanical contrivance, as the description is rather technical and difficult to understand without an accompanying diagram. However, for Tomeo it would go beyond the scope of his commentary to supply diagrams of machines other than the ones described by the author of the *Mechanics*. For a picture of this specific machine, he refers to the excellent illustrations by the architect Giovanni Giocondo in his 1511 edition of Vitruvius' *On Architecture*. This shows again that Tomeo's aim was to produce a translation and commentary that focussed on the philological aspects of the text; for those readers with an interest in the more technical details of machines, he refers to Vitruvius' work on architecture at several points throughout his commentary.

We have seen that some of Tomeo's diagrams can be understood as an extension of and commentary on the text, as in problem 25 on the cording of beds in which additional diagrams were included to explain the principle that remained incomplete in the original text. In the next section I will look more closely at the text-image relationship in the different sources and take problems 3–5 of the *Mechanics* as a case study to illustrate Tomeo's translation project.

Ships and Levers

The third problem of the *Mechanics* asks why it is that small forces can move great weights with the help of a lever. In the author's explanation of this question, we find an early formulation of the law of the lever (850a39–850b2): 'ὃ οὖν τὸ κινούμενον βάρος πρὸς τὸ κινοῦν, τὸ μῆκος πρὸς τὸ μῆκος ἀντιπέπονθεν' [Therefore, as the weight moved is to the weight moving it, so, inversely, is the length of the one arm to the length of the other]. This explanation of the working of the lever is a key problem in the treatise, since many other mechanical devices, such as the oar in problem 4 and the rudder in problem 5, are expounded by the author while referring back to the lever principle. The end of the third problem includes a short description of a diagram, by which all elements of the lever are identified. The *Quaestiones Mechanicae* present a literal translation of the Greek text. Certain emendations and errors in the two Latin versions in codex *Vat.Reg.Lat.* 1291 again illustrate that the second version contains the final text for the printed translation. In the first version of the Latin manuscript, Tomeo translated lines 850a37–39 from the Greek as follows: 'sint autem tria circa vectem, hypomochlion quidem spartum et centrum, duo vero pondera, quod movet, et quod movetur' [and there are three elements in the lever, the fulcrum, viz. the cord and center, and the two weights, that which moves, and that which is moved]. As Table 1 shows, part of these lines were unintentionally omitted by Tomeo from the second version, and are therefore missing from the print as well. The table further illustrates that the translation *onus* for the Greek word for weight, βάρος, has been replaced a number of times by *pondus*. This revision was also carried out in Tomeo's diagram accompanying problem 3. Tomeo did not only use the letter labels from the text in his diagram, but also added Latin terms to indicate the respective parts of the lever: the fulcrum (*hypomochlion*), the moving force (*movens*), and the weight are identified. In the first diagram in *Vat.Reg.Lat.* 1291, the weight is named *onus*, whereas in the diagram in the second version and the printed edition it has been emended in *pondus* [Fig. 7.3]. In his comments accompanying this problem, Tomeo makes several linguistic remarks. When describing a lever as an instrument to move weights more easily, Tomeo notes for instance that the shorter end of a lever is called a *lingua vectis*. Next he remarks that the Latin equivalent of the Greek term *hypomochlion* is *pressio*. We see here that Tomeo chooses deliberately to retain the Greek terminology in his Latin translation by using transliterations, and that only in his comments he refers to the corresponding Latin concepts.

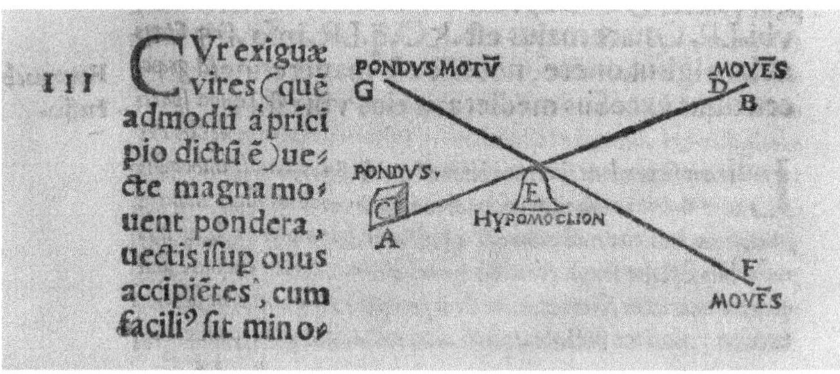

FIGURE 7.3 *Niccolò Leonico Tomeo*, "Diagram of a lever", *in* Quaestiones Mechanicae (*Venice, Bernardino Vitali: 1525*), *fol. 3rv. Woodcut. Berlin, Library of the Max Planck Institute for the History of Science.*

In the next short question the author of the *Mechanics* addresses a ship-problem which can be related to the previous problem of the lever. Problem 4 asks why it is that the rowers amidships contribute most to its movement. The reason for this, according to the author of the text, is that the oar works like a lever. The position where the distance between the rower and the fulcrum, i.e. the pin to which the oar is attached, is greatest, the ship is moved forward most. This occurs precisely in the middle of the ship, since here the ship is widest and a greater part of the oar is contained inside the ship when compared with the prow or stern. Tomeo's Latin text offers an accurate translation of the Greek and displays some revisions in preparation for the print, for instance at lines 850b13 and 850b14 (see Table 1). The comments give again evidence of Tomeo's linguistic interests when he asserts that the rowers on the different benches were distinguished by a particular name, depending on their position in the ship: *thranitae, thalamii* and *zeugitae*, for those rowers at the stern, prow or middle of the ship. In this context Tomeo does not remark on the varying lengths of oars of the different rowers, a fact that could be related to the author's question as to which group of rowers brings the ship forward most.

The fifth problem of the text is also concerned with seafaring and inquires about the working of a ship's rudder. Just as in the case of an oar, the rudder can be explained by the lever principle: the rudder is the lever, the point where it is attached to the ship the fulcrum, the sea the weight, and the steersman the moving force. The author of the *Mechanics* contrasts the principle of the rudder with the oar, in that they differ in the direction in which they work. An oar acts upon the side of a ship and contributes to its forward movement, whereas a rudder is placed at the end of a ship and helps in turning it. One reason why the rudder is attached to the end of the ship, the author continues, is that it can cause a greater deflection from that position. He gives a kind of mathematical argument for it, namely that the same angle faces a larger base, and the lines containing it are greater. We will see that in his commentary Tomeo reflects on this statement. A description of the device in terms of a diagram is included in the next lines (851a17–28), which I reconstructed as follows:

> ἔστω γὰρ ἡ ΑΒ κώπη, τὸ δὲ Γ ὁ σκαλμός, τὸ δὲ Α τὸ ἐν τῷ πλοίῳ, ἡ ἀρχὴ τῆς κώπης, τὸ δὲ Β τὸ ἐν τῇ θαλάττῃ. εἰ δὴ τὸ Α οὗ τὸ Δ μετακεκίνηται, τὸ Β οὐκ ἔσται οὗ τὸ Ε· ἴση γὰρ ἡ ΒΕ τῇ ΑΔ. ἴσον γοῦν[23] μετακεχωρηκὸς ἔσται. αλλ' ἦν ἔλαττον. ἔσται δὴ οὗ τὸ Ζ. τὸ Θ ἄρα τέμνει[24] τὴν ΑΒ, καὶ οὐχ ἡ τὸ Γ, καὶ κάτωθεν. ἐλάττων γὰρ ἡ ΒΖ τῆς ΑΔ, ὥστε καὶ ἡ ΘΖ τῆς ΔΘ· ὅμοια γὰρ

23 οὖν *Bern.* 402 and editions.

24 Ζ. τὸ Θ ἄρα τέμνει *Par.gr.* 2115; Ζ ἢ τὸ Θ. ἄρα τοίνυν *cett.*

τὰ τρίγωνα. μεθεστηκὸς[25] δὲ ἔσται καὶ τὸ μέσον, τὸ ἐφ' οὗ Γ· εἰς τοὐναντίον γὰρ τῷ ἐν τῇ θαλάττῃ ἄκρῳ τῷ Β μεταχωρεῖ, ᾗπερ τὸ ἐν τῷ πλοίῳ ἄκρον τὸ Α μετεχώρει δὲ τὸ Α[26] οὗ τὸ Δ. ὥστε μετακινηθήσεται τὸ πλοῖον, καὶ ἐκεῖ οὗ ἡ ἀρχὴ τῆς κώπης μεταφέρεται.

For let AB be the oar, Γ the pin, and A the end of the oar inside the ship, while B is the end in the sea. Then if A be moved to Δ, B will not be at E; for BE is equal to AΔ, and its displacement would be the same. But it is smaller, and it will be at Z. Θ then cuts AB not at Γ but below it. For BZ is less than AΔ, so that ΘZ is less than ΔΘ; for the triangles are similar. The center Γ will also be displaced; for it moves in a contrary direction to B, the end of the oar in the sea, and it moves in the same direction as A, the end in the ship, and A will be at Δ. So the ship will also be moved there where the end of the oar is displaced.

There are several corruptions in the Greek text, which led to variations in the manuscripts. Tomeo translated the description of the diagram in his *Quaestiones Mechanicae* as follows:

Sit enim AB remus, C vero scalmus; A autem in navigio sit remi principium, B vero in mari palmula. Si igitur A ubi D translatum est, non erit B ubi E. Aequalis enim BE ipsi AD. Aequale igitur translatum erit. Sed erat minus. Erit igitur ubi est F. [om. 851a22–23]. Minor enim BF ipsa AD, quare ipsa GF ipsa DG: similes enim sunt trianguli. Stans autem erit medium, ubi est C. In contrarium enim ipsi quod in mari est, extremo videlicet B procedit, ubi extremum in navigio est A. Non procederet autem ubi est D, nisi commoveretur navigium et eo transferretur ubi remi est principium.[27]

We notice some differences in Tomeo's text when compared with the Greek and the English translation. As visible from footnote 24, the Greek manuscript tradition contains an incomprehensible reading at lines 851a22–23. The scribe of *Par.gr.* 2115 proposed a sound emendation by replacing the particle τοίνυν [therefore] with the verb form τέμνει [it cuts]. Tomeo did not find this emendation in his source for *Bern.* 402, but only the variant τοίνυν of all other manu-

25 καθεστηκὸς Pachymeres and editions.

26 μετεχώρει δὲ τὸ A *Par.gr.* 2115; μὴ ἐχώρει *cett.*

27 Leonico Tomeo Niccolò, *Opuscula nuper in lucem aedita* [...] (Venice, Bernardino Vitali: 1525), fol. 33v.

TABLE 1 *A selection of variant readings in problems 3–5*

Source	*Bern.* 402	*Reg.Lat.* 1291 – 1	*Reg.Lat.* 1291 – 2	*Print* (1525)
850a30, 37, 39, 850b1, 7, 9	βάρος	onus	pondus	pondus
850a37–39	ἔστι δὲ τρία τὰ περὶ τὸν μοχλόν, τὸ μὲν ὑπομόχλιον, σπάρτον καὶ κέντρον	sint autem tria circa vectem, hypomochlion quidem spartum et centrum	om.	om.
850b12, 14	βάρος	onus	pondus	pondus
850b13	ναύτης	nauta	remex	remex
850b14	πλέον	plus	magis	magis
850b15	βάρος	onus	id	id
850b30	δυνάμεως	potentiā	viribus	viribus
850b30	ἠρεμαίας	quiescitis	modice utentis	modice utentis
850b32–34	om.	om.	om.	om.
851a22–23	ἢ τὸ Θ ἄρα τοίνυν τὴν AB, καὶ οὐχ ἡ τὸ Γ, καὶ κάτωθεν; remark in margin that he considers it a scholium	om.	om.	om.
851a28–29	τὸ αὐτὸ	idem	id ipsum	id ipsum

scripts. Possibly due to the corruption of this passage, Tomeo added in the margin of fol.104r that he considered these lines a scholium and consequently omitted them from his Latin translation.[28] At line 851a24 an emendation was proposed to read καθεστηκὸς [to stay in place], for μεθεστηκὸς [to change

28 See Table 1 for the incomprehensible reading at lines 851a 22–23 in *Bern.* 402 and the omission of this passage in Tomeo's Latin texts. Another omission in problem 5, which was already missing from codex *Bern.* 402, concerns lines 850b 32–34.

position]. This emendation does not make sense, since it is important in the argument of the text that the fulcrum changes its position as well. Therefore, it is all the more surprising to see that Tomeo adopted the emendation and translated it in his Latin translation as *stans* [to stand still]. As becomes apparent from the Bern manuscript, Tomeo corrected the reading μεθεστηκὸς in his text by replacing the first letter with a *kappa*.

The diagram does not explain the working of a ship's rudder, but rather expounds the previous question on oars. After his description the author simply remarks that a rudder works in the same way, except that it acts upon the end of the ship, thereby changing its position. While the Greek manuscript tradition contains one diagram based on the description in lines 851a17–28, Tomeo included three diagrams to this problem in *Bern.* 402, as Figure 7.4 illustrates. All three diagrams were copied into the first version of *Vat.Reg.Lat.* 1291. The first of them depicts an isosceles triangle standing on its top. It is shown that when both legs are extended the base will grow larger as well. The letter labels in the diagram do not refer to a description in the text, but as we will see, Tomeo gives an explanation of the diagram in the comment section of his Latin translation. The second diagram of a rudder complies with the author's description. The fulcrum remains at the same position in the diagram in accordance with the reading καθεστηκὸς, which shows that Tomeo drew the diagram after his revision of the Greek text. There is no reference in the text to the third diagram, it only depicts the prow and stern of the ship, in *Bern.* 402 with the words πρῷρα and πρύμνα, in *Vat.Reg.Lat.* 1291 with the corresponding Latin terms *prora* and *puppis*.

In the second version of the manuscript in the Vatican Library, Tomeo made some further modifications in the diagrams. Now a total of five diagrams are included to this problem, all of which were adopted in the printed translation. The first diagram does not have an explanatory function at all; it depicts an isosceles triangle on its top, and, since there are no letter labels included in the diagram, it remains unclear what it represents exactly. Figure 7.5 shows the next two diagrams from the 1525 *Quaestiones Mechanicae*. These are contained to the description in lines 851a17–28, the first is similar to the diagram in Tomeo's Greek manuscript [Fig. 7.4: second diagram] and the first version of his Latin translation, the second presents an alternative diagram to the same text passage. Apart from the different orientation of the two diagrams, we notice one significant difference: in the first diagram point C, the pin of the oar which acts as the fulcrum of the lever, stays in the same position, whereas in the second diagram it has moved to G. These variants may be related to the different readings in the Greek text. The first diagram corresponds to the emendation of μεθεστηκὸς in καθεστηκὸς that was adopted by Tomeo in *Bern.* 402 and

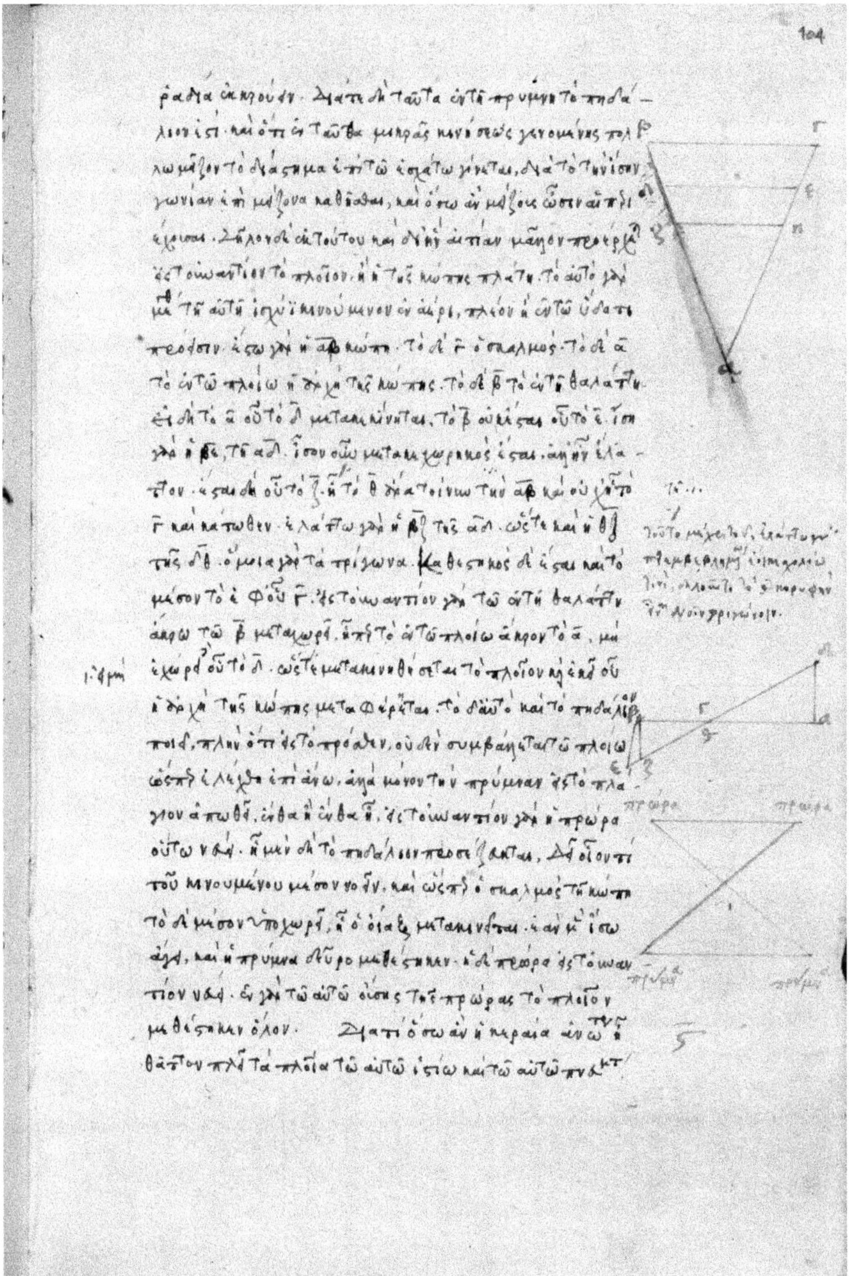

FIGURE 7.4 *Niccolò Leonico Tomeo,* "Diagrams of a ship's rudder" *(15th century, before 1497). Pen on paper. Burgerbibliothek Bern, Cod. 402, fol. 104r.*

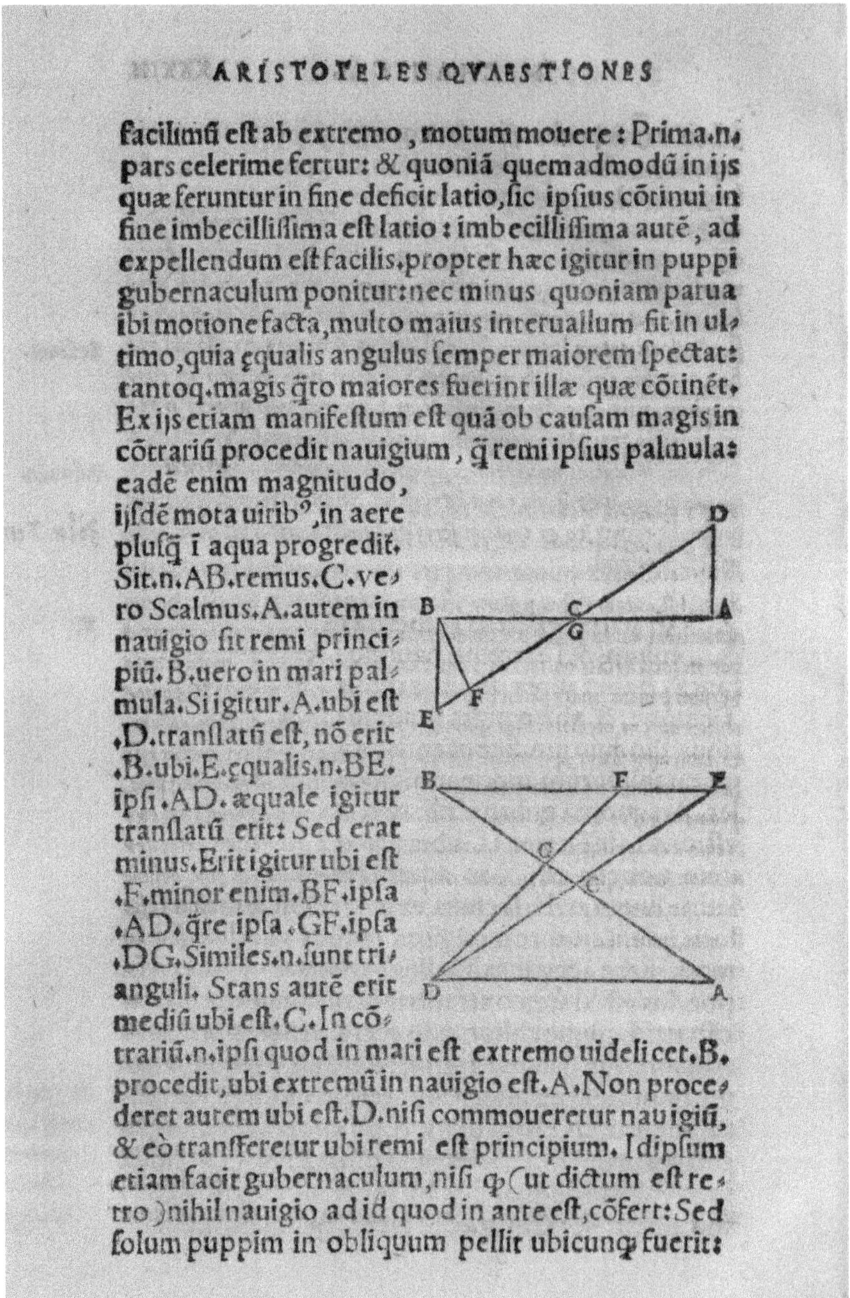

ARISTOTELES QVAESTIONES

facilimũ est ab extremo, motum mouere: Prima.n.
pars celerime fertur: & quoniã quemadmodũ in ijs
quæ feruntur in fine deficit latio, sic ipsius cõtinui in
fine imbecillissima est latio: imbecillissima autẽ, ad
expellendum est facilis, propter hæc igitur in puppi
gubernaculum ponitur: nec minus quoniam parua
ibi motione facta, multo maius interuallum fit in ul-
timo, quia æqualis angulus semper maiorem spectat:
tantoq. magis q̃to maiores fuerint illæ quæ cõtinẽt.
Ex ijs etiam manifestum est quã ob causam magis in
cõtrariũ procedit nauigium, q̃ remi ipsius palmula:
eadẽ enim magnitudo,
ijsdẽ mota uirib⁹, in aere
plusq̃ ĩ aqua progredit.
Sit.n.AB.remus.C.ve-
ro Scalmus.A.autem in
nauigio sit remi princi-
piũ. B.uero in mari pal-
mula. Si igitur.A.ubi est
.D.translatũ est, nõ erit
.B.ubi.E.æqualis.n.BE.
ipsi.AD.æquale igitur
translatũ erit: Sed erat
minus. Erit igitur ubi est
.F.minor enim.BF.ipsa
.AD.q̃re ipsa.GF.ipsa
.DG.Similes.n.sunt tri-
anguli. Stans autẽ erit
mediũ ubi est.C.In cõ-
trariũ.n.ipsi quod in mari est extremo uidelicet.B.
procedit, ubi extremũ in nauigio est.A.Non proce-
deret autem ubi est.D.nisi commoueretur nauigiũ,
& eò transferetur ubi remi est principium. Idipsum
etiam facit gubernaculum, nisi q (ut dictum est re-
tro) nihil nauigio ad id quod in ante est, cõfert: Sed
solum puppim in obliquum pellit ubicunq; fuerit:

FIGURE 7.5 *Niccolò Leonico Tomeo,* "Diagrams of a ship's rudder", *in* Quaestiones Mechanicae *(Venice, Bernardino Vitali: 1525) fol. 33v. Woodcut. Berlin, Library of the Max Planck Institute for the History of Science.*

translated into Latin (*stans*) on the one hand and, on the other, the second diagram represents the original reading from the Greek manuscript tradition. The latter is more consistent with the author's intentions, then, only when point C is moved as well, the triangles appear to be alike (line 851a24). From these alternative diagrams we notice again Tomeo's philological considerations: in his search for a sound text, Tomeo did not only pay attention to different readings, but also examined what these variants would imply for the diagrams. His philological activities, therefore, also permeate his diagrammatic practices.

With the other two diagrams to the problem of the rudder, Tomeo carried his translation project of the Aristotelian *Mechanics* a step further and included also diagrams in his commentaries. We find a diagram depicting the *prora* and *puppis* of the ship, that, different from the previously mentioned diagram, now shows that a small movement of the rudder at the stern leads to a large displacement of the prow. The other diagram is supplied with letter labels and can be understood in the same context. We have seen that the author of the *Mechanics* did not explain the working of a rudder in the diagram, but expanded his previous question on oars. Rather than criticizing the author for this deficit, Tomeo takes up the remark at 851a13–14 that the same angle faces a larger base and that the lines containing it are greater, and elaborates on it in his comments. After stating that a small motion on one side of the ship results in a large displacement on the other side, Tomeo adds the following observation:

> Sit enim trigonum (verbi causa) ABC, cuius angulus sit A, latera vero illum complectentia AB, et AC, coniuncta per hypotenusam BC. Extendatur autem latus AB usque ad BD. Similique modo AC aequaliter usque ad CE: iunganturque ipsa DE. Certum enim est quod idem angulus A qui prius ē regione spectabat hypotenusam BC, laterum productione maiorem spectat hypotenusam DE.[29]

> Let there be a triangle (for example) ABC, of which A is the angle enclosed by the sides AB and AC that are connected by the hypothenuse BC. Now let the side AB be extended to BD and in the same way also AC to CE. And let DE be connected. It is certain that the same angle A which first faced directly the hypothenuse BC, now after the sides have been produced, faces the larger hypothenuse DE.

In this description Tomeo employs the same letter labels as in the diagram that was already present in *Bern.* 402 [Fig. 7.4: first diagram]. Apparently, he

29 Leonico Tomeo, *Opuscula*, fol. 33v.

included the diagram in the Greek manuscript with these comments in mind. Tomeo here gives his own interpretation of lines 851a13–14 by adding a diagram in the commentary section of his translation, which forms an exceptional case in the *Quaestiones Mechanicae*. He concludes his comments by connecting problems 4 and 5 in that he argues that both ask the same question, only the direction of the motion is different for an oar or a rudder. However, although the diagrams and comments added by Tomeo are a nice illustration to show that a small motion on one side of the ship can bring about a large motion on the other side, it is merely an observation and does not offer a demonstration of a rudder. While referring to the law of the lever, Tomeo at his time still lacked the principles to fully explain the working of a rudder.

Tomeo's other notes to the rudder problem emphasize his philological interests. He notes for instance that there are two words in Latin for the Greek οἴαξ [handle of a rudder], namely *temo* and *ansa*, the first of which is used by Tomeo himself in his translation. This type of linguistic comments, as well as the fact that the diagrams to problem 5 reflect different textual variants, shows the importance of textual criticism for Tomeo's translation project. Tomeo displays an identifiably humanist approach in his *Quaestiones Mechanicae*, which is very different from several later commentators, as, for example, Bernardino Baldi who presents himself as a mathematically inclined reader of the text.

Conclusion

Tomeo's translation project of the Aristotelian *Mechanics* is characterized by its focus on textual criticism. In his aim to reconstruct the original Greek text, Tomeo paid attention to variant readings and proposed emendations of corrupted passages. The Latin translation offers a faithful rendering of the Greek text, which is supported by the use of transliterations of Greek terms instead of representing these by their Latin equivalents. Revisions in all different sources—the Greek codex *Bern.* 402, both versions of the Latin translation in codex *Vat.Reg.Lat.* 1291 and the 1525 printed edition—underline that for Tomeo an interpretation of the text can only proceed from a sound philological basis. Tomeo's philological interests also permeate the commentary part of his work, which contains many notes on the meaning and use of Greek and Latin words. It is in the diagrams to the *Quaestiones Mechanicae* that we notice that Tomeo also commented upon the mechanical problems. Different diagrammatic practices were consciously employed by Tomeo in his translation project. Several diagrams have lost their relevance in relation to the text—they no longer form part of a geometrical proof, for example, on the functioning of a balance but are very loosely connected with the surrounding text, now only illustrating a certain type of balance. Such illustrations show that Tomeo did not aim to provide a mathematical analysis of the treatise, but wrote for a humanistically-oriented audience. Early modern commentators could use different modes of visualization depending on the functions of their texts and their main audiences. A study of the different diagrams therewith sheds light on the contexts in which mechanical knowledge was spread in the early modern period. Tomeo's most important contribution to the circulation of the Aristotelian *Mechanics* was precisely to provide the much needed philological basis for later generations of commentators on the treatise.

Selective Bibliography

Primary Sources

Baldi Bernardino, *In mechanica Aristotelis problemata exercitationes* (Mainz, Johannes Albinus: 1621).
Bekker I. (ed.), Aristotelis Opera II (Berlin: 1831).
Bern, Burgerbibliothek, Cod. 402.
Leonico Tomeo Niccolò, *Opuscula nuper in lucem aedita* [...] (Venice, Bernardino Vitali: 1525).
Piccolomini Alessandro, *In mechanicas quaestiones Aristotelis paraphrasis paulo quidem plenior* (2nd ed., Venice, Traianus Curtius: 1565).
Vatican, Biblioteca Apostolica Vaticana, Cod. Reg.Lat. 1291.

Scholarly Literature

Andrist P., *Les manuscrits grecs conservés à la Bibliothèque de la Bourgeoisie de Berne—Burgerbibliothek Bern: catalogue et histoire de la collection* (Zurich: 2007).
Burnikel W., *Textgeschichtliche Untersuchungen zu neun Opuscula Theophrasts* (Wiesbaden: 1974).
De Bellis D., "Niccolò Leonico Tomeo interprete di Aristotele naturalista", *Physis* 17 (1975) 71–93.
Gamba E., "Un nuovo manoscritto copiato da Niccolò Leonico Tomeo (Par.gr. 1833). Appunti per la ricostruzione della sua biblioteca", *EIKASMOS* XXV (2014) 329–360.
Geanakoplos D.J., "The Career of the Little-known Renaissance Greek Scholar Nicholas Leonicus Tomaeus and the Ascendancy of Greco-Byzantine Aristotelianism at Padua University (1497)", *Byzantina* 13 (1985) 355–372.
Nobis H.M., "Über zwei Handschriften zur frühneuzeitlichen Mechanik in italienischen Bibliotheken", *Sudhoffs Archiv* 53, 3 (1969) 326–332.

Perfetti S., "Three Different Ways of Interpreting Aristotle's *De Partibus Animalium*: Pietro Pomponazzi, Niccolò Leonico Tomeo and Agostino Nifo", in Steel C. – Guldentops G. – Beullens P. (eds.), *Aristotle's Animals in the Middle Ages and Renaissance* (Louvain: 1999) 297–316.

Rose P.L. – Drake S., "The Pseudo-Aristotelian Questions of Mechanics in Renaissance Culture", *Studies in the Renaissance* 18 (1971) 65–104.

Sicherl M., *Griechische Erstausgaben des Aldus Manutius: Druckvorlagen, Stellenwert, kultureller Hintergrund* (Paderborn: 1997).

Valleriani M., "The Transformation of Aristotle's *Mechanical Questions*: A Bridge Between the Italian Renaissance Architects and Galileo's First New Science", *Annals of Science* 66, 2 (2009) 183–208.

Van Leeuwen J., "Thinking and Learning from Diagrams in the Aristotelian *Mechanics*", *Nuncius* 29 (2014) 53–87.

Van Leeuwen J., *The Aristotelian Mechanics: Text and Diagrams* (Dordrecht: 2016).

Vendruscolo F., "Manoscritti greci copiati dall'umanista e filosofo Niccolò Leonico Tomeo", in Funghi M.S. (ed.), *ΟΔΟΙ ΔΙΖΗΣΙΟΣ. Le vie della ricerca. Studi in onori di Francesco Adorno* (Florence: 1996) 543–555.

CHAPTER 8

'Secrets of Industry' for 'Common Men': Charles de Bovelles and Early French Readerships of Technical Print*

Richard J. Oosterhoff

Where Charles de Bovelles has a reputation at all, it is as a highly innovative philosopher in the intellectual mold of Nicolas of Cusa, Giovanni Pico della Mirandola, or perhaps Giordano Bruno.[1] But beyond being possessed of a mathematical curiosity and turn of imagination, the Picard canon was also deeply invested in the early sixteenth-century efforts to rework French as a language with a distinctive cultural heritage.[2] He experimented with arithmetical and geometrical theory in French and wrote studies of the language itself, such as a collection of French proverbs and a short study of French's origins, via the

* Besides thanks to Sietske Fransen and Niall Hodson for inviting me and for their tireless diligence, I owe gratitude to Robert Goulding for overseeing this work at an early stage. I should also thank Pascal Brioist for sharing his forthcoming work on Bovelles. The final version of this chapter was partially funded by the European Research Council under the European Union's Seventh Framework Programme (fp7/2007–2013)/erc grant agreement no. 617391.

1 Ernst Cassirer revived interest in Bovelles in *Das Erkenntnisproblem in der Philosophie und Wissenschaft der neueren Zeit*, 2 vols. (Berlin: 1920), vol. 1, 61–72, and especially his edition of Bovelles' *De sapiente* appended to *Individuum und Kosmos in der Philosophie der Renaissance* (Leipzig – Berlin: 1927). Bovelles's key insight included a Pican confidence in the intellectual powers of man to perfect and even co-create himself, a reading powerfully extended by Emmanuel Faye in *Philosophie et perfection de l'homme: De la Renaissance à Descartes* (Paris: 1998). Fundamental bibliography includes: Victor J.M., *Charles de Bovelles, 1479–1553: An Intellectual Biography* (Geneva: 1978); Trédaniel G. (ed.), *Charles de Bovelles en son cinquième centenaire, 1479–1979* (Paris: 1982); Ferrari M. – Albertini T. (eds.), *Charles de Bovelle's* Liber de sapiente, special issue of *Intellectual History Review* (2011); Klinger-Dollé A.-H., *Le* De sensu *de Charles de Bovelles (1511). Conception philosophique des sens et figuration de la pensée. Suivi du texte latin du* De sensu*, traduit et annoté* (Droz, Geneva: 2016); and Klinger-Dollé A.-H. – Faye E. – Sfez J. (eds.), *Bovelles philosophe et pédagogue* (Paris: *forthcoming*). An especially rich account of Bovelles's life can be gleaned from Margolin J.-Cl., *Lettres et poèmes de Charles de Bovelles* (Paris: 2002).

2 To sense this enormous moment in the formation of French literature, see DellaNeva J.A., *Unlikely Exemplars: Reading and Imitating beyond the Italian Canon in French Renaissance Poetry* (University of Delaware Press, Newark: 2009).

 | DOI 10.1163/9789004349261_010

ancient Druids, in Greek—like many other such theorists, he composed these theoretical studies of the vernacular in Latin.[3]

Bovelles also wrote three vernacular manuals of geometry.[4] He published the *Geometrie Francois* in 1511, the same year that the first illustrated edition of Vitruvius's ten books on architecture was published in Venice. Unlike Vitruvius, however, Bovelles claimed to have written his book not for elite, Latinate readers, but for those he called 'common' [*plebes*] workmen. Today, this book exists only in few copies, and it was not reprinted.[5] But there are more copies of a similar book Bovelles published in 1542, the *Livre singulier et utile touchant l'art et practique de geometrie*, again claiming a readership of craftsmen. Demand ensured a revised edition in 1547 (with the new title *Geometrie practique*), which was republished at least five times in French, besides Dutch and Latin

3 Bovelles Charles de, *Proverbiorum vulgarium libri tres* (Paris, Galliatus Pratensis: 1531); Bovelles Charles de, *Liber de differentia vulgarium linguarum* (Paris, Robert Estienne: 1533).

4 On vernacular geometry in a German context see also the article by Thomas Morel in this volume.

5 Bibliothèque municipale de Rouen, shelfmark Leber 1159; Bibliothèque de Gand, Centrale Bibliotheek, shelfmark A 11066(2); Bibliothèque municipale de Blois, shelfmark I 958, digitized at <http://www.bvh.univ-tours.fr/Consult/index.asp?numfiche=715>, last accessed 6 January 2017. Jean-Claude Margolin, in an important article of 1976, placed Bovelles at the beginning of an emerging tradition of French mathematical teaching, with a focus not on Bovelles' earlier theoretical treatises, but more on the vernacular handbook that became popular in 1542. Margolin J.-Cl., "L'enseignement des mathématiques en France (1540–70): Charles de Bovelles, Fine, Peletier, Ramus", in *French Renaissance Studies, 1540–70: Humanism and the Encyclopedia*, ed. P. Sharratt (Edinburgh: 1976) 109–155. In this article, Margolin claimed that Bovelles's 1511 *Geometrie Francoys* was the first printed vernacular mathematics, a point René Taton repeated, calling the volume 'the direct heir of the commentaries of Boethius and Bradwardine which formed the basis of Paris university education at the end of the fifteenth century'. Taton R., "Bovelles et les premiers traités de géométrie en langue française", in *Charles de Bovelles en son cinquième centenaire, 1479–1979: actes du colloque international tenu à Noyon, les 14–15–16 septembre 1979* (Paris: 1982) 196. In contrast, Taton judged the 1542 geometry to be a 'confused mixture' of mathematics, esotericism, and natural philosophy, all of which set him outside the lineage of modern science. When Margolin responded in 1993, he emphasized that Bovelles's 1542 geometry fit a growing trend to write learned literature in the vernacular. Moreover, even though Bovelles's rigor left much to be desired—this was by no means cutting edge mathematics—Margolin identified the work's contribution not in mathematics but precisely in its 'composite character'. What Taton called a 'confused mixture' Margolin saw as an important effort to popularize his anthropological and 'cosmo-theological' Latin theory by dressing it in practical garb. Margolin J.-Cl., "Une Géométrie fort singulière: la Géométrie pratique de Charles de Bovelles (Paris, S. de Colines, 1542)", in *Verum et Factum. Beiträge zur Geistesgeschichte und Philosophie der Renaissance zum 60. Geburtstag von Stephan Otto* (Frankfurt am Main: 1993) 445.

translations, into the early seventeenth century.[6] French mathematics had found a broader readership. Bovelles's practical geometry is a key case study because it self-consciously claims to be a practical text—furthermore, it is one of the first French practical geometries to be printed, and thus sheds light on the sources and aims of newly popular 'practical' genres that have long occupied historians of science and technology.[7]

Although previous historians have assumed that the advertised artisanal audience was merely a trope, Pascal Brioist has recently refocussed study on the question of Bovelles's relation to artisanal practice.[8] Reading the text in the light of sixteenth-century architectural and military practical manuals, Brioist extends an observation made by René Taton and Jean-Claude Margolin: that Bovelles's language regularly refers to material and physical conditions of figures, implying that this geometry belongs to craftsmen, not scholars. Brioist also cites places where Bovelles drew on his journeys through Germany and the Low Countries to give concrete examples, such as the difference between German tables (usually square) and French ones (usually rectangular). This is a much different picture than that given by Taton and Margolin; Brioist shows Bovelles not only trying to anticipate what information might be useful in practice, but attentively noting and assembling regional differences in artisanal practice.

---

6 Bovelles Charles de, *Geometrie en françoys. Cy commence le Livre de l'art et Science de Geometrie, avecques les figures sur chascune rigle au long declarees par lesquelles on peut entendre et facillement comprendre ledit art et science de Geometrie* (Paris, Henri Estienne: 1511); Bovelles Charles de, *Livre singulier et utile, touchant l'art et practique de Geometrie, composé nouvellement en Francoys* (Paris, Simon de Colines: 1542); Bovelles Charles de, *Geometrie practique* [...] *nouvellement par luy reveue, augmentee et grandement enrichie* (Paris, Reginald Chauderon: 1547). The 1547 edition was printed again in 1551, 1555 (twice), 1557, 1566, and 1608. See Appendix A of Oosterhoff R.J., *Mathematical Culture in Renaissance Paris: University, Print, and the Circle of Lefèvre d'Étaples*, Ph.D. dissertation (University of Notre Dame: 2013).

7 The developing relationship between craft and print is a key theme in Kusukawa S. – Maclean I. (eds.), *Transmitting Knowledge: Words, Images, and Instruments in Early Modern Europe* (Oxford: 2006). A useful review of the historiography and overview of various new technical genres is Long P.O., *Artisan/Practitioners and the Rise of the New Sciences, 1400–1600* (Corvallis, OR: 2011). The cultural significance of new French technical works was already noted by Davis N.Z., "Sixteenth-Century French Arithmetics on the Business Life", *Journal of the History of Ideas* 21 (1960) 18–48; for most recent studies of French in particular, see now Tura A., *Fra Giocondo et les textes français de géométrie pratique* (Geneva: 2008).

8 I am grateful to Pascal Brioist for sharing his paper prior to publication: "Les singularités de la géométrie pratique de Charles Bovelles", forthcoming in *Bovelles philosophe et pédagogue*. This paper provides the closest analysis of Bovelles's mathematical practice and artisanal use.

My own focus will be Bovelles's successive revisions of his practical geometries, which help us see the difficulties of two forms of translation during the crucial early sixteenth century, when Joachim du Bellay, Jacques Peletier du Mans, and others in Bovelles's circles were reimagining French as a literary and technical language. These geometries were translations in the usual sense of rendering Latinate texts into French. But they also attempted to translate expertise from one sphere to another, from library to workshop. By claiming (in Latin) to write for the *vulgari*, Bovelles proffered his Latin knowledge to a French audience of workmen.

The difficulties of translating expertise come into view when we ask: who was the readership for Bovelles's French geometry? Was it the early Republic of Letters, or was it the rising class of artisans who were engaging with the published word? In an effort to answer this question, I consider how Bovelles presented practical geometry in 1511, 1542 and 1547. By 1547, we see him imagine a growing public which was increasingly interested in technical books and the language of practical secrets as entertainment.

Imagining Mathematical Publics

Bovelles belonged to the Parisian circle of university humanists around Jacques Lefèvre d'Étaples. With Josse Clichtove, Bovelles was one of Lefèvre's closest collaborators at the Collège du Cardinal Lemoine, and was interested in what we might call the popularization—or, in their terms, 'vulgarization'—of learning. Around 1500, the circle produced textbooks and introductions used to simplify the Latin learning of the university; by the 1530s, they had also presented the Bible in French and composed Latin-French grammars. They formed the core of Marguerite de Navarre's network, out of which grew the diverse vernacular literary projects of the du Bellays, Scève, Ronsard and Jacques Peletier, the generation which reinvented French as a literary language.[9] In this context, writing in the vernacular was a potent experiment.

Bovelles and his Paris circles, I would argue, also fostered an emerging reading public for mathematics, first in Latin and increasingly in French. By 1526 the instrument maker, designer of engravings, and teacher of mathematics Oronce Fine could claim that he was publishing an *aequatorium*, an instrument for calculating the locations of planets, for the use of a 'mathematical

9 Reid J.A., *King's Sister-Queen of Dissent: Marguerite of Navarre (1492–1549) and Her Evangelical Network*, 2 vols. (Leiden: 2009).

republic'.[10] But how do we get at these publics? The most rigorous tool at our disposal, perhaps, is the history of reading, seeing who a text's readers were and what they made of these texts. This tool is not available in this case, since I am aware of only three exemplars of the first book—none annotated.[11]

Another line of evidence lets us say something about this reading public. In the early sixteenth century, we find a growing *expectation* or *promise* of utility surrounding mathematics, injected into a developing sense of 'public' in early modern Europe.[12] Lefèvre and others intended their books to be patronized, bought, and used by a public that was not necessarily mathematical, but found university mathematics useful for public goals. The Greek émigré George Hermonymus convinced Lefèvre to restore the discipline by noting (as Plato had) that mathematics is 'of the greatest importance not only to the republic of letters, but also to the civil republic'.[13] In another letter, Lefèvre related the insights of the philosophers to the immediate practical benefits of mathematics: 'Therefore, take away numbers and their learning and you will leave laws unkept, justice will be left blind, there will be no rules of [musical] modulation found, no entry to the contemplation of the heavens, and the mysteries of sacred letters [i.e. Scripture] will be obscured—as indeed will be the universal philosophy which includes the understanding of both human and divine things'.[14] Some years later, Oronce Fine's student would repeat this claim to a new royal patron, Henry II of France: mathematics 'provides the

10 Fine claimed that he devised his aequatorium for the benefit of the 'respublica mathematica'. Fine Oronce, *Aequatorium planetarum, unico instrumento conprehensum, omnium antehac excogitatorum et intellectu et usu facillimum* (Paris, Nicolas Calceolarius: 1526), fol. a2v. See Oosterhoff R.J., "Lovers in Paratexts: Oronce Fine's Republic of Mathematics", *Nuncius* 31 (2016) 549–583.

11 See note 5.

12 Foundational works on the early modern origins of publics include Habermas J., *The Structural Transformation of the Public Sphere: An Inquiry into a Category of Bourgeois Society*, trans. T. Burger (Cambridge, MA: 1962; 1991); Anderson B., *Imagined Communities: Reflections on the Origin and Spread of Nationalism* (London: 1992).

13 Lefèvre d'Étaples Jacques, *Textus de sphera Johannis de Sacrobosco, cum additione (quantum necessarium est) adiecta, novo commentario nuper edito ad utilitatem studentium philosophice parisiensis academie, illustratus* (Paris, Wolfgang Hopyl: 1495), fol. a1v: 'non modo reipublicae litterariae sed et civili momentum habent maximum'.

14 Ibidem, fol. a4r. (= Rice E.F. (ed.), *The Prefatory Epistles of Jacques Lefèvre d'Étaples and Related Texts* (New York: 1972), ep. 5, 18.) 'Tolle igitur numeros numerorum disciplinam, leges imperficis, iustitia caeca relinquitur, nulla modulationum reperietur regula, nullus caelestium contemplationum aditus, sacrarum litterarum delitebunt mysteria, immo et universa philosophia, qua pariter humanorum divinorumque cognitio describitur'. This section of the note had wider currency, for example excerpted by Caesarius Johannes

sweetest fruit for the use of the community of the kingdom, and the safe care of the republic'.[15] Of course, such promises are cheaply made to prospective patrons. Nevertheless, the fact that such promises *were* made indicates that these patrons—high-ranking officials and royalty—could be openly held accountable to some notional public, or what Benedict Anderson has called an 'imagined community'.[16] Those making such promises believed, at the very least, that there was enough of a public that such appeals would matter to their prospective patrons. In the absence of annotations and other evidence about a public, we can usefully consider what community printers and authors *imagined* in their works, recalibrating as successive editions failed and succeeded.

Failure (1511)

The *Geometrie en Francoys* that Bovelles first published in 1511 was considerably reimagined for the *Geometrie practique* thirty years later. The long gap suggests that Bovelles first misjudged his public.

In the preface of 1511, he invoked the trope that friends had begged him to publish a geometry in French:

> a plerisque amicorum instigati, hanc vernacula lingua Geometriam cudimus, in qua partim speculari, precipue vero operari et singula perficere edocemus. In hac enim magis rei utilitati ac usui, quam sermonis honestati studuimus. [...] Haud ergo latinis aut speculativis, sed factivis plebeisque viribus, hoc gallico sermone conscriptum exhibemus opusculum.[17]

(ed.), *Introductio Jacobi Fabri Stapulensis in Arithmeticam; Ars supputandi Clichtovei; Epitome rerum geometricarum Bovilli* (Deventer, Richard Pafraet: 1507), fol. A2v.

15 Fine Oronce, *De rebus mathematicis hactenus desideratis libri IIII* (Paris, Michel Vascosan: 1556), fol. *2v. See also Oronce Fine, *Protomathesis* (Paris, G. Morhius: 1532), fol. AA3r. Here Fine promised that the recovery of pure mathematics would help theologians, philosophers, physicians, judges, and indeed all aspects of civil order.

16 Anderson, *Imagined Communities*. See especially chapter 3, "The Origins of National Consciousness", for an argument about print-as-commodity at the origins of publics. For further reflection on this theme, see Watts J., "The Pressure of the Public on Later Medieval Politics", in Clark L. – Carpenter C. (eds.), *Political Culture in Late Medieval Britain* (Woodbridge: 2004) 159–180.

17 Bovelles, *Geometrie en françoys* (1511), fol. [1]v.

> I was instigated by many friends to print this Geometry in the vernacular tongue, in which I teach partly to speculate, but mostly to work and to construct each thing. For in this I sought the use and utility of the matter rather than the integrity of words. [...] I therefore offer this little book, written in French, not to Latin or speculative men, but to constructive and common men.

Yet it was no straightforward choice for an arts master to write in French—Bovelles wrote this preface in Latin. The tension between Bovelles's learned context and his popular aims was already apparent on the title page, which announced the subject as 'ledit art et science de Geometrie' [this art and science of geometry], but offered as sole ornament a woodblock depicting souls in the cosmos speculating on the zodiac, a figure more fitting to heavenly theory than earthly practice.[18]

And so, despite its claim of a common readership, the book mixed registers in both format and contents. Though printed in quarto, it used the Gothic typeface that Henri Estienne normally used for Latin, mostly prestige texts. One might contrast another early printed practical geometry in French, by Pierre Verney around 1530,[19] which employed the 'dagger' version of 'batard' most often used to print and copy vernacular texts.[20] Even after the title page,

18 In this case, the printer seems to have reused a woodcut that was used (more appropriately) in Bovelles' magnum opus, published that same year: Bovelles Charles de, *Liber de intellectu; Liber de sensu; Liber de nichilo; Ars oppositorum; Liber de generatione; Liber de sapiente; Liber de duodecim numeris; Epistole complures. Insuper mathematicum opus quadripartitum: De numeris perfectis; De mathematicis rosis; De geometricis corporibus; De geometricis supplementis* (Paris, Henri Estienne: 1511), fol. 29v.

19 Verney Pierre, *Succinte, briefve et compendieuse Collection Geometrale* (Metz, Jehan Pelluti: [c. 1530]). Verney may also be the author of some prognostications, first in Latin but later published in French in Lyon (1539). The only published information on Verney I have found is Tura, *Fra Giocondo et les textes français de géométrie pratique* 55, 71. The book follows closely the division of medieval French geometries into *altimetrie, planimetrie,* and *solimetrie* (i.e. the study of heights, surfaces, and solids).

20 Henri Estienne did not commonly print in the vernacular, so the shop may not have invested in appropriate type. See examples in Renouard A.A., *Annales de l'imprimerie des Estienne; ou, Histoire de la famille des Estienne et de ses editions* (Paris: 1843). The distinction between Gothic and *batard* should not be overstated, and many counter-examples should be expected. But during the fifteenth century and the first part of the sixteenth, it appears that printers did generally try to distinguish the type they used for Latin and vernacular books. For example, Antoine Vérard, who printed much more in the vernacular than did Estienne, distinguished quite clearly between the typefaces used in his Latin and his vernacular books.

the visual program of Bovelles's book advertised to a learned public, with few accommodations to the artisanal audience it claimed in the Latin preface. In particular, Bovelles's images are a curious mixture of theoretical and practical conventions. Consider drawings illustrating the same kind of operation, measuring the volume held by a cylindric hollow. Where Verney's woodcuts illustrate a man squinting through the instrument to measure a tower or well [Fig. 8.1a], Bovelles's images include only the figure in question, forcing the reader to focus on the mathematical abstraction more than the material context [Fig. 8.1b]. This choice was not consistent. In other places Bovelles's images admitted the materiality of the tasks they illustrated by shading in objects that, in the text, Bovelles identified as wood or stone.

The book's textual contents matched its jumbled visual format, mingling practical and theoretical concerns. Again, first consider Verney's text, which closely follows the medieval French tradition of practical geometry. The first text actually known as *Geometrica practica* is thought to be by Hugh of St Victor in the twelfth century, who divided geometry into the measurement of heights, surfaces, and volumes (*altimetria, planimetria, cosmimetria*). French geometries as early as 1275 took on the same division, focusing on the use of astrolabes or quadrants in measurement.[21] Verney's printed French geometry of c. 1530 followed this tradition closely, simply presenting a series of problems in *altimetrie, planimetrie*, and *solimetrie* (the last term used synonymously with stereometry).[22] Such texts showed little concern for mathematical demonstration, but focussed on practical construction. For example, they listed the steps necessary accurately to deploy a Jacob's Staff when measuring the height of a tower.

Bovelles's *Geometrie en francoys* tried to chart a course between theoretical geometry and this artisanal, 'constructive' tradition. Bovelles offered a comprehensive overview of geometrical objects, which he called the 'principles' (points, lines, surfaces, bodies).[23] In contrast, Verney simply began *in medias res*, with instructions for basic problems of measuring heights, areas, and

21 Shelby L.R., "Geometry", in Wagner D.L. (ed.), *The Seven Liberal Arts in the Middle Ages* (Bloomington: 1983) 203. The first French *Pratike de geometrie* again was heavily influenced by the Latin tradition of *agrimensores*, put in iconic form by Hugh of St Victor; see Victor S.K., *Practical Geometry in the High Middle Ages*, Artis cuiuslibet consummatio *and the* Pratike de Geometrie (Philadelphia: 1979).

22 Verney, *Collection Geometrale.*

23 The motivation behind this language of 'principles' can be glimpsed in Bovelles's expansion on the them in 1542, where he described these principles as the geometrical analogues to the integers 1, 2, 3, 4 in Pythagorean number theory. Bovelles, *Geometrie practique* (1542), fols. 3v–4r (preface).

vaisseau coulõnaire et le pallellogrãme a.g.h.b. signi
fie et repñte la cõcauite dud vaisseau car la li-
gne g.h. repñte le fõs leql est vng vray cercle
et les deux lignes a.g. et b.h. ppẽdiculaires
sur g.h. repñtẽt la rõde supfice de la rotũdite
et cõcauite du vaisseau ie dis dõcqs q̃ ce vais
seau coulõnaire a.g.h.b/ sera double au p̃mi
er vaisseau speriq̃ a d.b/ car p les rigles deuãt
dictes la coulõne a.g.h.b. est egale a la spere
entiere dõt la moitie est le vaisseau a.d.b/ p quoy se tu as le vaisseau cou
lõnaire double au p̃mier tu auras incõtinẽt le vaisseau a luy egal / car il
ne fault q̃ copper la moitie de dess⁹ ⁊ le vaisseau q̃ demourera sera aussy
bien coulõnaire q̃ lentier et sera egal au vaisseau speriq̃ p̃mier fait/ ou se
tu ne veux riẽs copper il te fault cauer la pierre en forme coulõnaire seu
lemẽt selõ la moitie de la q̃rte ptie d la circũferẽce cõme
selõ les lignes a.i/ ⁊ b.k/ ⁊ tu auras le vaisseau cou
lõnaire q̃ tu demãde leq̃l sera d egale ptinẽce ⁊ ca
pacite au p̃mier vaisseau speriq̃ a.d.b. cõe en la
pñte figure est demõstre: et pour cõposer ⁊ pfaire
le tiers vaisseau cubiq̃ il te fault diuiser le dyame
tre du cercle a.c.f.b. en huyt pties egales et puis
pr̃edre vne aultre ligne ayãt dix de ces pties/ car
cõe deuãt est dit ceste ligne de dix sera la dyametre du vray cube egal a
la spe du cercle a.c.f.b. par quoy il te fault p̃mieremẽt sur la tierce pier

FIGURE 8.1 *Two methods for calculating a hollow cylinder.* (*a*) *Verney*, Collection géométrale (*c. 1530*), *fol. C4r, detail;* (*b*) *Bovelles,* Geometry en francoys (*Paris, Henri Estienne: 1511*), *fol. 35v, detail.* Bibliothèque Municipale de Blois, *fonds ancien, Cote : 1 958, and the* Bibliothèques Virtuelles Humanistes, *CESR, Tours; by permission.*

volumes, notably using the ancient instrument of the Jacob's Staff. Bovelles began by identifying geometrical objects: points, lines, surfaces, and bodies. He proceeded in a Euclidean fashion, by giving propositions that were to build up into a larger mathematical narrative. He served his practical goals by reformulating the propositions and demonstrations of Euclidean geometry into constructive elements: *rules* and *problems*. The rules in each book showed how to construct a particular shape, such as how to 'enlarge a given square in any proportion';[24] problems or questions tended to be 'chiefly useful to carpenters and masons', such as an example in the third chapter showed how to translate spheres into a column of the same volume.[25] In a word, Bovelles's book was hybrid. It offered the systematicity of an academic tract, but of constructive geometry instead of demonstrative.

Whether simply because of the book's austere visual program, or more because of its apparently theoretical text, craftsmen did not flock to bookstalls. Bovelles's book, I suspect, fell between the needs of two readerships: craftsmen found it unnecessary to learn the conceptual underpinnings of practical techniques they probably already knew, while few literary elites were yet interested in the mechanical arts.

Finding a Public (1542)

By 1542, a mathematical public appears to have emerged. In that year Bovelles published a new practical geometry, titled *Livre singulier et utile, touchant l'art et practique de Geometrie* [A singular and useful book, concerning the art and practice of geometry]. Once again, Bovelles claimed an artisanal readership, citing 'certain craftsmen and manual laborers' who had requested that he write the book for them 'in the vulgar language', even though he was unaccustomed to writing in his mother tongue. Nevertheless, the vulgar tongue did not guarantee popularity—Bovelles complained that printers had promised 'mountains of gold' but in the end only 'gave birth to a mouse'—apparently they were hesitant to fully engage Bovelles's project.[26]

24 *Geometrie en francoys*, fol. 12v.

25 Ibidem, fol. 32v: 'Sensuyvent aucunes questiones en la pratique de Geometrie pour la reduction de la spere pyramide, cube, et colonne a equalite. Et sont ces choses utiles principallement aux charpentiers et massons'.

26 Bovelles, *Livre singulier* (1542), fol a2r: 'et quidam ex Parisiensibus Chalcographis, in illius excussione aureos polliciti montes, ridiculum murem peperissent'. The reference is to Horace, *Ars poetica* 139.

In the end, the book was finally published with the support of Oronce Fine, who by this time had been the royal professor of mathematics in Paris for over a decade.[27] Fine brought the project both a popular audience and special skills, as Bovelles acknowledged in the preface. On hearing that the work needed a printer, he remembered, Fine had promised two things:

> Duo protinus ingenue spopondit: se quidem cum primis daturum operam, ut aereis typis invulgata, plurimis esset usui; figurarum quoque, quas ibidem frequentius inscripsi, futurum ligneis in tabellis pictorem. Necnon (quod praecipuum est) adversum mendas observaturum vigiles praeli excubiat.[28]

> that he would himself give the work to printers, to be made popular in print, so that it could be used by many; and also that the figures which I had everywhere drawn, he would shape on woodblocks, and (what is most important) he would take vigilant care that errors be corrected in press.

In other words, Bovelles not only needed Fine's support to ensure relations with Paris printers and for correcting the proofs in press, but he also required his technical expertise for designing woodcuts.

Fine's expertise as a craftsman, particularly in designing adequate woodcuts to accompany the text, should not be overlooked. By this time Fine was well known as a mathematical practitioner, also for his own craftsmanship. He had designed important frontispieces for Lefèvre's circle in the 1520s, and he became widely known for crafting instruments. Antoine Mizauld, his student, later recalled that Fine employed craftsmen to work out of his house, which was always full of bishops, courtiers, and important Parisians who came to see the marvelous instruments Fine made with his own hands.[29]

Bovelles advertised that the book had been 'composé nouvellement' [newly composed], and indeed it was very different from the 1511 Geometrie. Some of this was simply due to updated print conventions introduced by the printer Simon de Colines: a more elegant italic typeface, foliated capitals, and Fine's distinctive, elegant woodcuts [Fig. 8.2]. But it was not just on the strength of better production values that the book succeeded. The first page includes a

---

27 Bibliography on Fine's career can be found in Marr A. (ed.), *The Worlds of Oronce Fine. Mathematics, Instruments and Print in Renaissance France* (Donington: 2009).

28 Bovelles, *Livre singulier* (1542), fol. a2r–v.

29 Mizauld Antoine, "Vita Orontii", in Fine, *De rebus mathematicis hactenus desideratis libri IIII*, fol. *6r.

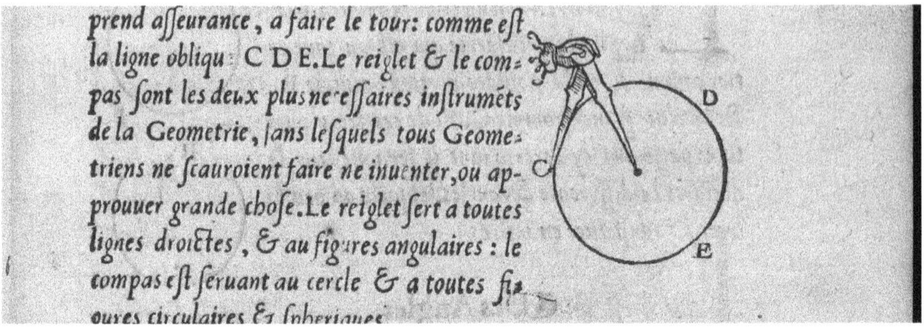
prend aſſeurance, a faire le tour: comme eſt la ligne obliqu: C D E. Le reiglet & le compas ſont les deux plus neceſſaires inſtrumẽts de la Geometrie, ſans leſquels tous Geometriens ne ſcauroient faire ne inuenter, ou approuuer grande choſe. Le reiglet ſert a toutes lignes droictes, & au figures angulaires: le compas eſt ſeruant au cercle & a toutes figures circulaires & ſpheriques

FIGURE 8.2 *Bovelles,* Geometrie practique *(Paris, Regnauld Chaudière: 1551),* CUL *Syn.5.55.7, fol. 6r, detail typical of Fine's decoration from 1542. This edition reused woodcuts from the editions of 1542 and 1547 (reproduced by kind permission of the Syndics of the Cambridge University Library).*

poem to the reader, promising not only understanding of measurement, but also the 'secretz d'industrie'; the poem then enjoined the reader to deploy the geometer's instruments, the square, rule, and compass, with illustrative woodcut below. Images progressed from simple lines to actual objects; these included vessels complete with handles, towers with pyramidal roofs, and even the geometrical structures of carts.[30]

The contents of the 1542 practical geometry also shifted. Bovelles did not give up the effort to present systematically the basics of geometrical objects, from point to solid. But he compressed the introduction to concepts, getting quicker to the construction of useful figures. While explaining constructions, he vacillated between the kind of description that belonged to theoretical treatises, and careful attention to material figures, as when teaching how to use a compass: 'The curve is produced by means of the compass, which steadies the hand to make the turn'.[31] Moreover, Fine's woodcuts—perhaps especially significant for a practical audience—changed the book's emphasis. The stronger visual program of concrete objects (instead of abstractions) went beyond the 1511 geometry, beyond mere cubic and circular vessels. The seventh chapter in particular addressed how to put bells in harmony, how the four legs of a horse conform to geometrical norms (nature sans cause riens ne fait) [nature does nothing without a cause]; the geometrical proportions of wagon loads, and the

30 Bovelles, *Geometrie practique* (1542), fols. 47v, 48v, and 51r.

31 Ibidem, fol. 6r: 'La ligne oblique, se produyt par le moyen du compas, par lequel la main prent asseurance, à faire le tour'.

TABLE 1 *Organization by book of early French geometries*

Verney, c. 1530	Bovelles, *Geometrie en francoys*, 1511	Bovelles, *Livre singulier*, 1542	Bovelles, *Géométrie practique* (2nd ed. of *Livre singulier*), 1547
1. Altimetrie	1. Lines	1. Principles, dimensions, circles	
2. Planimetrie	2. Surfaces	2. Angular shapes	
3. Cosmimetrie	3. Bodies	3. Figures inscribing and circumscribing circles	
		4. Quadrature of the circle	4. *Two new propositions*
		5. Dimensions of bodies	
		6. Cubing of the sphere	
		7. Explanations of bells, horses, carts, etc.	7. *Twenty-one new propositions*
			8. On the utilities and excellence of Geometry

equal height and arm extension of a human body, as well as the symmetrical arrangement of the organs of sight, smell, and hearing.[32]

Bovelles offered little more than a description of regularities to be found in nature, but in a couple of places he hinted at deeper reasons for such regularities, alleging that

> Ainsi appert que la Goemetrie n'est de petite utilité, par laquelle on peust cognoistre plusieurs choses dignes de scavoir. Et n'est aucunement possible, que l'engin humain puist bien profiter en la philosophie et science des choses naturelles, sans l'aide des arts mathematiques, esquelles sont contenues plusieurs mystiques, sur lesquelles se sont fondez et reiglez les anciens philosophes, pour inventer et descrire les occultes proprietez de toutes choses naturelles.[33]

32 Bovelles had a longstanding interest in figuring the senses, with and without mathematics. See Klinger-Dollé A.-H., *Le* De sensu *de Charles de Bovelles (1511)*.

33 Bovelles, *Geometrie practique* (1542), fol. 56r.

> thus it appears that geometry has no little utility, by which one can know many things worth understanding. It is thus impossible that human ingenuity [l'engin humain] benefit in philosophy and the sciences of natural things, without the aid of mathematical arts, in which are contained many mysteries, on which ancient philosophers based and directed themselves in order to discover and describe the hidden properties of all natural things.

Most interestingly, his claim is not that geometry is useful to explain nature, but only that it allows one 'to discover and describe' (inventer et descripre) the secrets of nature—thus asserting without unveiling the causal mystery. Here 'practical' geometry was a tool for practical discovery. It helped one see geometry in nature, and so manipulate nature.

The renovated *Livre singulier* of 1542 sold much better, judging by the only (albeit crude) measure we have available: surviving copies, new editions and translations.[34] One part of the explanation must be that Bovelles, with the help of Oronce Fine, had ordered some of the jumble that confounded the work of 1511. But we see this new order as a response to what Bovelles and Fine thought their public wanted, the success also tells us about that projected audience. In particular, one of the successful shifts Bovelles made between 1511 and 1542 was to add language about mathematics and the secrets of nature. The edition of 1547 only accentuated this language.

Secrets of Nature, Secrets of Industry (1547)

Most of Bovelles's additions to the 1547 edition of the *Geometrie practique* fall into the category of 'secrets of nature'. Pascal Brioist was the first to point out this language which existed already in the 1542 edition. In the prefatory poem, Oronce Fine alludes to '*secretz d'industrie*' as he exhorts 'all artisans and Mercurial people who want to find out new secrets' to adopt practical mathematics. Bovelles did offer some geometrical constructions from the artisanal tradition, such as a method for finding what 'common folk and mechanics call the lost centre' of a circle.[35]

Here we observe the difficulty of translating expertise between artisanal and learned spheres. Did such acquaintance with artisanal language mean Bovelles

---

34 See note on publishing history above (n. 6).

35 Brioist, "Les singularités de la géométrie pratique de Charles Bovelles".

was writing in the tradition of artisanal secrets—secret because they belonged to the unwritten education of guilds—and so served an artisanal audience?[36] To suppose so is artificially to limit the 'secrets' tradition to just those practical recipes which belong squarely in the middle of the 'maker's knowledge tradition'.[37] Technical books were also read for entertainment; secrets were increasingly meant to delight the *peuple moyen*, the growing public that these books meant to inform and entertain simultaneously. Bovelles's *Geometrie practique* signals the growth of an intermediate literature neither theoretically rigorous nor the unvarnished fruit of practice.

Brioist rightly observes that Bovelles meant 'to show that the immaterial mathematical ideas govern the universe of forms'.[38] One might further remark the notion of analogy that governs Bovelles's examples, highlighting how human art imitates nature. When he first describes the geometrical motion of four-legged beasts, he does so in order to point out that, in nature, rear legs are longer; likewise, wagons are best designed with larger wheels at the back.[39] He conceptualizes rivers as flowing from sources on the outside of a circle, flowing to the low point at the middle, in order to talk about 'la grande encyclie du monde universel' [the grand circle of the whole world] with its arrangement of heavy earth at the center and fire beyond the outermost sphere. In his additions to the 1547 edition, Bovelles describes the order of nature, in which the wind normally blows from east to west, moving the upper sails of a windmill in that direction, while water 'according to the order of nature' moves below an undershot waterwheel (unless human art arranges for a flume to oppose the order of nature [Fig. 8.3]).

Throughout, Bovelles is eager to take everyday experiences, and then show how these either reflect nature's mathematical reasons—or a human intervention that relies on the same mathematical principles. But at no point does he rigorously account for those principles. Throughout the book he referenced 'common people' and 'children' and deployed French proverbs such as one that played on the opposition of 'sharp' and 'round':

36 On this tradition, see Eamon W., *Science and the Secrets of Nature: Books of Secrets in Medieval and Early Modern Culture* (Princeton: 1994); Leong E. – Rankin A. (eds.), *Secrets and Knowledge in Medicine and Science, 1500–1800* (Farnham: 2011).

37 For an account of 'maker's knowledge' in early modern Europe, see Pérez-Ramos A., *Francis Bacon's Idea of Science and the Maker's Knowledge Tradition* (Oxford: 1989).

38 Brioist, "Les singularités de la géométrie pratique de Charles Bovelles".

39 Bovelles, *Livre singulier* (1542), fols. 50v–52r.

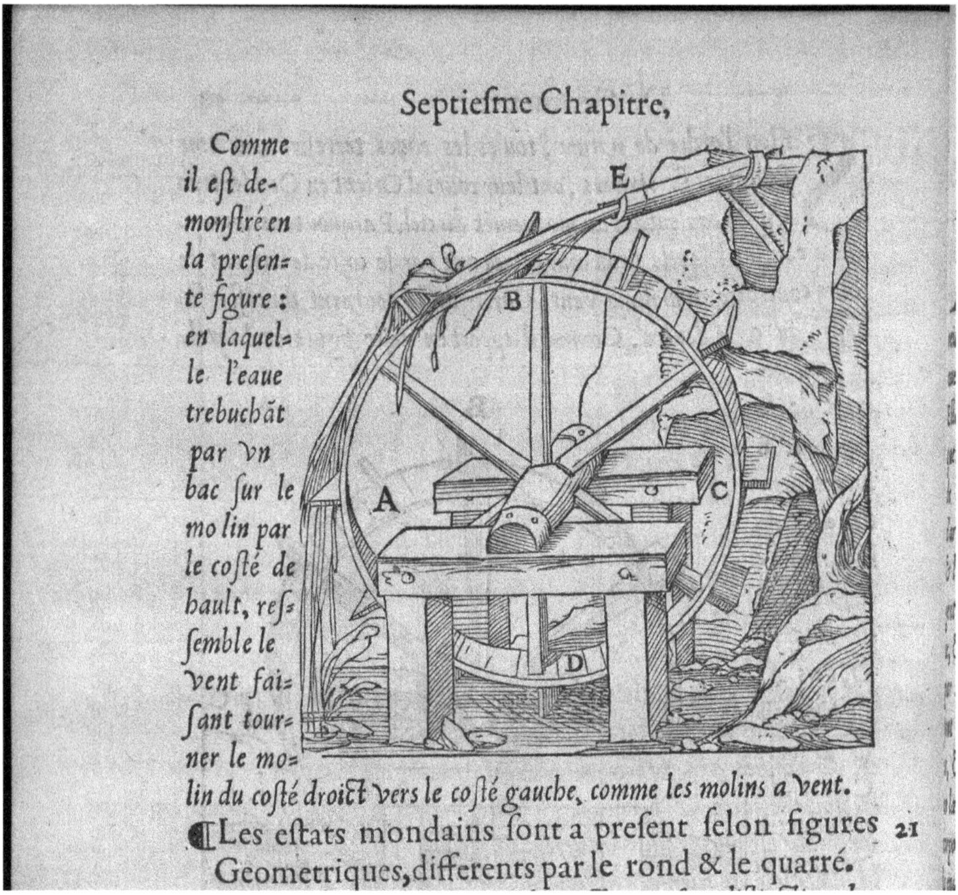

FIGURE 8.3 *Bovelles,* Geometrie practique *(Paris, Regnauld Chaudière: 1551),* CUL *Syn.5.55.7, fol. 58v, detail of a woodcut introduced in 1547 (reproduced by kind permission of the Syndics of the Cambridge University Library).*

Ronde memoire, agu entendement,
Fait l'homme habil, discret, sage, & prudent.
Memoire ague, & ronde engin,
Rend l'homme simple, & non fort fin.[40]

Round memory, sharp understanding,
Make man skilful, careful, wise, and sensible.
Sharp memory and round wit
Turn a man simple, and not so smart.

40 Bovelles, *Geometrie practique* (1547), fol. 65v.

Occasionally he sent his reader to experts, if they wished to learn more on the mysteries he noted. Discussing the arrangement of a living space to benefit from the most healthful winds, he advised that 'on this point one should consult the philosophers or physicians, who understand the disposition of the air and the differences of the four winds that come from the four directions'.[41]

But the *Geometrie practique* offer no explicit recipes for practice. Even if Bovelles thought that the first edition might help artisans learn better their own business, the material he added in 1547 confirms that he mostly meant to foster delight and wonder over such secrets rather than explain how actually to accomplish them. His first addition to the new edition was a discussion of perpetual motion. 'Each art possesses in itself some difficulty, not in transcending the power of Nature, but only the capacity and subtlety of our ingenuity'.[42] Perpetual motion thus was naturally possible, but someone had not yet discovered its secret. Bovelles added a caution: the would-be inventor should fear kings and princes, who would persecute such a discovery just as Domitian, fearing it would devalue gold, suppressed the discovery of a fabled unbreakable glass.

Bovelles's explanation of how windmills work shows that his readership could not have been artisans or practitioners. He claimed that the wind always acts on the top sails of the mill, driving the top sail around and allowing the bottom to move in reverse.[43] The explanation that the mill turns because wind above is more vigorous than wind near the ground seems to be the explanation of an observer, rather than the knowledge of artisans. Windmills of the Low Countries used sails slightly angled away from the plane of movement [Fig. 8.4]. By using angled sails, windmill makers evidently accounted for wind pressure directed perpendicularly to sail's plane of motion; millers had to trim their sails as often as the wind changed. Bovelles's explanation suggests that the most efficient kind of sail would have sails angled perpendicular to their plane of motion (like the vanes of a waterwheel), a construction artisans did not in fact use.[44] Based on these reasonable assumptions, there seems to be a curious disconnect between Bovelles's explanations and how artisans actually used nature. Similarly, in another passage Bovelles pointed out that one

41 Ibidem, fol. 64v.

42 Ibidem, fol. 56v.

43 Bovelles may have intended to explain by analogy with Vitruvius's observation that sails higher on a ship's mast are more effective than lower ones. *De architectura* x, 5–6. Hero of Alexandria's *Pneumatika* also described a windmill driving an organ, but made no comment on how the wind moved the sails.

44 On the construction of such 'post mills' see Lucas A., *Wind, Water, Work: Ancient and Medieval Milling Technology* (Leiden – Boston: 2006) 114–121.

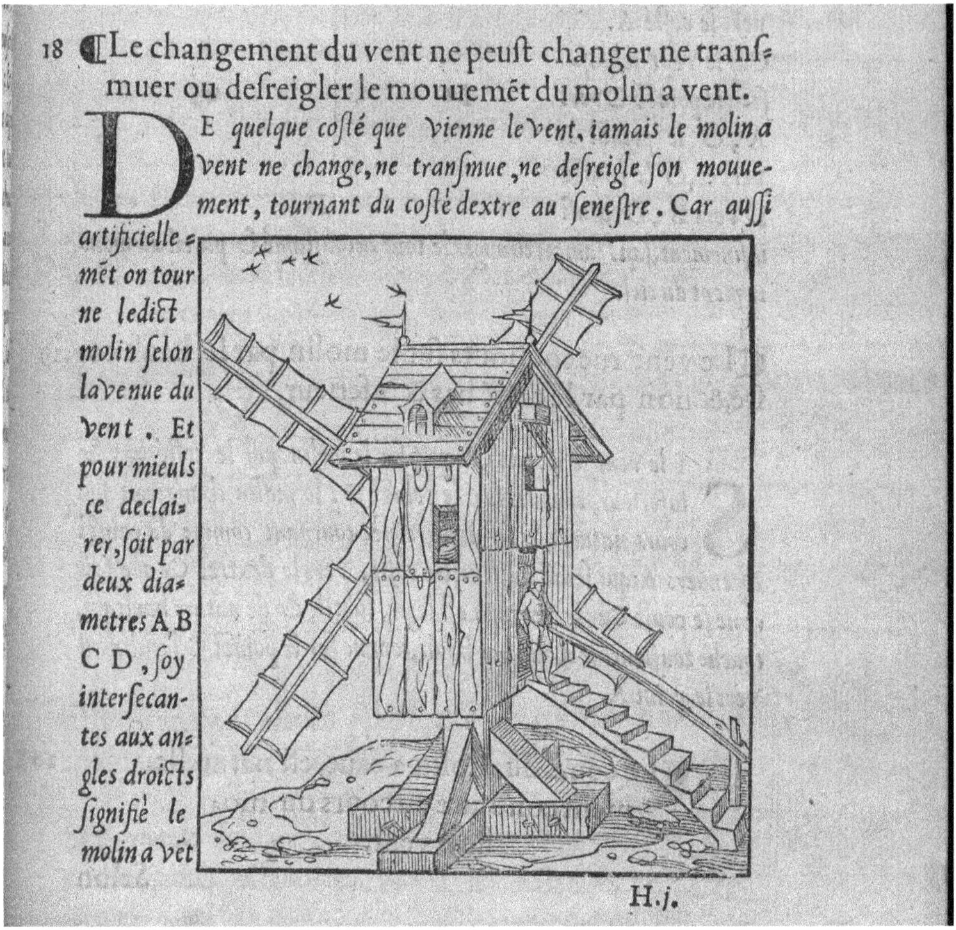
18 ❡Le changement du vent ne peuſt changer ne tranſ-
muer ou deſreigler le mouuemēt du molin a vent.

DE quelque coſté que vienne le vent, iamais le molin a vent ne change, ne tranſmue, ne deſreigle ſon mouuement, tournant du coſté dextre au ſeneſtre. Car auſſi artificiellemēt on tourne ledict molin ſelon la venue du vent. Et pour mieuls ce declairer, ſoit par deux diametres A B C D, ſoy interſecantes aux angles droicts ſignifié le molin a vēt

H.j.

FIGURE 8.4 *Bovelles,* Geometrie practique *(Paris, Regnauld Chaudière: 1551),* CUL *Syn.5.55.7, fol. 57r, detail of a woodcut added in 1547 (reproduced by kind permission of the Syndics of the Cambridge University Library).*

could use lines or circles as the basis for constructing typefaces—something that had to be well known to any reader of Pacioli, Dürer, or Geoffroy Tory, all widely known and available to artisans in the book trade.[45]

45 Geometry is used to design letters for type in Pacioli Luca, *Divina proportione* (Venice, Alessandro Paganini et Paganino I Paganini: 1509); Dürer Albrecht, *Underweysung der Messung mit Zirckel und Richtscheyt in Linien, Ebnen, und gantzen Corporen* (Nuremberg, Hieronymus Andreae: 1525); Tory Geoffroy, *Champfleury auquel est contenu l'art et science de la deue et vraye proportion des lettres attiques* (Paris, Olivier Mallard, for Geoffroy Tory and Gilles de Gourmant: 1529). These potential influences are mentioned by Margolin, "Une Géométrie fort singulière" 440.

Therefore, besides presenting geometry as useful as a kind of natural theology for recognizing the divine wisdom of numbers behind everyday objects and experiences, Bovelles meant his book to be enjoyed. He had written the practical geometry as a 'diversion', he stressed in the dedicatory letter of 1511. Both Margolin and Brioist have pointed out that in the 1542 edition Bovelles indulged in a Rabelaisian play on the microcosmic image of man, as he correlates the three lower holes in the human body (anus, genital, navel) emit elements similar to the cosmic elements (earth, water, air)—the heart correlates to fire, and appropriately remains secret, as fire apparently ought. A poem plays on the image of a broom to represent corporal punishment as the means of justice: the green bundle of twigs represent the switch for correcting youths; the thicker staff for older public offenders; while the string binding the fibers to the staff should be warning that those beyond correction may earn the hangman's noose. This collection of observations may seem harsh to modern sensibilities, but in sixteenth-century schoolyards likely served as humour.[46]

The eighth chapter was explicitly composed for useful recreation. As the largest single addition to the 1547 edition, this chapter sings the praises of geometry's 'utilities and excellencies'. Here Bovelles elucidated geometry's place among the quadrivium and walked his reader through ways geometry permitted one to deduce the size of stars and the distances of planets, as well as observe the nobility of the sun as the only planet without an epicycle—a fit image of human reason (while the other planets befit the wandering senses). Bovelles presented this as 'a little digression' to show the subaltern dependence of astronomy on geometry, as with perspective and the science of weights.[47] The mention of the mixed science of weights permitted Bovelles once again to digress into a list of secrets of nature: the diverse weights of kinds of earth and metals, the rare wood Gaiac (the remedy for syphilis), and the relative weights of food such as bread and cheese (the sort the Spanish called 'fermage' because it closes both meal and stomach). Bovelles designated this string of remarkable observations 'a joyous digression, in order to entertain and please the reader'.[48]

---

46 I go with caution here. After Huizinga J., *The Waning of the Middle Ages* (Toronto: 1954), historians have worried that the association of late medieval/Renaissance with macabre has been overplayed; but literary historians have still found the macabre a present feature of early sixteenth-century intellectual life. A *locus classicus* on the humorous elements of the macabre, extending Huizinga's own reflections on *homo ludens*, is also Bakhtin M., *Rabelais and His World*, trans. H. Iswolsky (Bloomington, 2009), here at 51.

47 Bovelles, *Geometrie practique* (1547), fol. 68r: 'Ici avons faict une petite evagation'.

48 Ibidem, fol. 68v: 'une joieuse evagation, pour recreer et resiouir le lecteur'.

Conclusion

Bovelles's practical geometry illuminates the source and aims of a genre that historians of science and technology have observed became 'popular' and widely diverse throughout Europe, from cosmographies and mapmaking manuals to instruction pamphlets sold with instruments.[49]

Bovelles (and Fine) begin to imagine a vernacular public in these books, I have suggested. It is extremely difficult to circumscribe a vernacular public, and scholars have long wrestled with the problem of who read books of secrets and other practical manuals. Like most, the *Geometrie practique* would have been of strictly limited use to actual craftsmen.[50] But it does not follow that artisans would have lacked the means to buy the books, or—more importantly—been uninterested in them. The very limited studies we have of sixteenth-century artisanal book ownership tell us that vernacular book ownership did rise considerably around mid-century.[51] Some form of literacy was quite widespread among the *menu peuple*; Natalie Zemon Davis judged that about half of males at the level of textile- and leatherworkers had a 'medium' literacy.[52]

This fits with the picture I have drawn. Bovelles's book captures the formation of a reading community of middling socio-economic status. Both supply

---

49 See studies cited in note 7. Cf. exhibition catalogues of instruments: Bennett J.A., *The Measurers: A Flemish Image of Mathematics in the Sixteenth Century* (Oxford: 1995); Korey M., *The Geometry of Power: Mathematical Instruments and Princely Mechanics Around 1600* (Berlin: 2007); Gerbino A. – Johnston S., *Compass and Rule: Architecture as Mathematical Practice in England, 1500–1750* (New Haven: 2009); Dackerman S. (ed.), *Prints and the Pursuit of Knowledge in Early Modern Europe* (New Haven: 2011).

50 A good example is the apparently practical ship-building manual by Michael of Rhodes in the late fifteenth century, which in fact omits measurements that could only be supplied by craftsmen who already knew what the book taught. Moreover, the book describes ships that were no longer being built in his time; it could not have been intended to guide the building of new ships. Long P.O. – McGee D. – Stahl A.M. (eds.), *The Book of Michael of Rhodes: A Fifteenth-Century Maritime Manuscript, Vol. 3: Studies* (Cambridge, MA: 2009). See also Tura A., *Fra Giocondo et les textes français de géométrie pratique* (Geneva: 2008) 103. More generally, see useful studies in Damm H. – Thimann M. – Zittel C. (eds.), *The Artist as Reader. On Education and Non-Education of Early Modern Artists*, Intersections 27 (Leiden-Boston: 2013).

51 Hackenberg M.R., "Books in Artisan Homes of Sixteenth-Century Germany", *The Journal of Library History (1974–1987)* 21, 1 (1986) 72–91. Hannah Murphy kindly shared with me Hackenberg M.R., *Private Book Ownership in Sixteenth-Century German-Language Areas*, Ph.D. Dissertation (University of California, Berkeley: 1983).

52 Davis N.Z., "Printing and the People", in *Society and Culture in Early Modern France* (Stanford: 1985) 210.

and demand create this market. The 1511 edition of his *Geometrie en francoys* failed to find a market—perhaps, I would conjecture, because it was aimed primarily at the bottom rung. For it succeeded when it reached higher, to a readership that was more literate, and had more money. It is this audience, I think, that is new. The *Geometrie practique* fits in between, as a work for mid-level elites who might never dream of participating in the higher reaches of the Republic of Letters.

There is no reason to reduce this vernacular, and indeed popular, readership to something manipulated by literary elites. The demand here is not either for mechanical utility or for theoretical abstraction, but a mix of both.[53] In his ground-breaking study of early modern printed books of secrets, William Eamon sharply distinguished between 'high' contemplative understanding of nature's *arcana* and 'low' recipes based on empirical, artisanal use of nature.[54] Yet in books such as the *Geometrie practique*, these extremes mingle. In this sense, they quite naturally fit the tradition of literature on wonders, which joined the extremes of the contemplative 'high' and empirical 'low' appreciation of nature, by viewing the experience of wonders as an entrance into the secret operations of nature.[55] Although these books were often written by university-trained natural philosophers, the phenomena described in this tradition were commonplaces of the broader swathe of culture. What is worthy of note, then, is that mathematical topics too could become part of this public, vernacular culture. Neither the mathematics nor the secrets exposed in the *Geometrie practique* are deep, rigorously argued, or particularly novel. Yet the genre was new in bringing abstract mathematics together with this empirical tradition of 'secrets', for a popular readership.

In particular, I should like to highlight the fact that Bovelles's French geometries were most successful when intended for entertainment; such secrets were increasingly meant to delight the *peuple moyen*, the growing public these books aimed to inform and entertain at once.[56] Pamela Smith and Allison Kavey have

53 This mixed quality of experimental, new genres of books at the time can be seen in Horodisch A., "Die Geburt eines Kinderbuches im 16. Jahrhundert", *Gutenberg-Jahrbuch* (1960) 211–222.

54 Eamon also distinguished between a medieval, esoteric language of secrets and early modern 'popular' secrets in newly printed books. Contrast chapters two and three of Eamon W., *Science and the Secrets of Nature*.

55 The basic studies are Céard J., *La nature et les prodiges* (Geneva: 1996); Daston L. – Park K., *Wonders and the Order of Nature, 1150–1750* (New York: 1998); Evans R.J.W. – Marr A. (eds.), *Curiosity and Wonder from the Renaissance to the Enlightenment* (Aldershot: 2006).

56 On reading for entertainment, see Eamon W., "How to Read a Book of Secrets", in Leong – Rankin (eds.), *Secrets and Knowledge in Medicine and Science* 23–46. The popular and

both noted that readers gained from these books not some proxy for actual experience, but the sense that there were more kinds of secrets than technical ones, and that nature lay open to them if they would only look.[57] Even in the process of outlining natural theology for laymen, such books fed early modern cultures of curiosity.[58] By extension, with mathematical curiosities and entertainments, they also made mathematics into popular culture.

Selected Bibliography

Bovelles Charles de, *Geometrie en françoys. Cy commence le Livre de l'art et Science de Geometrie, avecques les figures sur chascune rigle au long declarees par lesquelles on peut entendre et facillement comprendre ledit art et science de Geometrie* (Paris, Henri Estienne: 1511).

Bovelles Charles de, *Proverbiorum vulgarium libri tres* (Paris, Galliatus Pratensis: 1531).

Bovelles Charles de, *Liber de differentia vulgarium linguarum* (Paris: Robert Estienne, 1533).

Bovelles Charles de, *Livre singulier et utile, touchant l'art et practique de Geometrie, composé nouvellement en Francoys* (Paris, Simon de Colines: 1542).

Bovelles Charles de, *Geometrie practique* [...] *nouvellement par luy reveue, augmentee et grandement enrichie* (Paris, Reginald Chauderon: 1547).

Brioist P., "Les singularités de la géométrie pratique de Charles Bovelles", in Klinger-Dollé A.-H. – Faye E. – Sfez J. (eds.), *Bovelles philosophe et pédagogue* (Paris: *forthcoming*).

Faye E., *Philosophie et perfection de l'homme: De la Renaissance à Descartes* (Paris: 1998).

Fine Oronce, *Aequatorium planetarum, unico instrumento conprehensum, omnium antehac excogitatorum et intellectu et usu facillimum* (Paris, Nicolas Calceolarius: 1526).

Hackenberg M., "Books in Artisan Homes of Sixteenth-Century Germany", *The Journal of Library History* (1974–1987) 21, 1 (1986) 72–91.

---

entertainment value of books of secrets is also a theme in Eamon, *Science and the Secrets of Nature* 234–266.

57 Kavey A., *Books of Secrets: Natural Philosophy in England, 1550–1600* (Champaign, IL: 2007); Smith P.H., "What Is a Secret? Secrets and Craft Knowledge in Early Modern Europe", in Leong – Rankin (eds.), *Secrets and Knowledge in Medicine and Science* 52–54.

58 The rise of genres of books for the curious is a recurring theme in Kenny N., *The Uses of Curiosity in Early Modern France and Germany* (Oxford: 2004); Evans and Marr, *Curiosity and Wonder*.

Lefèvre d'Étaples Jacques, *Textus de sphera Johannis de Sacrobosco, cum additione (quantum necessarium est) adiecta, novo commentario nuper edito ad utilitatem studentium philosophice parisiensis academie, illustratus* (Paris, Wolfgang Hopyl: 1495).

Margolin J.-Cl., *Lettres et poèmes de Charles de Bovelles* (Paris: 2002).

Margolin J.-Cl., "Une Géométrie fort singulière: la Géométrie pratique de Charles de Bovelles (Paris, S. de Colines, 1542)", in *Verum et Factum. Beiträge zur Geistesgeschichte und Philosophie der Renaissance zum 60. Geburtstag von Stephan Otto* (Frankfurt am Main: 1993) 437–451.

Marr A. (ed.), *The Worlds of Oronce Fine. Mathematics, Instruments and Print in Renaissance France* (Donington: 2009).

Oosterhoff R.J., Mathematical Culture in Renaissance Paris: University, Print, and the Circle of Lefèvre d'Étaples, Ph.D. dissertation (University of Notre Dame: 2013).

Taton R., "Bovelles et les premiers traités de géométrie en langue française", in *Charles de Bovelles en son cinquième centenaire, 1479–1979: actes du colloque international tenu à Noyon, les 14–15–16 septembre 1979* (Paris: 1982).

Verney Pierre, *Succinte, briefve et compendieuse Collection Geometrale* (Metz, Jehan Pelluti: [c. 1530]).

Victor S.K., *Practical Geometry in the High Middle Ages,* Artis cuiuslibet consummatio *and the* Pratike de Geometrie (Philadelphia: 1979).

PART 3

Translating Philosophical Knowledge

CHAPTER 9

Taming Epicurus: Gassendi, Charleton, and the Translation of Epicurus' Natural Philosophy in the Seventeenth Century

Rodolfo Garau

Along with other early modern translations, a crucial characteristics of the translation of ancient philosophical texts, or portions thereof, was that they rarely appear as self-standing editorial pieces, but were consistently published along with, or even integrated in, substantial commentaries, notes, or discussions.[1] The increasingly outmoded but still quite vivid connection with the tradition of the *commentaria* of Aristotle's philosophy represents well such phenomenon.[2] Similarly, works such as Justus Lipsius' *Manuductionis ad Stoicam Philosophiam*, despite not being translations in the traditional sense of the term, result in fact in a long commentary, along with the translation of a number of Greek passages from Epictetus, Laërtius, Plutarch, and others, merged in Lipsius' more general reconstruction of Stoicism.[3]

As in other disciplines, early modern translations and commentaries of ancient philosophical texts 'were not primarily or exclusively focused on explaining (or reconstructing) the supposedly authentic meaning of the works of the past in a historical sense'. Rather, '[t]he primary concern was about the

1 On the early modern tradition of commentaries, see Enenkel K.A.E. – Nellen H. (eds.), *Neo-Latin Commentaries and the Management of Knowledge in the Late Middle Ages and the Early Modern Period (1400–1700)* (Leuven: 2013); and Enenkel K.A.E. (ed.), *Transformations of the Classics Via Early Modern Commentaries* (Leiden – Boston: 2013). Enenkel points out that '[e]arly modern intellectuals rarely read classical authors in a simple and "direct" form, but generally via intermediary paratexts: dedications, prefaces, and other introductory texts; *argumenta*; indices, illustrations; and above all, all kinds of commentaries—*annotationes, notae, commenta, commentaria, commentariola, animadversiones, paraphrases*, etc. These intermediary texts presented the classical text to modern readers in certain ways that determined and guided the reader's perception of the text being commented upon', in Enenkel K.A.E., "Introduction", in Enenkel, *Transformation of the Classics* 1.

2 On this, see Lohr C., "The Social Situation of the Study of Aristotelian Natural Philosophy in the Sixteenth and Early Seventeenth Centuries", in Leijenhorst C.H. et al. (eds.), *The Dynamics of Aristotelian Natural Philosophy from Antiquity to the Seventeenth Century* (Leiden: 2002).

3 See Lipsius Justus, *Physiologiae Stoicorum Libri Tres* (Antwerp, Joannes Moretus: 1610).

 | DOI 10.1163/9789004349261_011

present-day use (*usus*) and application of antiquity's writings in every possible sense'.[4] In focusing on the contemporary usefulness of classical tradition, early modern translations and commentaries displayed a broad range of different functions—from means of authorization to educational tools, from encyclopaedias of learning to textual criticism.[5] For instance, scholastic *commentaria* (save some exceptions) had mostly the educational aim to explain and teach Aristotle's philosophy, bringing light on potentially obscure passages.[6]

However, the early modern period is also an era in which the relationship with the philosophical classicity becomes more ambiguous and complicated. On the one hand, while certain aspects of Aristotle's natural philosophy (as for instance his *Parva naturalia*) still deeply influenced and inspired the natural-philosophical production, the discontent with Aristotelian-Scholasticism triggered the spread of a pervading rhetoric of the novelty or of the 'new beginning' for the philosophical enterprise, as testified for instance by the very title of Bacon's *New Organon*, or by Descartes' search for a new foundation of the whole edifice of knowledge. On the other hand, the search for alternatives to Scholasticism sparked a new interest for a diverse, non-Aristotelian philosophical past. The revival of Scepticism (often employed as an intellectual camouflage for radical anti-Scholastic stances); of Stoicism (especially in political and moral philosophy); and finally of Epicureanism marked the end of the sixteenth and the beginning of the seventeenth century, informing and influencing the following philosophical productions.

In this frame, the early modern translators of non-Aristotelian philosophical texts were often exposed to a twofold pressure. First (as in the case of Stoicism and Epicureanism), they were often dealing with historically controversial texts, which required, in addition to very refined linguistic and philological skills, also the ability to emend potentially impious passages in order to show how they could fit an intellectual framework characterized by intense

4 Enenkel, *Transformations of the Classics Via Early Modern Commentaries* 4; See also Enenkel K.A.E. – Nellen H., "Introduction", in Enenkel – Nellen, *Neo-Latin Commentaries and the Management of Knowledge* 3: 'Commentaries were mainly studied—in various intellectual settings, to be sure—in order to acquire knowledge and skills. A 'pure' understanding of the source text was not the exclusive goal. This radically different way of approaching both objects is exemplified by the fact that more often than not the commentary was consulted without reference being made to the text itself. Commentaries were seen as very useful tools for gathering and processing information from entirely different disciplines'.

5 Enenkel – Nellen, "Introduction", in *Neo-Latin Commentaries and the Management of Knowledge* 3.

6 For a discussion of a translation of and commentarary on an Aristotelian mathematical text, see the chapter by Joyce van Leeuwen in this volume.

censorship. Second, in an era characterized by the above-described rhetoric of the novelty or of the 'new beginning', they often had to push back the criticism of being mere 'historian of philosophy', and to show the relevance of their texts vis-à-vis the contemporary philosophical, scientific, or political discussions. As a consequence, such kind of early modern translators of philosophical texts had to adopt different strategies to meet both these two potential criticisms.

In this paper, I focus in particular on the seventeenth-century 'Epicurus Renaissance',[7] with particular regard to the complex relationship that the translation of Epicurean texts entertained with the early modern natural-philosophical and scientific discourse on the one hand, and with the Christian faith on the other.[8] Indeed, amongst the attempts to revive the philosophical past, the one concerning Epicureanism was by far the most complicated and controversial. In the seventeenth century, the name of Epicurus was still associated with the infamous reputation of an atheist philosopher, advocate of a lascivious ethics, denier of the immortality of the soul, and propounder of a materialistic physics. The world was conceived by Epicurus as deriving from mere chance; the human soul as a material entity that was destined to die with death; death in turn, as the cessation of sense perception, excluded any eternal reward and punishment. Moreover, as in the case of his follower Lucretius, the

7 The rediscovery of the manuscript of Lucretius' Epicurean poem *De rerum natura*, found in 1414 by the Italian humanist Poggio Bracciolini, (probably) in the Benedictine library at Fulda, Germany, stimulated new studies on the figure of Epicurus. In the Renaissance, scholars such as Jacopo Zabarella, Lorenzo Valla, Francesco Filefo, Cristoforo Landino, Cosma Raimondi, and Leonardi Bruni paid particular attention to Epicurus' theory of pleasure, offering fundamental, and philologically more accurate, revisions of the widespread misconceptions circulating in the Christian world since the late antiquity. Although their works were generally not apologetic (for instance, Valla's *De voluptate* uses Epicureanism mostly as a polemic tool against Stoic ethics, then discarding both of them in favour of the Christian moral), they had the fundamental role of reintroducing Epicurean philosophy in the intellectual discourse. Despite this, the Renaissance interest in Epicureanism was far from a systematic reappraisal; at the same time, while discussing the theory of pleasure, Epicurus's physics was relatively left aside from such discourse. On this, see for instance Jones H., *The Epicurean Tradition* (New York: 1989); Joy L., "Epicureanism in Renaissance Moral and Natural Philosophy", *Journal of the History of Ideas* 53, 4 (1992) 573–583; Joy L., *Gassendi the Atomist: Advocate of History in an Age of Science* (Cambridge: 2002); Butterfield D., *The Early Textual History of Lucretius' De Rerum Natura* (Cambridge, 2013); Wilson C., *Epicureanism at the Origins of Modernity* (Oxford – New York: 2008).

8 On the importance and pervasiveness of Epicurean thought in the early modern period, see Wilson, *Epicureanism at the Origins of Modernity*.

figure of Epicurus was slandered as that of an unlearned and uneducated man, an enemy of erudition and true knowledge.[9]

Representing a radical alternative to Aristotelian hylomorphism and a form of radical empiricism, Epicurus' atomistic philosophy constituted a very appealing theoretical frame for early modern inquirers. But since Epicureanism was still seen as the doorway to atheism, the temptation of Epicureanism was often an unavowable one. The famous repression of the thirteen anti-Aristotelian theses of Etienne de Clave, Jean Bitault and Antoine de Villon by the Parliament of the Sorbonne in 1624, as well as the harsh repression of its proponents, testifies the atmosphere of relative hostility towards atomistic themes.[10] In this complex frame, while Epicurean motives were tacitly adopted by many, if not by most part of early modern inquirers, an explicit and systematic re-evaluation of Epicurean philosophy was still a bold desideratum.

In this paper, I offer a case study of such complicated attempt to restore this ancient philosophical school, and to show the 'strategies of domestication' (to use Peter Burke's expression) that two of the protagonists of the seventeenth-century revival of Epicureanism – Pierre Gassendi (1592–1655) and Walter Charleton (1619–1707)—enacted in order to reach such goal.[11] I shall focus in particular on two texts. First, I pay attention to Gassendi's *Animadversiones in decimum librum Diogenis Laertii* (1649)—a long and complex text comprising a translation, from ancient Greek to Latin, of the tenth book of Diogenes Laërtius' *Lives of Eminent Philosophers*, devoted to Epicurus, and Gassendi's commentaries on the text. In these commentaries, all printed as appendices, all doctrines of Epicurus are reconstructed, while referencing additional sources. A first series of appendixes is not completely related to the topic (*De facto novo circa Inane Experimento; De Sestertiorum moneta nostra expressorum Abaco; De Eclipsi dice Novembr. Ultimae 1648; Exscriptum ex tractatu de Globo Telluris; De nutritione animalium; De pulsu et nutritione animalium*), however,

9 Joy, *Gassendi the Atomist* 67.

10 Scholars have suggested that this hostility was motivated, in addition to its strong connection with Epicurean (and therefore atheist) philosophy, also by the specific theological framework of Scholasticism. The doctrine of the forms was the basis of the explanation of the transubstantiation of the bread and the wine in the mystical body and blood of Christ, as well as of the idea of the resurrection of the human soul. On the contrary, atomism, with its anti-hylomorphistic understanding of matter, paved the way to the doctrine of the first and secondary qualities, and, with it, to the idea that different forms are just the by-product of our peculiar way to perceive complex arrangements of matter. On this, see for instance Redondi P., *Galileo eretico* (Turin: 2004).

11 See Burke P., "Cultures of Translation in Early Modern Europe", in Burke P. – Po-chia Hsia R. (eds.), *Cultural Translation in Early Modern Europe* (Cambridge – New York: 2007).

a long appendix is, abridging in a schematic fashion Epicurus' philosophy, the *Syntagma philosophiae Epicuri*. I then focus on Charleton's *Physiologia* (1654), largely consisting of an English paraphrase of selected parts of Gassendi's *Animadversiones*, with the addition of comments.[12]

Both actively engaged, though with different roles and importance, in the Epicurean revival of the seventeenth century, as well as in various scientific practices, Gassendi and Charleton complemented their works with a critical apparatus of comments and notes with two aims in mind. First, that of emending Epicurean philosophy from those aspects that made it glaringly incompatible with the Christian faith, or to show the lack of real contrast between the two. Second, the introduction of experimental supports, as well as the comparison or contrast with modern sources, show the relevance of Epicurus's system of natural philosophy vis-à-vis the emerging scientific sensibility. All the while this 'baptism of Epicurus', to use an expression employed by Margaret J. Osler, made acceptable the use and discussion of Epicurus's philosophy (and in particular his natural philosophy).[13]

Within this framework, I shall show that the combination of translations and commentaries are part of an integrated strategy that allowed Gassendi and Charleton to reach a threefold goal. First (and at least in the case of Gassendi), the translation becomes an instrument to open a discourse on Epicurean philosophy. The translation of Laërtius' *Tenth Book* made it possible for Gassendi to address openly the possibility of Epicurean philosophy as a viable alternative to Scholastic Aristotelism. Second, the *apparatus* of notes and commentaries makes it possible to show the compatibility of Epicurean philosophy with the Christian faith by emending the most incompatible parts of his doctrine. At the same time, by integrating and testing the Epicurean framework with experiments and observations, it shows its usefulness for replacing Scholasticism as a theoretical framework of early modern experimental science.

The treatment of Charleton's *Physiologia* will make it possible to address another key-issue of this delicate work of domestication of Epicurean philosophy: its target. Gassendi still speaks to a world of literati, and uses philology

12 Gassendi Pierre, *Animadversiones in decimum librum Diogenis Laertii, qui est de vita, moribus, placitisque EPICURI* (Lyons, Guillaume Barbier: 1649); Charleton Walter, *Physiologia Epicuro-Gassendo-Charltonia or a Fabrick of Science Natural Upon the Hypothesis of Atoms Founded: Epicuruius, Repaired: Petrus Gassendus, Augmented: Walter Charlton, Dr. In Medicine, and Physician to the Late Charles, Monarch of Great Britain* (London, Newcomb: 1654).

13 Osler M.J., *Divine Will and the Mechanical Philosophy: Gassendi and Descartes on Contingency and Necessity in the Created World* (Cambridge: 1994).

and erudition to justify his broader design of vindication and restoration of Epicureanism. On the contrary, Charleton—though, I shall show, motivated by the same need of emending Epicurus' philosophy from his potentially anti-Christian contents while at the same time showing its resourcefulness—speaks to a different intellectual context—that of the English-speaking British Isles. This motivates not only his use of the vernacular, but also informs his style of exposition as well as the contents of his work.

Gassendi's Translation of Laërtius' Tenth Book in the *Animadversiones*

Gassendi's *Animadversiones*, besides being also a translation, *stricto sensu*, of the cryptic tenth book of Diogenes Laërtius' *Lives of Eminent Philosophers*, also represented the most systematic attempt at 'domesticating' ancient Epicureanism with regard to the Christian religion and early modern science. By 1649, year of publication of the *Animadversiones*, Gassendi had already established himself as one of the key intellectual figures of his time. Then professor of mathematics at the prestigious Collège Royale of Paris, he had published *Exercitationes paradoxicae adversus Aristoteleos* [Paradoxical Exercises Against the Aristotelians] (first volume appeared in 1624), based on public disputations he held as professor of philosophy at the University of Aix-en-Provence, which had gained him the reputation of an anti-Aristotelian philosopher. Confirming some of the predictions made by Kepler, his astronomical observations (such as his *Mercurivs in sole visus, et Venus invisa Parisiis, anno 1631: pro voto, et admonitione Keppleri*, 1632), spread all over the European continent and contributed to the success of the Keplerian model. His writings on the Galilean law of free-fall (*De motu impresso a motore translato*, 1642, and *De proportione, qua gravia decidentia accelerantur*, 1646) reporting on the first performance of Galileo's 'mental experiment' of the fall of a cannon ball dropping from the mast of a moving ship, and applying Galileo's theory of free fall to one of the first modern accounts of gravitation, are credited to have provided the first account of rectilinear inertia.[14] What is more, they opened a much larger debate on Galileo's physics in French culture.[15] His polemic with René Descartes, which started with the sixth set of objections to the *Meditations*, and then published in a more extended version under the title *Disquisitio metaphysica. seu,*

14 See Koyré A., *Galileo Studies*, trans. J. Mepham (Atlantic Highlands, NJ: 1978) 249.

15 See Galluzzi P., "Gassendi and l'Affaire Galilée of the Laws of Motion", *Science in Context* 14, 1 (2001) 239.

Dubitationes et instantiae adversus Renati Cartesii Metaphysicam, et responsa (1644), had made him the exemplar advocate of empiricism against Cartesian rationalism. His intellectual network included key figures of the early modern philosophy and science such as Galileo, Kepler, Descartes, Hobbes, Mersenne, de Sorbière, and Schickard.

But besides his astronomical observations, his fierce polemic with Descartes and with the Aristotelians, Gassendi had been working for over 20 years to a more ambitious project: that of providing a comprehensive reconstruction and reappraisal of philosophy of Epicurus. Although it is hard to reconstruct exactly the origin of Gassendi's interest in Epicureanism, we know that Lucretius was part of his readings as early as 1621 (letter to Faur de Pibrac, 8 of April).[16] The letters to the Dutch humanist and philologist Eric Van de Putte (also known as Erycius Puteanus) of 1628, and to the scholar and astrologer Jacques Gaffarel of march 1629, show that Gassendi initially intended to publish his research on Epicurus as a part of his *Exercitationes paradoxicae adversus Aristoteleos*, whose original project extended far beyond the two books published respectively in 1624 and in the *Opera Omnia* of 1658.[17] In the *Exercitationes*, Gassendi makes extensive use of arguments derived from ancient Scepticism in order to uncover the shaky foundations of Scholastic philosophy. This effective rhetorical move carried with itself the not banal consequence of casting a dark shadow on the very possibility of attaining any sound knowledge of the external world whatsoever. The letter to Puteanus makes it clear how Gassendi understood Epicurus' philosophy as both the alternative to Scholastic Aristotelianism and Scepticism:

> Vera loquor, ex quo tempore tuum illud de Epicuro, cum evulgata ipsius effigie pellegi Eulogium, nescio quo tacito affectu occoeperim te et suspicere etamare. Scilicet ego tanto viro paravi Apologiam, destinato ipsius doctrinae volumine integro, quod Paradoxicarum Exercitationum adversus Aristoteleos volumini, ciuius ideam primumque librum faci iam iuris publici, attextatur.[18]

> [...] after I read your *Eulogy of Epicurus*, accompanied by the publication of his portrait, I have started to admire and love you with a silent feeling. So I have prepared for this man a great Apology, planning to devote

16 Gassendi Pierre, *Opera omnia*, 6 vols. (Lyons, Anisson – Devenet: 1658; reprint, Stuttgart: 1964) vol. 6, 1–2.

17 Ibidem 11–12.

18 Ibidem 11. All translations are mine unless otherwise stated.

> a whole volume to his doctrine with the idea of attaching it to my volume of the *Exercitationes paradoxicae* against the Aristotelians of which I have already published the first volume.

Gassendi's letter to Puteanus is particularly significant as it shows that, as early as 1629, Gassendi individuated in the philosophy of Epicurus as both the alternative to Scholasticism and the answer to the threat of Scepticism. However, in order to do so, Gassendi had to fight the widespread prejudice the figure and philosophy of Epicurus still carried with themselves. In this respect, the letter to Gaffarel is particularly telling:

> Addis ad calcem aliquid de emittendis illis meis in Epicuri philosophiam elucubrationibus. Id habeo sane gratissimum, quando et mea tibi studia innotuerunt, neque vereor, ut quos liber tuus lectores aequos nanscitur, mihi propterea succenseant ob quaesitam illam Epicuro apud vulgus infamias. Facilissime enim nosse potuerunt, quod Seneca habet, Epicurum sacra et recta praecipere; eiusque Sectam male quidem, sed immerito audire, quod sola voluptatis vox fronti praefixa et prave accepta fecerit locum fabulae.[19]
>
> I add something concerning the publication of my research on Epicurus. I am very thankful that you know my studies and I do not fear that the readers who have received favourably your book will be outraged with me because of the infamy that people attributes to Epicurus. For they will be easily learn that Epicurus proposed holy and right doctrines, as Seneca defined them, and it is unjust that his sect has a bad reputation, because the term pleasure that have attribute to him was misinterpreted.

At the centre of Gassendi's concerns there is Epicurus' doctrine of pleasure, often interpreted, as propounding unrestrained hedonism and moral lasciviousness. Gassendi is therefore convinced that he has to correct this widespread prejudice, that he sees as the result of a slander, in order to make possible a treatment of Epicurus's philosophy at large. This clearly shows that, since from the beginning of his engagement with Epicurean philosophy, Gassendi had clearly in mind that only through reappraisal of the reputation of Epicurus, and not only of his doctrine, would make it possible to open a more comprehensive discourse on Epicureanism in general.

---

19 Ibidem 16.

This widespread prejudice surrounding the figure of Epicurus probably motivated Gassendi's reluctance to publish anything at all on his figure and philosophy. Through a letter to Beeckman (September 14, 1629) we know that as early as 1629 Gassendi thought that the completion of his draft of the philosophy of Epicurus was at hand.[20] However, it was only in 1649—and arguably not entirely by choice—that Gassendi started publishing the outcomes of his twenty-years research.

Indeed, in 1649, Gassendi's translation and edition of Diogenes Laërtius's tenth book of the *Lives of Eminent Philosophers*, devoted to Epicurus, was published in Lyon by the editor Guillaume Barbier with the title *Animadversiones in decimum librum Diogenis Laertii*. With the translation Gassendi published a dense set of notes, an epitome of the Epicurean doctrines on epistemology, physics, natural philosophy, and ethics, and the report of some natural experiments and observations aimed at corroborating the validity of Epicurus' theories. The bizarre circumstances of its publication are recapitulated in the dedicatory letter to the Lyonnais lawyer and philologist François Barancy (dating September 1646) by its author:

> Obtinisse te abs meo LULLERIO, quam ad eum ante duodecim annos vitae, ac morum Epicuri Apologiam transmiseram, grave mihi haudquaquam fuit; quando illi per me fueram integrum, ut quibuscum vellet, eam comunicaret. At vero te eo incubuisse, ut etiam typis mandaretur, ea praesumptione, quod facilius ex me impetraturus facti veniam, quam faciendi licentiam fores; id non potuit esse non grave.[21]

> It was not at all distressing that you had obtained from my Lulllier the Apology on the life and morality of Epicurus which I had sent him twelve years before; for he could certainly share it, I believe, with whomever he wanted. On the contrary, that you had exerted yourself so that it was sent to the print, under the presumption that it would have been easier to obtain my forgiveness than my permit—this was necessarily distressful.

Indeed, Barancy had obtained from Gassendi's patron and friend, the scholar Claude-Emmanuel Luillier, a copy of Gassendi's draft of a biography of

20 Though from this letter it is impossible to understand whether it is the translation of Laërtius that Gassendi appoints as 'my *Epicurus*', or a biography of the philosopher, or even a early version of the *Syntagma philosophiae Epicuri* (which was published along with the *Animadversiones* in 1649, not to be confused with the *Syntagma philosophicum*).

21 Gassendi, *Animadversiones* 1.

Epicurus had published, without Gassendi's consent, in 1647, with the title *De vita et moribus Epicuri libri octo*—a writing that was clearly part of Gassendi's plan to re-establish Epicurus's reputation before dealing with his philosophy at large. Gassendi, notoriously reluctant to publish any work concerning Epicurus, suddenly saw his name printed at the title page of a biography that was clearly addressed to defend Epicurus from the false accusation of being morally lascivious and an enemy of culture and wisdom. In other words, an apology was published in his name, therefore publicly associating his name to that of someone who, in the early modern sensibility, was still associated with atheism and perversion. Although published with an introductive letter of Luillier that made clear the philological ground of Gassendi's work—that of re-establishing the truth concerning Epicurus' moral—the publication of *De vita et moribus Epicuri* surely pushed Gassendi well out of his comfort zone. This explains the unexpected reaction that Gassendi's had at Barancy's request to send him the drafts of the other Epicurean writings that he had been brooding over for more than twenty years:

> Quod subinde porro exoptasti, ut, dum id, quod supererat, operarum procederet, emendatum a me contextum Decimi libri Laërtiani, una cum interpretationem, Apologiae attexendum, ad te transiderem; dici non potest, quam repugnarim; idque conspirantibus licet ac pene convitio agentibus amicis illis, quibus, quid cepisses consilii, curas innotescere.[22]

> That then you desired that I transmitted you the structure of the Tenth book of Laërtius from me emended, to be combined with an interpretation and an Apology, so that what remained of the work would come out; it cannot be said that I opposed it.

Also convinced by his friends, whom Barancy had informed, Gassendi decided to finally publish his material on Epicurus. In this context, the choice of publishing the translation of the *Lives* is particularly telling. Indeed, translation becomes a way of corroborating the new image of Epicurus provided by the unauthorized publication of his biography with textual and philological evidence, and, at the same time, to directly address his philosophy in general. Gassendi himself, in the *Preface*, seems to insist on the sheer philological interest of his work. He refers to the French erudite and Latinist Gilles Ménage, as one of the friends who convinced him to come out with this publication. Ménage's own translation and emendation of Laërtius' *Tenth Book* would be

22 Gassendi, *Animadversiones* 1–2.

completed almost twenty years later and printed with the title *Observationes et emendationes in Diogenem Laertium* in Paris 1663, with reprints London in 1664 and Amsterdam in 1692. The connection between the two also offers an interesting cross-section of the mid-seventeenth century state of the art of the philological interest on Laërtius' *Vitae*. 'Aegidius Menagius, qui exhibere totum Laërtium locis plusquam mille feliciter castigatum aggressus' [Aegidius Ménage, who has started to edit the complete Laërtius, successfully emended in more than one thousand places], Gassendi writes, 'persensi tamen haerere se circa huiuscemodi librum' [found himself in difficulties in dealing with such a Book].[23] So that 'quod norit me non nihil meditatum Epicurea dogmata, de quibus in eo agitur, ac reputarit sibi ex meis qualibuscumque coniecturis affulgere posse lucis aliquid' [because he knew that I had reflected profoundly on the Epicurean dogmas, of which the Book treats, he thought that he could derive some sort of illumination from my conjectures of whatever quality they are].[24]

The *Preface* also makes clear how Gassendi's translation fits in his more general commitment to restore the figure, and the doctrines of Epicurus. As Gassendi recalls, the tenth book of the *Lives* was portrayed as a very obscure text—presenting crucial difficulties both in the individuation of the right *lectiones* and in contextualizing its text—so that they would be even more difficult than the enigmas to which Oedipus had been exposed.[25] Gassendi claims that such obscurity of Laërtius' text—an opinion reinforced by many authoritative philosophers such as Cicero, Plutach, and Sextus Empiricus—had dramatically affected Epicurus's reputation both as a person and as a philosopher.[26] As it emerges from the introduction the notion of pleasure is still at the core of Gassendi's concerns. In the only passage where he emphatically uses capital letters, Gassendi anticipates that from his translation of Laërtius it will emerge that Epicurus's pleasure is nothing but the condition we achieve 'quam NON DOLERE CORPORE, AC ANIMO NON PERTURBARI' [WHEN OUR BODY

23 Ibidem 2.

24 Ibidem.

25 Ibidem: 'eos grifos contuexterint, quibus Oedipus impar sit'.

26 As Joy nicely put it: 'Gassendi deplored the present state of scholarship on Book X, for it seemed to him that a combination of badly copied Greek manuscripts of the *Lives* and a few inadequate Latin translations of the texts contained in such manuscripts were chiefly responsible for Epicurus' reputation as an obscure and unimportant thinker'. Although he blamed scholars for having avoided book X: 'Still he placed most of the blame for their shortcomings on the poor state of the existing manuscripts, and he also recognized that it was the Hellenistic interpreters of Epicurus – Cicero, Plutarch, and Sextus Empiricus—who had led more recent scholars wrongly to assume that Epicurus' texts should be avoided because they were unreadable'. Joy, *Gassendi the Atomist* 74.

DOES NOT HURT, AND OUR MIND IS NOT UPSET].[27] Philology, in this frame, becomes a way to legitimize such interpretative line and to provide a basis for a more general treatment of Epicurus' doctrines.

Between Translation and Commentary: Taming Epicurus

The translation of Laërtius *Tenth Book* occupies the first sixty pages of the *Animadversiones*. The commitment to a rigorous philological work is testified by the presence of the Greek text, which is transcribed entirely next to Gassendi's Latin translation, and divided in numbered sub-chapters to provide a rigorous match between the two. Still, also in the translation Gassendi takes good care in emphasizing the most sensitive passages with the use of capital letters. This is the case, for instance, of the passage concerning the description of atoms.[28] But in particular, once again, the passages concerning pleasure, which are still at the core of his concern. In this way, Gassendi intended to attract, also in a visual way, the attention of the readers on the passages that could disprove their prejudices on Epicurus' morality.[29]

But Epicurus's natural philosophy emerged equally well in all its powerful, anti-teleological and godless nature. The imposing *Animadversiones* [commentaries] to Laërtius' *Tenth Book* (which spread from page 100 to page 751 of the first edition of the work) consists of a very dense commentary of all the aspects of Epicurus' philosophy. Commenting on everything from logic to ethics, from physics to astronomy and meteorology—combined with the quotation of other ancient sources (as for instance Plutarch, Lucretius, and Cicero), Gassendi's twofold goal emerges in his entirety: that of emending the dangerous aspects of Epicurus' philosophy, and yet showing at the same time how his readers' understanding of nature could benefit from such trade-off, showing its relevance vis-à-vis modern science. In this light, the commentary integrates the translation in a threefold way. First, some passages are meant

27 Gassendi, *Animadversiones* 11.

28 Gassendi, *Animadversiones* 17.

29 See for instance ibidem, 81: 'Cupiditatum aliae sunt naturales et necessariae; aliae naturales nec tamen necessariae; aliae neque naturales neque necessariae, sed ex opinione inani prognatae. NATURALEIS PORRO ATQUE NECESSARIAS EXISTIMAT EPICURUS EAS, QUAE DOLORES TOLLUNT, CUIUSMODI EST REI POTULENTAE, DONEC VIGET SITIS. NATURALEIS VERO ET NON NECESSARIAS ILLAS, QUAE NIHIL ALIUD QUAM VOLUPTATEM VARIANT, NON VERO *absolute* DOLORI DETRAHENDO, *quatenus sunt tales, necessariae* SUNT, QUAE EDULIORUM SUNT DELICATIORUM. *Postremo* NEQUE NATURALEIS, NEQUE NECESSARIAS, QUALES *verbi gratia* CORONARUM SUNT STATUARUMVE ERIGENDARUM'.

to *integrate* the translation of Laërtius' *Tenth Book* with a new set of sources, which provide a more coherent picture of Epicurus' doctrines at large. Second, other passages are meant to show the immediate usefulness and tenability of Epicurus' doctrine in respect to early modern science, and are usually integrated with experimental evidence that is meant to test the validity of the physical doctrines of Epicurus. Third, other passages complement the work of emendation of Epicurus' philosophy from the doctrines that show the greatest incompatibility with the Christian faith. Indeed, while the notion of pleasure was addressed philologically by proving its erroneous interpretation, other traits of Epicurus' philosophy were still to be emended in order to make available as a viable alternative to Scholasticism. In what follow, I will give two examples of these latter two 'techniques of domestication'.

The Existence of Vacuum: A Test for the Doctrines of Epicurus

One of the most powerful elements of appeal of Epicurean natural philosophy to early modern inquirers was his idea that nature is fundamentally composed by two kind of entities: bodies (in turn composed by unbreakable corporeal element or atoms), and vacuum. The existence of vacuum was a major point of contention among early modern scholars at least following the divulgation of Evangelista Torricelli's observations on the barometer. Whether the space left by mercury in a tube previously filled with the substance and then set vertically in a basin with other mercury was actual vacuum, or air, or another subtle substance, was a question that could potentially shake not only the intellectual edifice of Scholastic natural philosophy, but also that of Descartes'. His entire picture of the world (and therefore his explanation of motion, light, optics, and even animal motion) depended after all on the identification of space and matter, labelling vacuum as an absurdity. While Scholastic and Cartesian philosophers were forced to find implausible explanations to reconduct the result of Torricelli's observation under the umbrella of their natural-philosophical systems (sometimes even joining forces to contrast the common enemy), such vacuist interpretation of the 'Torricellian experiment' reflected what an experimenter without prejudice was more easily brought to infer.[30]

30 On this, see Grant E., *Much Ado about Nothing: Theories of Space and Vacuum from the Middle Ages to the Scientific Revolution* (Cambridge: 1981); Gemelli B., *Aspetti Dell'atomismo Classico Nella Filosofia Di Francis Bacon E Nel Seicento* (Florence, 1996); Lupoli A., *Nei Limiti Della Materia. Hobbes E Boyle: Materialismo Epistemologico, Filosofia Corpuscolare E 'dio Corporeo'* (Milano: 2006); Trevisani F., "La Teoria Corpuscolare in Cartesio", in Baldini U. et al. (eds.), *Ricerche Sull'atomismo Del Seicento* (Florence: 1977).

In addition to offering valid epistemological support to empirical-driven scientific research, Gassendi's Epicureanism also provided a cosmological model in which vacuum was a fundamental part of the system. He supplemented this with a discussion of the strengths of this model compared to that of Aristotle and vis-à-vis the latest advancement of scientific observations. The chapter of the *Animadversiones*, entitled "Dari praeter corpora etiam inane in rerum natura" [In the nature of things there is besides bodies also vacuum], opens with quotes from Lucretius; continues by expounding Aristotle's anti-vacuist position (the famous 'horror vacui', according to which the space left by an object is immediately filled with particles of the medium in which the object moves); and then presents the Epicurean view that no motion is possible in the universe without vacuum, i.e. atoms impenetrable. Epicurus's arguments, Gassendi explains, are based on the evidence that, were nature full of bodies, the medium would have no elastic properties, and therefore motion would be impossible. Here, the reference to experiments and observations results/is? particularly important. A first set of experiments is quoted in reference to the phenomena of rarefaction and condensation. The most important experiment refers to the so-called 'ball of Heron', 'Heron's engine', or 'aeolipile', consisting in a ball of metal with two breathers pointing in the same direction and connected to a brazier filled with water through two horizontally pivoted pipes.

Once heated up, the water in the brazier would evaporate, making breathers propel the ball in the opposite direction. This experiment (that could be found

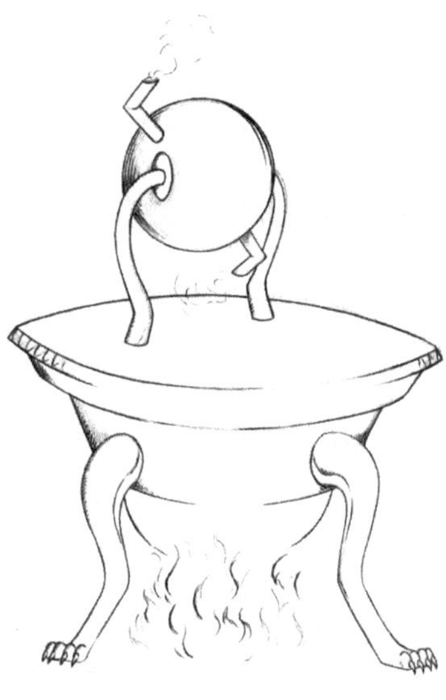

FIGURE 9.1
A representation of Hero's aeolipile (also known as Hero's engine) in a drawing by Marco Discalzo (2017).

in Vitruvius's *De architettura* and in Hero's *Pneumatica*) seems to corroborate the Epicurean claim that air is filled with minute empty *interstitia* [gaps]: indeed, 'Si putes nihil posse spatiorum intra ipsam inane fieri, quaeso ubi ad ignem apponitur, & incalescens aqua in vaporem rarescit, ac osculo illo exilissimo erumpit summo impetu; qui potest eadem capacitas semper oppleta perseverare?' [if you think that no empty space could be between them, how could it possibly be that, as soon as it is put over the fire, and the heated water rarefies in vapour, and bursts out with great violence from that tiny hole, it keeps its capacity always full?][31]

A second set of experiments refer to the property of solvability of salt crystals into water:

> noram pridem non posse aquam quantitate salis quantamlibet exsolvere ac in se transfusam veluti concipere, sed ad certam solum mensuram, adeo ut satiata, quicquid superest, inexsolutum relinquat [...]. [...] abeunte sale in particulas minutissimas deberent esse intra aquam consimilia spatiola ipsis excipiendis capacia, quibus repletis exsolutio conceptioque illa cessaret [...].[32]
>
> You might have noticed sometimes that water cannot solve whatever quantity of salt, and, so to say, receives it by transferring it in itself; but if a certain measure is surpassed, the salt remains unsolved, as the water was satiated. *This must be due to the fact that* the salt being absent, there must be in the water analogous minute spaces capable of receiving it in very small particles, whose faculty and capacity ends once those spaces are filled.

Lastly, Gassendi devotes an entire appendix to the third and last set of experiments in support of the existence of vacuum. This appendix, entitled *De facto novo circa Inane Experimento* (from page iii to x) constitutes the connection between Epicureanism and modern science that could prove how the former could provide a reasonable natural-philosophical basis of the latter.[33] Here, Gassendi talks extensively about Pascal's experiment (who starting from 1646 had begun to test Torricelli's experiment) and its implications for what concerned the theories on the nature of vacuum he had just presented in his *Animadversiones*. While Epicurus is not quoted directly, Gassendi's aim

31 Gassendi, *Animadversiones* 173.

32 Ibidem, 174.

33 Gassendi, *Animadversiones* iii–x.

appears to be clear: that is, to show the relevance of Epicurus' doctrines with respect to contemporary advances in experimental science. Once again, the commentary, as well as the translation, aim to their immediate application to the contemporary world.

'Contra Epicurum'

While the whole edifice of Epicurus physics is expounded without emendations, the parts dealing with more sensitive topic, such as the nature of the soul, the creation of the world, the nature of divine providence, are labelled as 'contra Epicurum' in the *Animadversiones*. They are object of a careful emendation in which Gassendi does not only take distance from their content, but also shows how they could be changed leaving the useful part of Epicurus's natural philosophy substantially/for the most part? intact. In particular, a set of three chapters is set to dispel the most controversial aspects of Epicurus's cosmology. The first, *Esse deum autorem Mundi, contra Epicurum* [God is the author of the world, against Epicurus], contrasts Epicurus's notion of the creation of the world from fortuitous collision of atoms. The basis of Gassendi's argument is that Epicurus is certainly right in arguing that every thing can be made out of atoms, as a flea is built from the same matter as an elephant.[34] However, the symmetry and regularity shown in the world presupposes the work of an author. Since the world needs an author, the second objection against Epicurus' cosmology, "Esse deum rectorem Mundi, contra Epicurum", [God is the governor of the world, against Epicurus] is answered in the following manner: if something is created by an author, it must possess a certain structure and regularity as in fact we experience in the world; and if we think that the creator is perfect, it is impossible to admit, as Epicurus does, that the world is not hold by his providence, and that everything is contingent.[35] Finally, in the last of the three chapters, *Gererem Deum hominum curam, contra Epicurum* [God takes care if men, against Epicurus], Gassendi defends the notion of special providence ('providentia specialis'), that is, the care that God has for the greatest of his creature, man.

34 Gassendi, *Animadversiones* 712: 'Neque vero moles operis eum terruit; quod visum illi fuerit, oportet solum atomorum numerum augere, quae suppeditari ex illa infinitate poterant; ac non debere molem aductam videri improbabilem, nisi quis putet pulices tantum et non elephantes ex eadem communi materia prodire'.

35 Ibidem 727.

These three chapters conclude the *Animadversiones* as if they symbolized that the rest of Epicurus's natural philosophy could maintain its validity. Through these strategies, Gassendi had provided a 'domestication' of Epicurus's philosophy showing at the same time its heuristic power vis-à-vis the new sciences, and its emendability vis-à-vis Christianity. While the translation of *Lives*, book X provided the possibility of a direct interrogation of the sources, the commentary based thereon complemented the work of domestication of Epicurus's philosophy and made possible its integration in the scientific and philosophical discourse.

Epicurus Paraphrased into Vernacular: Charleton's Physiologia

In a dedicatory letter dating from 1654 and addressed to the wife of his patron Robert Villiers, Elisabeth, Walter Charleton, 'Dr. In Medicine, and Physician to the late Charles, Monarch of Great-Britain', wrote:[36]

> [...] having the Honor of so great a Trust, as that of your most precious Lives committed unto me; it highly concern'd me, to study and pursue all ways of Demonstrating myself not altogether incapable thereof, and more *especially this of Natural Philosophy, which being the Grounds, is also the Measure of a Good Physician.*[37]

The natural philosophy 'ground and measure' of his activity as physician that Charleton is propounding in the book dedicated to Elisabeth Villiers is in fact Gassendi's Epicureanism, that he likely knew through the reading of the *Animadversiones*. The full title of the work is long and at the same time revealing: *Physiologia Epicuro-Gassendo-Charltoniana or a Fabrick of Science Natural, Upon the Hypothesis of Atoms, Founded by Epicurus, Repaired by Petrus Gassendus, Augmented by Walter Charleton*. The mentioning of Gassendi in the title of the book is the only acknowledgement of his philological and philosophical efforts: Gassendi is not mentioned anywhere else in the work. While the text of *Physiologia* is not a word-by-word translation of Gassendi's *Animadversiones*, the reader derives from its reading the lasting impression that it largely consists of a paraphrase (or sometimes a selective translation) of passages of Gassendi's *Animadversiones*. Its exclusive focus is on its natural-philosophical sections, relieved from the philological rigour and the sometimes haunting erudition that characterizes the *Animadversiones*. While

36 On Charleton's science and natural philosophy, see Booth E., *A Subtle and Mysterious Machine: The Medical World of Walter Charleton (1619–1707)* (Dordrecht: 2006).

37 Charleton, *Physiologia* 4. Italics mine.

the prose of Gassendi's *Animadversiones* is characterized by the constant quotation of authoritative sources, by eloquent prose, and by the constant display of erudition and frills, Charleton's prose, in addition to be in the vernacular, is stripped from any unnecessary quotation (including Gassendi's). To the extent that, while it appears largely inspired by the *Animadversiones*, it looks at the same time like a completely different work. Indeed, as also the dedicatory letter shows, Charleton targets a whole different readership than Gassendi. Gassendi spoke to a world of cosmopolitan intellectuals who knew Latin well and could appreciate his philological rigour and erudition. His project of proposing the philosophical doctrines of Epicurus as a theoretical basis for the emerging sciences is never disjoined from his humanistic sensibility for the accuracy of the sources and the rigour of the style. On the contrary, by translating, paraphrazing, summarizing, and complementing the long and complex text of the *Animadversiones*, Charleton intended to introduce Gassendi's Epicureanism to a vast readership of English intellectuals. Among his intended readers, there were certainly those who shared his interests for chemistry and his medical training, and could be fascinated by Gassendi's proto-experimental epistemology and atomistic matter theory.[38] But arguably Charleton's intended readership was even broader. Indeed, Gassendi's Epicureanism also offered a coherent system of thought that could be used to gain credit as a scientist and a savant. In France, a bit later in the century, many Cartesian scholars—who arguably struggled to find their place in still a conservative educational system—were looking at the bourgeois salons to popularize their science, in turn establishing their social position as scholars.[39] The dedicatory letter to Elisabeth Villiers offers a similar cross-section on Charleton's relation with his patrons. The work, Charleton writes, was written in their house and answered the need (or at least this is what Charleton wanted them to think) of satisfying their curiosity:[40]

> [...] The book comes not into your hands, to Informe, but only Remember you of many of those Discourses of Nature, which your Noble Husband and your self have often suffered me to entertain (would to God, I might have said, satisfy) your eager Curiosity withal, at those hours your indus-

38 On the culture of experimenting in seventeenth-century England, see for instance Gaukroger S., *The Emergence of a Scientific Culture: Science and the Shaping of Modernity 1210–1685* (Oxford: 2006).

39 See Roux S., "An Empire Divided: French Natural Philosophy (1670–1690)", in Garber D. – Roux S. (eds.), *The Mechanization of Natural Philosophy* (Dordrecht: 2013).

40 Charleton, *Physiologia* 4.

> trious Minds required Relaxation from the bent of more grave and advantageous Thoughts.[41]

Not only curing, but also to establishing his position as a learned man was at the centre of Charleton's concern, and especially, as we have seen, in respect to 'this of Natural Philosophy, which being the Grounds, is also the Measure of a Good Physician'.

Besides affecting the style of exposition, Charleton's target affects, as I shall show, also the content of his work. Indeed, *Physiologia* integrates the content of the *Animadversiones* with the addition of some experimental reports aimed to corroborate the 'physical hypotheses' concerning the existence of atoms and vacuum. In doing this, it encounters some of the research interests that were mostly preponderant in seventeenth-century England, therefore showing the usefulness of Epicurus's atomism to such readership. Furthermore, Charleton provides some consistent comments that aimed to further emend some seemingly anti-Christian Epicurean doctrines that Gassendi had not considered potentially controversial; in this way, the *Physiologia* answers to specific anti-Epicurean concerns of early seventeenth-century England. Robert Kargon, emphasizing the importance of *Physiologia* in the development of seventeenth-century English corpuscularism and empiricism, has claimed that during the 1640s English philosophers (such as the Cambridge Platonist John Smith) still saw Epicureanism as the doorway to atheism, and that the publication of the *Animadversiones*, while emending Epicurus's claims on the materiality of the soul and on the possibility of a universe without artificer, did not cancel this impression completely.[42] Still depicting Epicureanism as an impious doctrine insofar as it described the universe as emerging from chaos without the intervention of God, Henry More's publication of the *Antidote Against Atheism* (1653) reflects the lasting pervasiveness of anti-Epicurean stands on the Island. In this sense, Charleton's selected translation of the *Animadversiones* required, and in fact provided, also 'a domestication' of the text to the needs and sensibility of the English intellectual context. This operation, according to Kargon, was threefold: 'he tried to demonstrate that modern Epicurean atomism was purged of the heresies which admittedly contaminated the pagan formulations, specifically that the soul is material and mortal, and that motion is inherent in matter'. Secondly, 'he attempted to dissociate atomic doctrine of Gassendi from classical atomism by joining the assault'. Thirdly, 'he showed

41 Ibidem 7.

42 Kargon R.H., "Walter Charleton, Robert Boyle, and the Acceptance of Epicurean Atomism in England", *Isis* 55, 2 (1964) 184–192.

how powerful this doctrine could be in promoting by demonstrating that it is a very effective proof of God'.[43] Anticipating this, Charleton then complemented *Physiologia* with works such as *The Darknes of Atheism Refuted by the Light of Nature* (1652) and *Epicurus's Morals* (1656).

In this respect, Chaleton's addition to the text are particularly telling, since they mostly consist in the integration of experimental support, and the emendation of Epicurean doctrines that Gassendi did not consider as potentially controversial.

We can find an example of the former in chapter IV of *Physiologia*, 'A Vacuum in Nature', that contains a long depository of experiments in support of the existence of vacuum in nature—both what Charlton names 'natural' (that is, the one that he believes to be normally found between different portion of corpuscles)—and a 'praeternatural' vacuum—that is, the one artificially created by man. While the discussion of praeternatural vacuum consists of a defence of the vacuist interpretation of Torricelli's experiment, of a discussion of Heron's ball, and on the property of solubility of liquids (examples that he could well have found in the *Animadversiones*), some of the observations in support of natural vacuum largely draw upon Charleton's own experience as a chemist and naturalist. Such experiments reveal Charleton's engagement with themes and research lines that characterized seventeenth-century English science at large, and his tendency to read Gassendi's version of the philosophy of Epicurus in such frame. Also, it shows his superficial understanding (or, more charitably, partial adhesion) to the explanatory rationale of mechanical philosophy. In particular, on of such experiments concerns the hypothesis around the presence of a thick layer of ice around the mouth of a very hot furnace during cold winter in the Worchestershire. Charleton speculates that the phenomenon was due to the following causes:

> That the ambient Aer, surcharged with too great a cloud of exhalations from the fire, was forced to a violent recession or retreat, and a freshly supply of aer as violently came on to give place to the receding, and maintain the reception of fresh exhalations; and so a third, fourth and continued relief succeeded: and that by this continuous and impetuous afflux, or stream of new aer, loaden with cold Atoms, the activity of the cold could not but be by so much the more intense at the mouth of the furnace, then abroad in the open aer, by how much the more violent the stream of cold aer was there then elsewhere.[44]

43 Ibidem 185–186.

44 Charleton, *Physiologia* 28.

To prove this hypothesis, Charleton fills two dishes with water and puts one of them close to the mouth of the furnace, and the other "sub Dio" [under the open sky]:

> I found that near the furnace so nimby creamed with Yce, as if that visibly-freezing Tramontane Wind, which the Italian calls Chirocco (sic), had blown there, and much sooner perfectly frozen than the other.[45]

From this, he deduces that the aim of air is that of being:

> [...] common RECEPTARY of Exhalations, and that for the satisfaction of this End, it doth of necessity contain a *Vacuum Disseminatum* in those minute insensible *Incontiguitis* or Intervals betwixt its atomically Particles; since Nature never knew such pros improvidence, as to ordain an *End*, without the codestination of the *Means* requisite to that End.[46]

This experiment reflects Charleton's engagement in themes and research, those on the properties of gas and of respiration, that were carried out in England in the seventeenth century. In a way, this experiment proves Charleton's effort to make Epicurus-Gassendian natural philosophy appealing to a English readership by showing its immediate usefulness in respect to some of the research that many English intellectuals were carrying out in these years. It also show Charleton's seemingly amateurish engagement with Epicureanism: he still relies on a teleological conception of nature that clearly does not reflect his commitment to the doctrines of Epicurus, neither in their classical form nor in that emended by Gassendi.

The second example of Charleton's integration addresses precisely Henry More's objections against Epicurus of the *Antidote Against Atheism* (1653, criticizing the idea of the universe as emerging from chaos without the intervention of God), therefore showing how Charleton translation also strives to situate Gassendi's Epicureanism within the English context. Indeed, after an excursus on the division of philosophical sects, Charleton addresses the issue of the plurality or infinity of worlds.[47] While the choice of the first topic to treat might seem bizarre, it allows Charleton to clarify, from the very beginning of the book, his position on the Epicurean idea that the universe was the outcome of fortuitous collisions between atoms—a doctrine that clearly excluded

45 Ibidem 29.

46 Ibidem.

47 Ibidem 11–15.

divine causation and providence and was therefore at the core of More's criticism. Indeed, Charleton criticizes Epicurus's and Democritus's thesis of the existence of an infinite number of worlds on the basis that 'albeit we readily concede that there is an *Infinite Inanity* or Ultramundan Space, yet there can it not follow of necessity, that there are *Infinite Atoms* contained in that Ultramundan Space [...] insomuch as it sounds much more concordant to reason, that there are no more Atoms, then those of which this single World was compacted'.[48] From this, Charleton switches promptly to the real object of his concern: the anti-providential creation of the world. He writes,

> They argue; *Since the vacuity or ultramundane Space is infinite in Magnitude or Capacity, necessary it is that the Abyss of Atoms included therein be also Infinite in Extent: because otherwise they could never have convened, and coalesced in that Form, which the World now Holds*: we admit their *Induction* for natural and legitimate, but detest their *supposition* as absurd and impossible. For, they take it for granted, that the Chaos of Atoms was not only eternal and *Increate*, but also that it disposed, and compacted it self into that Form, which constitutes the World, by the spontaneous motion inhaerent in Atoms, and their fortuitous coalescence in such and such respective Figures: when to a sober judgment it appears the highest *Impossibility* imaginable, that either the Chaos of Atoms could be eternal, self-principle, or increate, or dispose and fix in it self into so vast, so splendid, so symmetrical, so universally harmonical, or Analogical a structure, as thus if the World.[49]

The appeal to a principle of harmony in nature, and the idea that such harmony cannot be the outcome of a chaotic universe, is therefore used to polish Epicurus's natural philosophy, arguably, from More's fair accusation of atheism. Once done, Charleton returns to the theme of the plurality of worlds to show how the atomist model could work within a providential frame:

> And therefore, since it is most probable that Atoms were the *Material Prima*, or material Principle of the World; as we shall clearly enunciate in a singular Chapter subsequent: we may adventure to affirm, that God created exactly such a proportion of Atoms, as might be sufficient to the making up of so vast a Bulk, as this of the World, and that there remained no on superfluous. 'Tis unworthy a Philosopher to acknowledge any su-

48 Ibidem 11.

49 Ibidem 11–12.

> perfluity in Nature; and consequently a dangerous soloecism to say the God of Nature knowing not how to proportion the quantity of his materials to the model or platform of his structure, crated more Atoms, then were necessary, and left an infinite Residue to be perpetually hurried too and fro in the untramundane space.[50]

While the theme of the creation of the world was already present in the *Animadversiones*, here Charleton not only expands it considerably, but also put it at the very beginning of the work, while, in the *Animadversiones* it was at the very end. In other words, from the first page onward, Charleton wants his reader to know his considerations on the erroneousness of Epicurus' doctrine of the origin of the world, while the two other chapters of the tryptich of objections in the *Animadversiones* (*Esse deum autorem Mundi, contra Epicurum,* and *Gererem Deum hominum curam, contra Epicurum*) are not mentioned. This shows how, in translating, paraphrasing, and in turn commenting the *Animadversiones* into English, Charleton is also providing a local context? to the work in order to respond to the most pressing theological criticism against Epicurus from his contemporaries in England.

Conclusions

The cases of Gassendi and Charleton show two similar example of how early modern translator attempted to domesticate potentially controversial, though in their mind useful, texts in natural philosophy. In a period when not only potentially anti-Christian claims, but also history of philosophy at large, were looked at suspiciously, the translator had to handle carefully the text (s)he was translating. Translation, whether a legitimate philological work as in the case of Gassendi, or a rough paraphrase as in the case of Charleton, opens up the possibility of debating a potentially controversial subject, such as Epicurus's philosophy, and per se resolves some issues connected to the misrepresented conception of Epicurus's notion of pleasure for Gassendi, and some of Epicurus's most controversial arguments for Charleton.

The integration of translation and commentary, present in both authors, serves to fully accomplish the 'domestication' of the philosophical text. The elements that more openly contradict Christianity must be emended in order to divert the suspicions on the author, and, at the same time, make the most relevant elements emerge. As the case of Gassendi shows, in a period in which

50 Ibidem 13.

the figure of the 'historian of philosophy' was blamed for excessive pedantry and sterile erudition, the author has to show the relevance of the text for his contemporaries.

At the same time, the comparison of these two works also shows the importance of the intended readership in affecting the style of the translations and contents of the commentaries.

Selective Bibliography

Booth, E., *A Subtle and Mysterious Machine: The Medical World of Walter Charleton (1619–1707)* (Dordrecht: 2006).

Burke P. – Po-chia Hsia R. (eds.), *Cultural Translation in Early Modern Europe* (Cambridge – New York: 2007).

Butterfield D., *The Early Textual History of Lucretius' De Rerum Natura* (Cambridge: 2013).

Charleton Walter, *Physiologia Epicuro-Gassendo-Charltonia or a Fabrick of Science Natural Upon the Hypothesis of Atoms Founded: Epicuruius, Repaired: Petrus Gassendus, Augmented: Walter Charlton, Dr. In Medicine, and Physician to the Late Charles, Monarch of Great Britain* (London, Thomas Newcomb: 1654).

Enenkel K.A.E. (ed.), *Transformations of the Classics via Early Modern Commentaries* (Leiden – Boston: 2013).

Enenkel, K. – Nellen H. (eds.), *Neo-Latin Commentaries and the Management of Knowledge in the Late Middle Ages and the Early Modern Period (1400–1700)* (Leuven: 2013).

Galluzzi P., "Gassendi and l'Affaire Galilée of the Laws of Motion", *Science in Context* 14, 1 (2001).

Gassendi Pierre, *Animadversiones in decimum librum Diogenis Laertii, qui est de vita, moribus, placitisque EPICURI* (Lyons, Guillaume Barbier: 1649).

Gassendi Pierre, *Opera omnia*, 6 vols. (Lyons, Anisson – Devenet: 1658; reprint, Stuttgart: 1964).

Gaukroger S., *The Emergence of a Scientific Culture: Science and the Shaping of Modernity 1210–1685* (Oxford: 2006).

Gemelli B., *Aspetti Dell'atomismo Classico Nella Filosofia Di Francis Bacon E Nel Seicento* (Firenze: 1996).

Grant E., *Much Ado about Nothing: Theories of Space and Vacuum from the Middle Ages to the Scientific Revolution* (Cambridge: 1981).

Jones H., *The Epicurean Tradition* (New York: 1989).

Joy L., "Epicureanism in Renaissance Moral and Natural Philosophy", *Journal of the History of Ideas* 53, 4 (1992) 573–583.

Joy L., *Gassendi the Atomist: Advocate of History in an Age of Science* (Cambridge: 2002).
Koyré A., *Galileo Studies*, trans. J. Mepham (Atlantic Highlands, NJ: 1978).
Leijenhorst C.H. et al. (eds.), *The Dynamics of Aristotelian Natural Philosophy from Antiquity to the Seventeenth Century* (Leiden: 2002).
Lipsius, Justus, *Physiologiae Stoicorum Libri Tres* (Antwerp, Joannes Moretus: 1610).
Lupoli, A., *Nei Limiti Della Materia. Hobbes E Boyle: Materialismo Epistemologico, Filosofia Corpuscolare E 'dio Corporeo'* (Milano: 2006).
Osler M.J., *Divine Will and the Mechanical Philosophy: Gassendi and Descartes on Contingency and Necessity in the Created World* (Cambridge: 1994).
Redondi P., *Galileo eretico* (Turin: 2004).
Roux, S., "An Empire Divided: French Natural Philosophy (1670–1690)", in Garber D . – Roux S. (eds.), *The Mechanization of Natural Philosophy* (Dordrecht: 2013).
Trevisani F., "La Teoria Corpuscolare in Cartesio", in Baldini U. – Zanier G. – Farina P. – Trevisani F. (eds.), *Ricerche Sull'atomismo Del Seicento* (Florence: 1977).
Wilson C., *Epicureanism at the Origins of Modernity* (Oxford – New York: 2008).

CHAPTER 10

Ibrahim Müteferrika's Copernican Rhetoric

B. Harun Küçük

In 1732, Ibrahim Müteferrika (d. 1745), the ex-Socinian printer to the Ottoman Sultan, voiced a pointed critique of the Empire's conservative scholars:

> Ve lakin gaflet olunmaya ki emr-i mezbur umur-ı itikadiyeden ve levazım-ı hulkiyeden olmadığı miraren ve kiraren işaret ve bu mahalde irad olunmak lazıme-yi halden olmayub bil ke müstedrek ve müstehcen add olunur iken mücerred fırka-ı merkumenin reyi icmalen tenkih ve takrir ve kayd-ı tahrir olunub akval ve aralarını tefhim ve zahirde iraet ve müşahede mülahazalarıyla ecram ve ecsamın keyfiyetlerine dair tarh ve tastih eyledikleri süver ve eşkal dahi ilhak olundı. Ta kim meydan-ı bi-payan fazl-ı irfan şecileri ve sahra-yı fesihe el-feza-i ulum-ü-maarif aşinaları el-hasıl erbab-ı kemal-ü-dikkat ve ashab-ı hamiyet-ü-gayret olan fuzela-i asr bad el-ıtla bu reye itiraz ve hekimane vaz-ü-hareket-berle sevk-i berahin-ü-izahe ve ihzar-ı edille-i kaviye-berle bina-yı ara-yı na-fercamlarını hedme himmet buyurub kar-ı kitabda haric suturda vus'at üzre bu matlubun husulüne vesile olmak ümidiyle terk olunan beyaz yerlere mahal-be-mahal haşiyeler yazub mezheb-i evvele şöhret ve takviyet ve bu kitab-ı müstetaba zib-ü-ziynet vireler. Ve eğerçe bu mezheb-i atıle hükema-i İslam caniblerinden dahi itiraz bervakitde hali olmayub kavl-i na-şayestelerine vechle men-ü-redd senedleri ne şekil ve suret ile ibtal olundığı kütüb-i heyetde mestur ve musarrihdir. Ve lakin herçend ke atıl ve batıl ise de mezheb-i kadim ve rey-i atik-i hükema olmağla her asır be her zamanda mail-ü-ragıbı ve haris-ü-hamisi bulunub bu reyi teyyid sevdası ve takviye hevası gayretiyle kütüb-ü-resail-i vafire dahi telif ve tedvin olunmuşdur ... Kopernikuş heyeti denilmekle şöhret bulunmuşdur ve bilad-ı Avrupa'da olan müteahhirinin ekseri kabul itmişdir. Ba-husus ke müteahhirin-i mezkurenin eşheri ve eşbehi olan Kartejyuş nam hekim, felsefe-yi cedidesinde ihtira eylediği bazı kavaid-i cedide-ü-ceyyidesiyle bu reye takvit virdüğünden ötürü kendüsüne nisbet olunub halk arasında Kartejyuş heyeti ünvanıyla dahi şayi olmuşdur.[1]

1 Katip Çelebi, *Cihannüma*, ed. and introduced by İbrahim Müteferrika (Istanbul, Sultanic Press: 1732) 34. The most comprehensive study of Müteferrika's preface is Kalaycıoğulları İ.,

DOI 10.1163/9789004349261_012

> Principles of theoretical astronomy are neither doctrinal nor do they pertain to the creation. I have simply described the various theories [Aristotelian, Copernican and Tychonic] in order to express these opinions clearly, with demonstrations, and added the shapes and diagrams of the celestial and terrestrial objects upon which each opinion depends. If the *virtuosi* who know the vast expanses of learning, that is, those who are masters of diligence and perfection, those honorable and hardworking men, have disagreements, they may point out the weaknesses of these opinions by presenting clear and strong evidence and explanations. And, let them do so on the margins of this book, which has been left wide open for their use. Let them write commentaries and glosses. Let them gain fame and power as they add beauty and ornament to this volume. Muslim scholars of the past have also opposed the idle opinions of [Aristotle and Ptolemy]. What they accepted and rejected, and the reasons as to why they did so are clearly written in the books on cosmography. Although the ancient opinion is both idle and superstitious, it has always found supporters, commanded esteem and enjoyed protection as well as preservation. Some have even written new books to support and strengthen the ancient view [...] The most famous among the philosophers who follow the Copernican opinion is Descartes, who has proposed a new philosophy that presents new rules to support this opinion. This is called the cosmography of Descartes.

There is no doubt that Müteferrika took aim at contemporary Muslim scholars and associated Aristotle as well as Ptolemy with tradition, idleness and superstition. He also deployed many of the rhetorical strategies we find in other Copernican writings in Europe, especially those of Galileo: the truth of the few against the opinion of the many, inviting critique and controversy and, even speaking truth to power. What does it mean that this familiar European rhetoric emerged in the Ottoman Empire?

In this chapter, I wish to draw a large frame to show the complex mixture of biographical, political, religious and finally, intellectual factors that can help us explain the genealogy and the ramifications of this unusual passage. What made Müteferrika's unique were his use of two deliberately non-radical sources—Edmond Pourchot's *Institutiones philosophicae* and Al-Ghazali's *Tahafut al-falasifa*—to articulate the radical notion that science was completely separate from religion. My goal is to explain how bluntly orthodox texts

---

Katip Çelebi'nin Cihannüma Adlı Eserine İbrahim Müteferrika'nın Yaptığı Ekler Doğrultusunda Çağdaş Bilimlerin Türkiye'ye Girişi, Unpublished MA Thesis (Ankara University: 2003).

from early modern Catholic and medieval Islamic contexts were used to level an attack against the official custodians of Islamic orthodoxy in the Ottoman Empire.[2] The translation and transfer of texts, ideas and even heresies will feature as key elements of this discussion.

Ibrahim Müteferrika: A Brief Biography to 1730

Ibrahim Müteferrika is tiringly familiar, yet also utterly mysterious to Ottomanists. For many, he is an unmoved mover and represents a moment of rupture in Ottoman history. He is an agent of radical progress, for he is the man who established the very first Muslim printing press. To others, he is an outsider who sits uneasily in the general ebb and flow of Ottoman history.[3] Part of the problem with Müteferrika is ideological. To write about Ibrahim Müteferrika is to write about Turkish modernity, because it is hard to decide on what to do with him. Does he belong to 'our' history or is he just a Westerner doing Western things in an Eastern setting? Was he an Ottoman or was he a foreigner who held the Ottomans in contempt? Especially today, such questions easily turn into other questions. Is Turkey part of the West? How deep are the foundations of the modernity that culminated in the modern Turkish Republic? Or was the republic an anomaly in the normal course of things, a small wave in the vast ocean that is the history of 'Islamic civilization'? A properly contextualized analysis of his writings will help disentangle a lot of this complex emotional engagement with the man, and show just how responsive he was to intellectual and political developments he observed and participated in first hand.

Yet, neither of these notions about Müteferrika is based on his writings and translations and even less on how his writings fit into a landscape of other contemporary writings in Istanbul. In this paper, I will be expanding on some of the ideas that İnan Kalaycıoğulları, Marinos Sariyannis and Vefa Erginbaş have brought to scholarly attention.[4] And, I hope to show below, that Georg Simmel's 'stranger' is the best term for understanding Müteferrika's place in

2 For a discussion of the de-radicalisation of a classical text to an early modern Western European context, see the chapter by Rodolfo Garau in this volume.

3 On Müteferrika in historiography, see Sabev O., *Ibrahim Müteferrika ya da Ilk Osmanlı Matbaa Serüveni* (*1726–1746*) (Istanbul: 2006).

4 Kalaycıoğulları İ, *İbrahim Müteferrika'nın Yaptığı Ekler*; Sariyannis M. – Tuşalp Atiyas E., *Ottoman Political Thought up to the Tanzimat: A Concise History* (Rethymno: Foundation for Research and Technology—Hellas, 2015); Erginbaş V., "Enlightenment in Ottoman Context:

Ottoman history.[5] He was not 'of the group', but he was integrated into different cultures of naturalism in Istanbul. He came from the outside, but he came to stay. He was neither a born-and-raised Istanbulite Westernizer, nor was he an outsider. Setting aside larger problems arising from the assumption that the Western civilization or the Ottoman Empire were monolithic entities in the early eighteenth century or that 'Westernization' makes sense as a special category of historical development, I will simply focus on providing a broad context that may help explain this specific passage and show how these categories were emergent, fractured and malleable in the early eighteenth century.

Most of what we know about Müteferrika prior to his appearance as Ibrahim, Sultan's envoy to the exiled Hungarian prince Rákóczy Ferenc II (1676–1735), is a matter of speculation. Historians agree that he was born to Christian parents in the Transylvanian town of Cluj. His denomination remains unknown. Sources are conflicted and scholars have disputed whether he was a Calvinist or a Unitarian. Orlin Sabev's recent work, however, leaves little doubt that he must have been a Socinian.[6] Socinianism is well known for its radicalism and anti-trinitarianism among historians of the early Enlightenment. Jonathan Israel's *Enlightenment Contested* devoted an entire chapter to Socinian controversies in Western Europe at the end of the seventeenth century. Originally a movement that emerged on the Eastern edge of the continent, particularly Poland and Transylvania, it spread rapidly across Europe as part of a family of philosophical deisms, especially Spinozism.[7] Martin Mulsow has argued that 'around 1700 there were numerous members of the intellectual avantgarde [including John Locke, Jean LeClerc and Philipp van Limborch] who promoted various mixes of Socinian, Cartesian, Spinozistic and Lockean views'.[8]

Some Socinians also harboured sympathy towards Islam because of their shared disbelief in the Trinity. The combination of Islam's claim to being an uncorrupted monotheism, and the rise of Arabic scholarship in Europe, particularly studies of anti-Christian polemics in early Islam, contributed to what

---

İbrahim Müteferrika and His Intellectual Landscape", in Roper G. (ed.), *Historical Aspects of Publishing in Languages of the Middle East* (Leiden: 2013) 53–100.

5 Simmel G., "The Stranger", ed. and trans. K.H. Wolff, *The Sociology of Georg Simmel* (New York: 1950) 402–408.

6 Sabev O., "Portrait and Self-Portrait: Ibrahim Müteferrika's Mind Games", *Journal of Ottoman Studies* 44 (2014) 99–121.

7 Israel J.I., *Enlightenment Contested: Philosophy, Modernity, and the Emancipation of Man, 1670–1752* (Oxford: 2006) 115–134.

8 Mulsow M., "Socinianism, Islam and the Radical Uses of Arabic Scholarship", *Al-Qantara* 31, 2 (2010) 553.

Mulsow has called a 'monotheistic transfer of heresy'.[9] Scholars generally associate the anonymous Turkish treatise from around 1710 that presented a biblical argument based on Immanuel Tremellius (1510–1580) and Franciscus Junius's (1545–1602) reformed rendition of the Gospel of Matthew, in favour of the veracity of Islam with Ibrahim Müteferrika's conversion to Islam.[10] While we do not know if Ibrahim had converted out of radicalism and spite for his Habsburg overlords or because he fell prisoner of war to the Ottomans and faced the galley without a prospect of being ransomed likewise remains a mystery. This much we know: Ibrahim spent the vast majority of his life working, quite ardently and persistently, to make Islam victorious over Catholicism.

Ibrahim's network was broad and he enjoyed great fame among European *hommes éclairés*.[11] We also know that he was friends with Nikolaos Mavrocordatos (1670–1730), the erudite Istanbulite prince of Moldavia, which then was an Ottoman tributary state. His friendship with Mavrocordatos and the latter's extensive correspondence with Jean Le Clerc (1657–1736), another early Enlightenment radical, lends credence to his Socinian background.[12] Furthermore, Ibrahim spent a great deal of time with a group of projectors and disillusioned Huguenots who lived in the Ottoman town of Tekirdağ (Rodosto) alongside Rákóczy Ferenc II's exiled court.[13] As he went back and forth between Istanbul and Tekirdağ between the early 1710s and 1735, he was likely exposed to Rákóczy's ideas about religious tolerance and peoples' right to self-determination, to the anonymous Huguenot officer who proposed to set up a new artillery corps for the Sultan and to the famous Claude Alexander, Comte de Bonneval (1675–1747), the French commander who later converted to Islam and became Humbaracı Ahmet Paşa.[14] In other words, Ibrahim was, without

9 Mulsow, "Socinianism," p.554.

10 Müteferrika Ibrahim, *Risale-i Islamiyye*, ed. M.E. Coşan (Istanbul: 2010). Coşan does not identify the Latin quotations or the edition of the Bible.

11 See, for example, Heumann Christoph August, *Poecile; Sive epistolae miscellaneæ ad literatissimos aevi nostri viros. Accedit appendix exhibens dissertationes argumenti rarioris* (Halle, Johann Gottfried Renger: 1728), vol. 3, 295–307. Kundmann Johann Christian, *Rariora naturae et artis item in re medic*a (Breslau, Michael Hubert: 1737) 703–717.

12 Bouchard J., "Les relations épistolaires de Nicolas Mavrocordatos avec Jean Le Clerc et William Wake", *O Eranistes* 11 (1974) 62–92.

13 Berkes N., *The Development of Secularism in Turkey* (Montreal: 1964) 36.

14 On Rákóczy's republicanism, see Dreisziger N., *Church and Society in Hungary and in the Hungarian Diaspora*, (Toronto: 2016) 71–3. Unat F.R., "Ahmed III. Devrine Ait Bir Islahat Takriri: Muhayyl Bir Mülakatın Zabıtları", *Tarih Vesikaları* 1 (1941) 107–121. For a good biography of Bonneval, see Stockinger H.E., "Die Apostasiee des Pascha-Grafen Alexander von Bonneval (1675–1747) und europäische Stimmen zum 'Fall' Bonneval", in Klein D. –

doubt, militantly anti-Habsburg. He might have also been a republican, a deist and a skeptic.

He entered the limelight of Ottoman history with his proposal to establish a Sultanic press in 1726. Considered the first Turkish printing press in history, Ibrahim Müteferrika's operation was broadly modeled after the French Imprimerie Royale.[15] The founding document, arguably penned by Müteferrika himself, presents the case for the establishment of a secular press under Sultanic patronage. Printed books and maps, according to Müteferrika, would not only educate and civilize readers from all around the Empire, but also would cater to European book collectors whose appetite had almost completely drained Istanbul's book market.[16]

Other important actors involved in founding the press were Yirmisekiz Mehmed Çelebi (d.1731) and his son, Mehmed Said Efendi (d.1761). Yirmisekiz Mehmed, who had spent his career as the Master of the Ottoman Mint and the Master of the Ottoman Foundry, was imperial treasurer when the press went into operation. In 1720–21, Sultan Ahmed III sent Yirmisekiz to France as ambassador. The main factor in the Sultan's choice seems to have been Yirmisekiz's profound familiarity with natural philosophy and the mechanical arts. Both the embassy report and Yirmisekiz's earlier work on the material utility of natural philosophy seems to corroborate this line of reasoning.[17] Mehmed Said—allegedly one of the first Ottoman Freemasons—was likewise a naturalist who wrote treatises on *materia medica* and on the pulse and later served as Grand Vizier between 1755 and 1756.[18]

Platos B. (eds.), *Wahrnehmung des Islam zwischen Reformation und Aufklärung* (Munich: 2008) 116–122.

15 Yirmisekiz Mehmed Çelebi – Veinstein G., *Paradis des infideles: un ambassadeur ottoman en France sous la regence* (Paris: 1981) 226. Irepoğlu G., "Topkapı Sarayı Müzesi Hazine Kütüphanesindeki Batılı Kaynaklar Üzerine Düşünceler", *Topkapı Sarayı Müzesi Yıllığı* 1 (1986) 56–72, 174–197.

16 See Ahmed III's edict attached to El-Vani Mehmed b. Mustafa – Müteferrika İbrahim – Yirmisekizzade Mehmed Said, *Tercemetü's-Sıhahü'l-Cevheri* (İstanbul: Sultanic Press, 1729); also see Müteferrika İbrahim, *Vesiletü't-Tıbaa* [The Purpose of Printing], attached to the same work. For a discussion see, Sabev O., *Ibrahim Müteferrika*, 78–100.

17 Küçük B.H., "Science Studies and Early Modern Ottoman Science", *International Journal of Middle East Studies* 47, 3 (2015) 584–587. On Yirmisekiz Mehmed's embassy, see Göçek F.M., *East Encounters West: France and the Ottoman Empire in the Eighteenth Century* (Oxford: 1987).

18 Osmanzade Ta'ib – Ahmed Cavid – Abdülfettah Şefkat Bağdadî – Ağazade Ömer Vahid, *Hadikatü'l-Vüzera* (İstanbul: 1855) vol. 3, 84–85.

The Sultanic press printed its first map and first book in 1729 after a long period of negotiation and preparation. During the first two years of the press, the publications served, like so many cultural and administrative efforts of the late 1720s, the Sunni Ottomans in their conflict against the Shiite Safavids. The very first item off the press was a map of Iran, which the Sultan claimed to have conquered during the war that began in 1722. Another seminal publication was a Sunni history of Baghdad, one of the main theatres of military confrontation between the Safavids and the Ottomans.[19] Other publications included a history of Ottoman Egypt, a Turkish-Arabic dictionary, Jean-Baptiste Holdermann's *Grammaire turc*, and other works that sought to consolidate Ottoman power and hegemony in the provinces.

The Rebellion of 1730 and Jesuit Cartesianism in Istanbul

A major rebellion broke out in Istanbul in late 1730.[20] The leader of the rebellion was Patrona Halil, a private person with neither discernible rank nor known connections to established political households. Contemporary authors presented both long-term and short-term reasons for this explosive reaction: Ahmed III's ineffectiveness against the Safavids, the general moral corruption of Istanbul, excessive taxation, and the lack of provisions in the capital.[21] The insurgents made camp just outside of the palace and, in a relatively short period of time, managed to get Ahmed III off the throne, kill and mutilate Grand Vizier Damad Ibrahim Paşa and exile or execute the majority of the Ottoman administration. While previous rebellions in the capital—frequent and destructive during the seventeenth century—routinely reflected the power play between major Ottoman political households, the circumstances were somewhat different in 1730. Part of Ahmed's efforts to centralize imperial power were institutional and included mint and coinage reforms, the consolidation of the Orthodox Greek Church under the Patriarch of Constantinople, and the establishment of the press. Equally important was the establishing of a physical court and uniting the dynastic family with the empire's elites through intermarriage.[22] The private aspects of administrative centralization meant

19 Avcı N., "Gülşen-i Hulefa", *Erciyes Üniversitesi İlahiyat Fakültesi Dergisi* 10, 8 (1992) 245–257.

20 Karahasanoğlu S., *Politics and Governance in the Ottoman Empire: the Rebellion of 1730* (Cambridge, MA: 2009).

21 Aktepe M., *Patrona İsyanı* (Istanbul: 1958) 1–130.

22 Artan T., "18. Yüzyıl Başlarında Yönetici Elitin Saltanatın Meşruiyet Arayışına Katılımı", *Toplum ve Bilim* 83 (1999–2000) 292–322.

that the rebellion touched the majority of the existing power-holders, bringing unprecedented political instability to the capital. In 1730, there was not a single elite household that was strong—or willing—enough to put an end to the atrocities.

Ibrahim Müteferrika's press went into hiatus during the first year of the rebellion and we do not know if he was sympathetic to the rebels, whose actions were celebrated in two anonymous republican treatises that were printed in London and Amsterdam in 1737.[23] However, when Ibrahim started publishing again, there was noticeable change in both the language and the contents of his publications. The press put out, in rapid succession, three works that advocated various versions of Cartesianism. Ibrahim used Edmond Pourchot's Cartesian textbook, *Institutiones philosophicae*, in his *Füyuzat-ı Mıknatısiyye* [Magnetic Effluvia] and in his printer's preface to *Cihannüma* [Cosmorama]. The former treatise was about magnetic declination and used Cartesian particles to explain magnetism. The latter was an updated version of Kâtip Çelebi's (d.1657) seventeenth-century geographical masterpiece, which in turn drew on the atlases of Gerhard Mercator and Willem Janszoon Blaeu. Ibrahim also used Pourchot's *Institutiones philosophicae* in his preface, and advocated mathematics as the foundation of good reasoning. The third item in İbrahim's post-revolutionary publications, *Foundations of Government in Various Social Orders*, not only presented a political philosophy that drew on the notion of natural right, but also advocated mathematical training for military officers, clear chains of command and drilling for the army. Here, Ibrahim also clearly spoke against the rebellion because it led to disorder.

It is difficult to explain Müteferrika's rapid intellectual transformation—or perhaps, emergence. Given the magnitude of the rebellion's trauma, Ibrahim's strong reaction to the toppling of the entire regime that made him into what he was is hardly out of place. Somewhat more puzzling is the direction of this change. Do some revolts produce Cartesians, while others produce empiricists?[24] A number of interlocking narratives may explain Müteferrika's defense of Cartesio-Copernican views. One explanation might be that Ibrahim was already an avant-garde intellectual in 1710 and Cartesianism was par for the course. He might have found it expedient to remain silent on certain issues

---

23 Anon., *A particular account of the two rebellions, which happen'd at Constantinople, in the years MDCCXXX, and MDCCXXXI, at the deposition of Achmet the Third, and the elevation of Mahomet the Fifth: composed from the original memorials drawn up in Constantinople:* [...] *publish'd in French* [...] *translated into English* (London, G. Smith: 1737) 6.

24 Shapin S. – Schaffer S., *Leviathan and the Air-Pump: Hobbes, Boyle, and the Experimental Life* (Princeton: 1989).

while he worked for other causes that were compatible with his general intellectual and religious orientation. His fame and his network allow for such an interpretation. However, it is nearly impossible to make a more general case because many of his known associates, particularly Yirmisekiz Mehmed and Nikolaos Mavrocordatos were both dead by the end of the rebellion. The rebels exiled Yirmisekiz Mehmed to Cyprus, where he died in 1731 and Mavrocordatos died a natural death shortly before the rebellion. Mehmed Said, who survived the rebellion and went on to have a stellar career in the Ottoman administration, was virtually silent during this period.

A second line of argument comes from Ibrahim's own life experience. Given that Müteferrika spent several years as a printer and as a cartographer, and as an associate of Yirmisekiz Mehmed Çelebi, there is certainly an argument to be made for his propensity towards mechanical philosophy.[25] Here, it is useful to summon Jim Bennett's discussion of mechanical philosophy as the mechanics' philosophy.[26] The notion that extensive experience with maps and machines led to the early modern mechanical conception of the world rather than the other way around has had some traction among historians of science.[27]

Neither of these explains why Ibrahim chose *Institutiones philosophicae* as his textual source for Cartesian philosophy after 1730. Here, we may turn to two different explanations. One possible explanation is that Müteferrika's drew on Vikentios Damodos's (1679–1752) Greek paraphrase of the entirety of Pourchot's *Institutiones*. Damodos, who studied in Venice and at the University of Padua between 1713 and 1723, used this text at the seminary he established on the island of Cephalonia. The two authors shared anti-Catholic sentiments and their translation strategies were certainly similar. However, there are several problems with this explanation. Cephalonia never was an Ottoman territory and we do not currently know an intermediary between Damodos and Müteferrika. Nevertheless, there were plenty of Greeks in the Ottoman capital

---

25 For the European case, see Warntz W., "Newton, the Newtonians, and the Geographia Generalis Varenii", *Annals of the Association of American Geographers* 79 (1989) 165–191; Livingstone D.N. – Withers C.W.J., "On Geography and Revolution", in Livingstone D.N. – Withers C.W.J. (eds.), *Geography and Revolution* (Chicago: 2005) 12–13. For an explanation of Müteferrika's mechanicism from this perspective, see Küçük B.H., "Emulating Petrine Russia: Mechanics and Social Order in *The Foundations of Government*", in Stamatopoulos D. (ed.), *War and Revolution in European History: A Balkan Perspective* (London: *forthcoming*).

26 Bennett J., "The Mechanics' Philosophy and the Mechanical Philosophy", *History of Science* 24 (1986) 1–28.

27 Roberts L. – Schaffer S. – Dear P. (eds.), *The Mindful Hand: Inquiry and Invention form the Late Renaissance to Early Industrialization* (Amsterdam: 2007).

who held arts or medical degrees from Padua, but, aside from Methodios Anthrakites (1660–1748), a mystery figure whose writings have not been studied in any detail, there is nothing that approaches 'smoking gun' evidence.[28] A second problem has to do with the two authors' approach to physics. While Damodos believed that physics was subservient to theology, Müteferrika's argument was quite the opposite.[29] Finally, while Müteferrika explicitly mentioned Pourchot, Damodos's generous and uncredited use of the *Institutiones* remained unknown until 2010.[30]

Müteferrika's post-revolutionary works were printed at a pro-French moment in Ottoman history. Thus, a second explanation involves the Ottoman-French relations, particularly ambassadorial relations. The Ottoman administration had a fruitful and friendly relationship with France when Jean-Louis d'Usson, marquis de Bonnac, represented the French crown in Istanbul. This was the period when Yirmisekiz Mehmed went to France as ambassador. The two kingdoms shared a common enemy, the Habsburgs. However, the relationships were less than rosy after 1718. Although the Treaty of Passarowitz (1718) had guaranteed Ottoman Catholics the right to observe their religion freely, this was something of a pyrrhic victory for the Catholic powers of

28 Camariano-Cioran A., *Les académies princières de Bucarest et de Jassy et leurs professeurs* (Thessaloniki: 1974) 193.

29 Vlahakis G., "Dissemination and Development of Non-Aristotelian Physics in Aristotle's Land", in Lertora Mendoza Conicet C.A. – Nicolaidis E. – Vandermissen J. (eds.), *The Spread of the Scientific Revolution in the European Periphery, Latin America and East Asia: Proceedings of the XXth International Congress of History of Science (Liège, 20–26 July 1997)* (Turnhout: 2000) 45–52.

30 Pourchot's text enjoyed broader reception among Ottoman Greeks. Demetracopoulos J.A., "Purchotius Graecus I: Vikentios Damodos's *Concise Ethics*", *Verbum – Analecta Neolatina* 12, 1 (2010) 43: '[M]ost, if not all, of D[amodos]'s philosophical (and theological) "writings" are not originals, but latent yet close translations or adaptations of Latin writings by well-known European authors who lived shortly before or virtually at the same time as D., Edmond Pourchot (1651–1734), an Eclectic Catholic philo-Cartesianist philosopher who taught in Paris, being the principal among them'; on p. 58, Demetracopoulos also remarks how Damodos masked the Catholic provenance of the text: 'Now D., when rendering into Greek P.'s declaration that, in arranging the material in his *Ethics*, he followed the method of the IIa Pars of the *Summa theologiae* of Aquinas, substituted for Aquinas the name of Aristotle. Most probably, he did so in order to prevent his Orthodox audience from recalling the traditional Roman Catholic use of Aquinas' arguments for the *Filioque* against the Orthodox Church'. Also see Demetracopoulos J.A., "Purchotius Graecus II: Vikentios Damodos's *Concise Metaphysics*, Part I ("Ontology") and II ("Pneumatology"), *Studia Neoaristotelica* 11, 1 (2014) 5–63. Damodos also translated the entirety of Pourchot's third volume.

Europe. In practice, the treaty led to acute conflicts between Ottoman authorities on the one hand and, Dominicans, Franciscans and Jesuits on the other.[31] In 1722, Ahmed III issued a *ferman* that forbade Catholic missionary activity, most likely to appease the Greek Patriarch Jeremias, and thus to put a wedge between the clergy and France.[32] Part of this anti-Catholic episode that lasted until the Grand Vizierate of Topal Osman Pasha in 1731 was the prevention of church reparations. It was also with Topal Osman that we see the resumption of the personal friendship between an Ottoman Grand Vizier and the French Ambassador, this time Marquis de Villeneuve.[33] In a letter to the French Keeper of the Seals in 1730, Villeneuve said that the Ottoman court put great confidence in him in matters of science and medicine.[34]

Villeneuve's connection with the Jesuits on the one hand and his connection with Topal Osman Pasha on the other hand probably had something to do with Müteferrika's choice of text. Edmond Pourchot's *Institutiones* was the only Cartesian textbook that made it through Louis XIV's suppression of Descartes' philosophy. In late seventeenth- and early eighteenth-century France, Jansenism and Cartesianism usually went together, and Cartesian philosophy elsewhere had other undesirable associations.[35] Pourchot's text, on the other hand, followed a syncretistic route that did not completely dispense with Aristotelian philosophy and that did not advocate a radical political agenda. As a consequence, his textbook was wildly popular at the University of Paris and in Jesuit colleges. One Parisian Jesuit college, alternatively called *College de Quatre-nations* or *College Mazarin*, trained *jeunes des langues* who worked closely with the French embassy in Istanbul. Furthermore, Jean-Baptiste Holdermann, whose *Grammaire turc* came out of Müteferrika's press, was a professor at the same college.

Grand Vizier Topal Osman Pasha, the Governor of Rumelia (the highest provincial office in the Ottoman administration) was one of the figures who were responsible for restoring Istanbul's political order in 1731. Topal Osman was a spy in his youth and, like many Ottoman spies, operated in the Mediterranean.

---

31 Frazee C.A., *Catholics and Sultans: The Church and the Ottoman Empire, 1453–1923* (Cambridge: 1983) 175.

32 Frazee, *Catholics and Sultans* 156, 200.

33 Vandal A., *Une ambassade française en Orient sous Louis XV: La mission de Marquis de Villeneuve, 1728-1741* (Paris: 1887) 172.

34 Villeneuve Louis Saveur de, *Lettres au Garde de Sceaux* II, Paris, BNF Suppl. Fr. 2272b (1728–31) 44.

35 On the orthodox Calvinist opposition to Descartes, see Verbeek T., *Descartes and the Dutch: Early Reactions to Cartesian Philosophy, 1637–1650* (Carbondale: 1992).

He had been captured by corsairs on his way to Cairo. Vincent Arniaud, a French merchant, ransomed him in Malta trusting Topal Osman's word that he would pay him back. This led to a lifelong relationship between the two. In 1732 Topal Osman, now Grand Vizier, summoned Arniaud to the capital to thank him. During what appears to have been a very public event, Topal Osman extolled the virtues of French merchants, reserving the highest praise to Arniaud: 'J'ai été esclave comme eux ajouta-t'il, j'étois chargé de chaînes, percé de coups, couvet de blessures, voilà celui qui m'a racheté, qui m'a sauvé; voilà mon Patron: liberté, vie, fortune, je lui dois tout' [I have been committed to chains, cut, covered in wounds. And here is the man who freed me, who saved me. This is my patron. Liberty, life, fortune, I owe all to him].[36] Unsurprisingly, Topal Osman—now the head of government and İbrahim's patron—was pro-French in his policies, especially those pertaining to trade.[37] He also facilitated the instatement as commander and religious conversion of Comte de Bonneval/ Humbaracı Ahmed Paşa.

Translation and Language

Although Müteferrika translated copiously from Pourchot's *Institutiones* in two of his post-revolutionary works, *Magnetic Effluvia and* the 'Printer's Preface' to *Cosmorama*, it is difficult to assess the fidelity of Müteferrika's translations to Pourchot's text. For one, Pourchot's text ran through many editions and each edition displayed a specific configuration of censorship and political intervention.[38] We do not know which particular edition ended up in Müteferrika's hands. There is also a lot of work to be done in the study of not only Müteferrika's translation practices, but also his unique scientific idiom. The best I can do here is to provide a few general observations that may help future scholarship.

36 "Lettre de M. D.L.C. à M. D.L.R. sur quelques particularitez de la vie de Topal Osman Pacha, cy-devant Grand Visir de l'Empire Ottoman, et aujourd'hui Séraskier de l'Armée Turque en Perse. A Paris, ce 18 Janvier 1734", *Mercure de France* (January 1734) 88: For evidence collaborating this anonymous piece in *Mercure de France*, see Marquis de Villeneuve, *Lettres au Garde des Sceaux II (1728–1731)*, BNF Suppl. Fr. 2272b, 358v, 363v–364r, 373r. Also see Abdülkadir Özcan, "Topal Osman Paşa", *Türkiye Diyanet Vakfı İslam Ansiklopedisi* (1988-) 41, 246.

37 *Lettres au Garde de Sceaux II* 358v, 363v–364r, 373r.

38 Schmaltz, T.M., "A Tale of Two Condemnations: Two Cartesian Condemnations in Seventeenth-Century France", in Del Prete A. (ed.), *Descartes e i suoi Avversari: Incontri cartesiani II* (Florence: 2004) 203–221.

Pourchot's style was scholastic. He enumerated arguments, spelled out objections, cited his sources clearly and produced a highly readable neo-Latin text that would appeal to students. Müteferrika's style was very different from that of Pourchot. For one, his language was pleonastic and rhythmic. He imitated the Persianate poetic idiom—*inşa* prose- of eighteenth century Istanbul.[39] To go back to the opening quotation, a very literal translation—which I often avoid in order to improve the intelligibility of the text—of the following passage would be:

> Ve lakin herçend ke atıl ve batıl ise de mezheb-i kadim ve rey-i atik-i hükema olmağla her asır be her zamanda mail-ü-ragıbı ve haris-ü-hamisi bulunub bu reyi teyyid sevdası ve takviye hevası gayretiyle kütüb-ü-resail-i vafire dahi telif ve tedvin olunmuşdur [...].
>
> And, although these [the opinions of Aristotle and Ptolemy] are idle and superstitious, being the ancient opinion and the old view of the philosophers, it has been attractive and well regarded in all ages and at all times. And, they have always had aspirants and patrons. And, many books and tracts have been written and compiled out of a desire to verify and of a wish to support this view.

One obvious reason for the use of such language was Müteferrika'a audience. As he was addressing the privileged and well-educated group at and around the Ottoman palace, he was speaking up and thus, also 'translating up'. He was by no means the only person to do so. Other period physicians and philosophers included a fair amount of poetry in their treatises and some of them were accomplished poets and storytellers.[40] Nevertheless, Müteferrika's use of high-register language in combination with novel technical vocabulary also meant that common people who probably had no access to the poetry-laden scientific manuscripts but could buy or otherwise access printed books had a hard time with his idiom. Petros Baronyan, the Ottoman-Armenian interpreter

---

39 Sılay K., *Nedim and the Poetics of the Ottoman Court: Medieval Inheritance and the Need for Change* (Bloomington, IN: 1994).

40 See, for example, Bachour N., *Oswaldus Crollius und Daniel Sennert in frühneuzeitlichen Istanbul: Studien zur Rezeption des Paracelsismus im Werk des osmanischen Arztes Salih b. Nasrallah Ibn Sallum al-Halabi* (Freiburg: 2013) 37–38. On poetry in Ottoman scientific writings, see Brentjes S., "The Interplay of Science, Art and Literature in Islamic Societies before 1700," in Dev A. (ed.), *Science, Literature and Aesthetics*, History of Science, Philosophy and Culture in Indian Civilization XV, Part 3 (New Delhi: 2009) 453–484.

to the Dutch mission in Istanbul, said as much in yet another translation, this time of Jacques Robbe's *La méthode pour apprendre facilement la géographie*: 'This insignificant and impotent little book is intended to help the common people (*ahali*) and the students access the chapter headings and the index, and learn the superior vocabulary used in Ibrahim Müteferrika's edition of Katip Çelebi's *Cihannüma*'.[41] Baronyan was simply trying to give the common reader a way to cut through Müteferrika's high-register language, which remains inaccessible to Turkish readers even today.

A second feature of Müteferrika's translations was opaqueness, especially when it came to citing authors. For example, *Magnetic Effluvia*, which also included translations from Pourchot's *Institutiones*, omitted the parts where Pourchot discussed specific authors:[42]

> Böyle ki seng-i mıknatıs batın-ı arzda mestur ve mütevari olduğu halde, zatında hulkı ve cebeli niçe niçe menafız ve mecari vaki olub, batın-ı zeminden ruy-i zemine ihrac olundukda dahi füyuzat-ı cevheriyye-i merkume ol menafız ve mecariden duhul ve huruc ve cereyan iderler. Ve mihver-i arzda mütevazı hutut üzre hareket ile mıknatıs ol siyaka muvafık nizama vaz ve kutublarını arzın kutublarına teveccüh itdirirler. Ve ekseri bu reye sahib oldular.[43]

---

41 Süleymaniye MS Yazma Bağışlar 7530, 4a.

42 Terrestrial magnetism was operationalized, but not necessarily theorized in various Ottoman texts from the fifteenth century onwards. See, for example, Bican Ahmet, *Dürr-i Meknun* [The Hidden Pearl], ed. and trans. L. Kaptein (Asch: 2007); Seydi Ali Reis, *Kitabü'l-Muhit* [On of the Indian Ocean] (1554), transcribed H. Büke, "Seydi Ali Reis-Kitabü'l-Muhit", unpublished MA thesis (Pamukkale University, 2010).

43 Müteferrika İbrahim, *Füyuzat-ı Mıknatısiye* (İstanbul: Sultanic Press, 1732) 4a; Pourchot Edmond, *Institutiones philosophicae ad faciliorem veterum ac recentiorum philosophorum lectionem comparatae*, 3rd ed., 5 vols. (Leiden, Antonius Boudet: 1711), vol. 3, 234–235: 'Verum id plurimis incredibile videtur, tam eas materiae striatae cochleas quam canales seu tubulos, in quibus moventur, figuram suam striatam constanter retinere, nec cochlearum strias prominentes atteri aut abradi [...] Haec et alia id genus clarissimum Hugenium Hagensem Batavum impulerunt, ut aliam hypothesim proponeret, quam Cartesianae anteponendam multi arbitrantur, tametsi suis etiam incommodis iisque gravissimis sit obnoxia. Statuit igitur profluvia, quae continenter e terra per lineas illius axi parallelas erumpunt, unum e magnetis polis penetrare ac per lineas axi magnetis parallelas progressa per oppositum polum egredi: cumque facilius ea profluvia per magnetis quam per aeris meatus moveantur, vorticem aliquem circum magnetem sic efficere, ut uno tantum polo ipsum subeant et altero tantum ac opposito ex eo erumpant'.

> Although the lodestone is hidden and veiled inside the earth, by its very creation and nature, it includes many many holes and channels. When it is extracted from the inside of the crust to the outside of the crust, the effluvium of the aforementioned essence keeps going into and out of, and flowing through these holes and channels. And the motion of the lodestone, by means of the holes and channels located inside the terrestrial globe, is parallel to the axis of the earth and aligns its poles to the poles of the earth. This is the opinion the majority [of European philosophers] have followed.

Such omissions did not follow a discernible pattern, and seem to apply as well to the Coimbra Jesuits as they did to Huygens:

> Hükema-i garbiye [...] ibreyi kutb-ı şimali canibine teveccüh itdüren kutub yıldızıdur. Ve bazılar semanın bir kıtasıdır [...] Ve bazı mütedeyyinleri havass-ı mıknatıs bir hikmet-i hüdadır ki irade-i aliyye-i ezeliyye anı izhar idüb kemal-i kudretin isbat ve kavllerinden istiğrab olunmasını murad idüb hakikatde sırrını ihfa eyledi didiler.[44]
>
> Some Western philosophers have said that what points the magnetic needle towards the north is the North Star. And some have said that it is a part of the heavens. And some devout [philosophers] have said that properties of the magnet are known to God alone. The great and infinite will wanted to reveal it as a proof and an expression of his omnipotence, but has hidden its secrets.

The filtering of references may also have something to do with his audience—leaving in only the most famous authors, such as Descartes or Copernicus—, but this raises more questions than it answers. How did Müteferrika decide what the European canon was? And what made Descartes and Copernicus more important than Cardano, Huygens and also Newton for that matter? Again, while it is impossible to attempt a generalization without a full critical edition of Müteferrika's writings, that we can learn as much from what he did not translate as we can from what he did requires no further argument.

---

44 Müteferrika, *Füyuzat* 3b. Pourchot, *Institutiones* 3, 233: 'Eam vim repetit Cardanus a stella polari; Conimbricenses ab aliqua coeli parte non longe a polo remota. Vulgares scholastici qualitatem occultam causantur, *quam Deus mirari nos voluit, scire noluit*. Recentiores Physici ad profluvium aliquod substantiale e terra profectum confugiunt'.

Lastly, Müteferrika and Pourchot had fundamental disagreements about the relationship between science and religion. Thus, the opaqueness of the translations also helped Müteferrika express his own views. Pourchot, unlike Müteferrika, never said that matters of theoretical astronomy were not also matters of religious creed. How could he, as an author who wrote in a Catholic country at a time when the works of Copernicus and Descartes were on the papal index? He had also included a discussion of Galileo's condemnation at the end of his treatment of Copernicus under the heading, 'Responses to Objections'.[45]

The first objection was that the system was indefensible because 'it seemed to run counter to the authority of the Holy Scripture'. Pourchot cited Ecclesiastes 1:4 '[...] but the earth stands forever' and Psalm 103 (104) 'Who has founded the earth upon its own bases: it shall not be moved for ever and ever', and added that Galileo had defended this opinion against the Scripture and had been condemned twice: by the Inquisition in 1616 and by Pope Urban VIII himself in 1633. Pourchot's response was to propose that the Copernican system was not defensible as a thesis, but was defensible as a hypothesis. That is, it saved the physical and astronomical phenomena (*spectantia salverentur*).

This Catholic argument for 'weak Copernicanism' was not part of Ibrahim's translation, though he voiced his personal skepticism towards astronomical theories with a couplet as he finished his treatment of the three systems:[46]

Kimse hal eylemedi müşkil-raz feleği
Var ise akd, Süreyya kodı nokta-ı şek[47]

No one has been able to understand the troublingly mysterious heavens.
Even if there were a consensus, Pleiades would conclude it on a doubt.

Müteferrika's doubts did not prevent him from posing the Copernican system as a challenge, inviting concrete proofs and objections to the view. We can tell that his challenge was not simply talk, as many of the printed editions of

45 Pourchot, *Institutiones* 3, 31: 'Obiicies primo: illud systema defendi non potest, quod Scripturae sacrae auctoritati videtur oppositum [...]. Respondeo primo distingui posse maiorem hoc modo. Defendi non potest, ut thesis, concedo; ut hypothesis, nego'.

46 The absence of this passage might have something to do with the particular edition of Pourchot that he used. The weak, hypothetical argument was more pronounced in the Venetian editions from the 1720s than they were in French editions. See Mayaud P.-N., *La condamnation des livres coperniciens et sa revocation* (Rome: 1997) 125–130.

47 Katip Çelebi, *Cihannüma* 48.

the book had blank pages sewn in against these passages. He was genuinely expecting a response from those scholiasts who bought and read this book. And, the copies at the Süleymaniye Manuscript Library—I have not consulted the editions kept elsewhere—suggests that no one responded to this challenge.[48] Was it that people who could scrutinize Copernican astronomy never read Müteferrika's book? Was it that there were no theoretical astronomers in Istanbul? Was it that the libraries that kept these books did not allow the readers to jot down annotations? Or, perhaps, every single one of Müteferrika's readers found Copernican astronomy convincing and did not feel the need to respond. Any answer is bound to be speculative in the absence of evidence, but I can provide further context that shows Ibrahim's invitation was not out of place.

Müteferrika and Ottoman Naturalism

During the 1720s, there was a community of proponents of empirical knowledge in the Ottoman Empire: leading the pack were the adherents of 'new medicine', who drew inspiration from the European discussions of Paracelsian medicine, and for others, the magnetic compass would be the object of empirical interest.[49] There certainly were no known Muslim Cartesians, but an investigation of the views of Petros Baronyan, who was employed at the Dutch embassy, and of Methodios Anthrakites is likely to change our understanding of Cartesianism in Istanbul. While Descartes' views were not empirical in the sense that Boyle's were, Descartes was famous for offering intelligible explanations of observed phenomena. Furthermore, recent scholarship shows that the relationship between empirical practices and Cartesian philosophy was more complicated than previously assumed.[50]

---

48 Copies I have consulted at the Süleymaniye Manuscript Library are: MSS Ayasofya 2606, Çelebi Abdullah 259, Düğümlü Baba 468, Esad Efendi 2046, Fatih 4317, Halet Efendi 638, Hamidiye 931, Hekimoğlu 736, Hüsrev Paşa 271, Mihrişah Sultan 308, Pertevniyal 754, Pertev Paşa 373, Nuruosmaniye 3005, Ragıppaşa 1062, Hacı Selim Ağa 737, Hacı Selim Ağa 738, Hacı Ahmed Paşa 170, Murad Molla 1420, Reisülküttab 634.

49 Küçük B.H., "The Compass and the Astrolabe: Empiricism in the Ottoman Empire", in Somel A. – Kenan S. (eds.), *Realms of Transformation in Ottoman History, Essays in Honor of Metin Kunt* (London: *forthcoming*).

50 Dobre M. – Tammy N. (eds.), *Cartesian Empiricisms* (New York: 2013).

Since Müteferrika's post-revolutionary writings advocated empiricism, his ideas about the separation between science and religion were in line with what we find in empirical texts of the period. By the early eighteenth century, Ottoman discussions of new medicine had precipitated a debate about the relationship between medical theory and medical practice on the one hand, and between medicine and religious sciences on the other. The role of experience in the theories of medicine and the independence of medical knowledge from religious creed were central themes in many of these writings. Most physicians argued that religion and medicine were completely separate domains, while others went so far as to claim that medicine was superior both in its utility and in its epistemology.[51] The argument for new medicine rested by and large on the experienced efficacy of new drugs. However, these efficacious chemical drugs created a whole host of problems about the proper method for following medical recipes, about accurate dosage and about the right time for administering drugs. The result was the emergence of physicians who presented themselves as consummate—and composite—naturalists, who knew astronomy, physics and mathematics, in addition to pharmacology and medicine. In the 1720s, some were already arguing—in a manner directly opposed to Avicenna—that medicine could provide a complete, better and empirical substitute for traditional natural philosophy. Also in tow was an emerging crisis around terrestrial magnetism where brahim Müteferrika was also an actor—was the compass more reliable than astronomical observation for reckoning direction?

Müteferrika also sought to broach several different issues with his 'Printer's Preface'. First, he tried to present geography as a theoretical science that was on par with other parts of philosophy.[52] He shared this notion with Katip Çelebi, whose *Cihannüma* followed Müteferrika's preface. Secondly, he introduced a new structure for the mathematical sciences. The combination of geography, mathematical astronomy (*hey'et*) and physical astronomy (*nücum*) belonged under the science of 'cosmography', a novel name for Ottoman readers and one that Müteferrika did not Arabicize.[53] He described geography in such a way as to make it synonymous with the 'knowledge of the world'. As such, geography included not only cartography, but also historical geography and

51 Küçük B.H., "New Medicine and the *Hikmet-i Tabi'iyye* Problematic in Eighteenth-Century Istanbul", in Langermann T. – Morrison R. (eds.), *Texts in Transit in the Medieval Mediterranean* (College Park: 2016) 222–242. Physicians' self-perception seem to match the outsiders' view of medicine. For poet Nabi's view of medicine, see Kaplan, M., *Nabi, Hayriyye-i Nabi* (*İnceleme-Metin*) (Ankara: 2015) 305–309.

52 Müteferrika, *Cihannüma* 15.

53 Müteferrika, *Cihannüma* 14.

political geography. And finally, he argued that geography was the noblest and the most useful science for the statesman, a sentiment that he shared with Katip Çelebi.[54]

Müteferrika identified two different ways to consume geographical knowledge: sporadically consulting geography texts yielded some use when waging wars, drawing borders—itself a recent invention—and collecting taxes. Even greater, however, were the benefits of habitual engagement with maps and geographical texts. Katip Çelebi had said that those who studied geography led a peaceful life on 'four pillows', and 'loved detail, stood sure-footed and knew their location on earth'.[55] This was a Cartesian move, in the sense that the study of geography, which rested on '*mebadi-i yakiniye*' [certain first principles], provided rules for the direction of the mind.[56]

Frontispiece of the *Ledger of Fools*: Ibrahim Müteferrika on Science and Religion

Historians of Ottoman science and philosophy have oscillated between calling the seventeenth century a 'triumph of fanaticism' and proclaiming that there was no opposition between science and Islam in the seventeenth and the eighteenth centuries. What we lose in such binary views of Ottoman history is the wide range of views on Islam, on natural knowledge and on the relationship between the two. Not only did the Ottoman Empire have heretics who could call the Koran an 'Ottoman notebook', but there were substantial contingents within the lower strata of Ottoman society who lived in direct breach of the holy law.[57] Hasan Çelebi (d.1660), Chief Astrologer to Sultan Mehmed IV, was also known as Küfri or the Profane on account of his complete rejection of prayer and fasting.[58] Nükhet Varlık's work shows that there were also traditions of determinism that go back to the sixteenth century, to the Ottoman

54 Müteferrika, *Cihannüma* 17: 'Bu fenn, siyaset babında cümleden enfa olur. Bunu bilen aziz ve memduhtur'. I would like to thank Gottfried Hagen, who helped me distinguish between Katip Çelebi's words and those of Müteferrika.

55 Müteferrika, *Cihannüma* 16.

56 Müteferrika, *Cihannüma* 1–2.

57 Majer H.G., "'The Koran: An Ottoman Defter!' Ottoman Heretics of the Eighteenth Century", in Veinstein G. (ed.), *Syncrétismes et hérésies dans l'Orient seljoukide et ottoman (XIVe–XVIIIe siècles)* (Paris: 2005) 299–310; Kurz M., *Ways to Heaven, Gates to Hell: Fazlizade Ali's Struggle with the Diversity of Ottoman Islam* (Berlin: 2011).

58 Ekinci, R., "Türk Hiciv Edebiyatının Sıradışı Bir Şairi: Küfri-i Bahayi ve Eserlerinden Örnekler", *Türkiyat Mecmuası* 24 (2014) 33–58.

battle against the plague.[59] Ethan Menchinger's recent article also shows that by the eighteenth century, a naturalistic view of the world from politics to science was mainstream.[60] Some eighteenth-century scholars, such as Fazlızade Ali (fl.1740), have also viewed the Abbasid legacy, Sufism and even the late medieval Persian philosophy—bread and butter of the Ottoman colleges—as being heretical. In short, the relationship between science and religion in the Ottoman Empire was a vast and complex field ranging from Sunni ultra-orthodox views to outright heresy.[61] Ibrahim Müteferrika's position falls somewhere in the middle. While he professed to the veracity of Islam, he was dismissive of both the ultra-orthodox and of the Empire's scholiasts. We can read this through the selective translation from Al-Ghazali with the implicit claim that his view was the properly Islamic view.

Al-Ghazali is a pivotal figure in the history of Islam. Nineteenth and twentieth-century scholarship considered him to be the man who put an end to philosophy in the world of Islam.[62] Born in 1058, Ghazali was a jurist who worked at one of the very first *medreses* that the Seljukid vizier Nizam al-Mulk established. As the *medreses*, both Seljukid and Ottoman, sought to train judges (*kadıs*), Ghazali's views were largely formulated from a judicial perspective: namely, whether philosophers were heretics and whether their words and deeds could be reconciled with Islam. Ghazali's solution was metaphysical occasionalism. That is, only God could activate what otherwise seemed like a habitual cause and effect relationship. Recent scholarship has shown that the legacy of Ghazali's occasionalism was considerably more complex than simply opposing scientific and philosophical inquiry.[63]

We can find various versions of Ghazalian occasionalism in late medieval and early modern Islam. Ibn Taymiyya, sometimes considered the father of Islamic fundamentalism, routinely invoked similar arguments.[64] Occasionalism was

59 Varlık N., *Plague and Empire in the Early Modern Mediterranean World: The Ottoman Experience, 1347–1600* (Cambridge: 2015).

60 Menchinger E.L., "Free Will, Predestination and the Fate of the Ottoman Empire", *Journal of the History of Ideas* 77, 3 (2016) 445–466.

61 Kurz, M., *Ways to Heaven*, 24 and in passim.

62 Gutas D., "The Study of Arabic Philosophy in the Twentieth Century: An Essay on the Historiography of Arabic Philosophy", *British Journal of Middle Eastern Studies* 29, 1 (2002) 5–25.

63 For an overview, see Daiber H., "God versus Causality: Al-Ghazali's Solution and Its Historical Background", in Tamer G. (ed.), *Islam and Rationality: The Impact of al-Ghazali, Papers Collected on His 900th Anniversary* (Leiden: 2015) vol. 1, 1–22.

64 See, e.g. Michot Y., "Ibn Taymiyya on Astrology: Annotated Translation of Three Fatwas", *Journal of Islamic Studies* 11, 2 (2000) 165.

also one of the main elements of post-thirteenth century philosophical theology—a compromise position that held the various branches of learning together.[65] In the fifteenth century, as Mehmed II mulled his relationship with Sunni Islam, Ghazali's views received fresh attention, as they were key to positioning the dynasty vis-a-vis the three main schools of Islamic thought: philosophical theology, Greek philosophy and mysticism. Mehmed commissioned two prominent scholiasts, Hocazade Müslihüddin Mustafa (d.1488) and Alaaddin Ali al-Tusi (d. 1472) to revisit Ghazali's text in a disputation.[66] While neither of these scholars fully adhered to Ghazali's occasionalism, Alaaddin al-Tusi was clearly more accommodating towards philosophy, calling it both beneficial and probable.[67] The Ottoman scholar whose views were closest to those of Al-Ghazali was Mehmed Birgivi, whose legacy in the Ottoman Empire resembles that of Phillipp Melanchton in Germany and, exerted the greatest influence on seventeenth-century Ottoman Islam.[68] Birgivi was not only against philosophy but also against the *medrese* establishment as a whole. This well-known theological position, combined with European travellers' impressionistic views of Ottoman Islam has given rise to the clearly false notion that Ottomans were fatalists and ignoramuses.[69]

A second line of argument in favor of geography estimates what Sebastian Conrad has called 'globality'.[70] Müteferrika invoked the example of the man 'who did not know his neighbourhood'. Such men, he argued, ran the 'risk of fire and of mutiny'.[71] Consequently, men had to know what was happening outside their household and this was precisely what geography and cosmography offered. In 1732, when Müteferrika was penning these lines, and many more about different forms of government and about the magnetic compass, Istanbul had just come out of a two-year revolt, which came on the heels of a major fire.[72] Although the relationship between the revolt and the misfortunes

65 Sabra A.I., "Science and Philosophy in Medieval Islamic Theology: The Evidence of the Fourteenth Century", *Zeitschrift für Geschichte der Arabisch-Islamischen Wissenschaften* 9 (1994) 1–42.

66 Özervarli M.S., "Arbitrating between Ghazali and the Philosophers: The *Tahafut* Commentaries in the Ottoman Intellectual Context", in Tamer (ed.), *Islam and Rationality* vol. 1, 375–398.

67 Özervarli, "Arbitrating" 390.

68 Küçük, "The Compass and the Astrolabe".

69 Menchinger, "Free Will" 462.

70 Conrad S., "Enlightenment in Global History: A Historiographical Critique", *American Historical Review* 117 (2012) 999–1027.

71 Müteferrika, *Cihannüma* 17.

72 Müteferrika, *Cihannüma* 2.

of Istanbul is not entirely clear, Müteferrika was clearly thinking of the events as part of larger developments, including the war on the Eastern front and social unrest in Europe.

The third line of argument had to do with science and religion. Müteferrika argued that science and religion had to be separate matters. While this may appear to be a perennial theme in the history of science, Müteferrika had at least two proximate reasons: one was a point that physicians often raised by reference to an alleged saying of the prophet: 'Science, 'tis two: first comes the science of bodies, then comes the science of religions'.[73] Although it is not entirely clear what they precisely meant by this separation, we can glean at least two contexts where this was meaningful: for one, the custodians of Ottoman law were *medrese* graduates whose primary training was in *fiqh* or Islamic law. It was increasingly common for Ottoman subjects to bring up their grievances against physicians in courts of law. The court, therefore, was one site of confrontation between the lay physician and the religious judge. And, medico-legal works became a well-represented genre, again, in the late seventeenth century. A second, more pertinent issue was the over-emphasizing Sunni pedigree in all matters of knowledge. The seventeenth century, for example, was also the golden age of prophetic medicine in the Ottoman Empire.

Ekmeleddin Ihsanoğlu and Avner Ben-Zaken have argued that Copernican astronomy did not contradict with Islam in the same way it contradicted with the Tridentine interpretation of heliocentrism.[74] Müteferrika's defense of astronomy, which was a reproduction of Katip Çelebi's arguments, presents a different picture. In a fairly long citation and translation of Ghazali's second introduction in *The Incoherence of Philosophers*, Katip Çelebi had presented an argument for why natural philosophy and astronomy were separate from creedal matters which immediately followed Müteferrika's own arguments:

> İmam Gazali, Tehafüt Felasife'de bi-ibaretihi bu mahalde irad ve tercüme olunır ki demüşdür.
>
> ليعلم ان الخلاف بينهم و بين غيرهم من الفرق ثلاثه اقسام.

73 Küçük, "New Medicine" 236.

74 İhsanoğlu E., "Introduction of Western Science to the Ottoman World: A Case Study of Modern Astronomy (1660–1860)", in İhsanoğlu E. (ed.), *Transfer of Modern Science and Technology to the Muslim World* (Istanbul: 1992) 67–120; Ben-Zaken A., "The Heavens of the Sky and the Heavens of the Heart: the Ottoman Cultural Context for the Introduction of Post-Copernican Astronomy", *British Journal for the History of Science* 37, 1 (2004) 1–28.

Malum olsun ki hükema ile fırak-ı enam beyninde olan hilaf üç kısımdır …

ما لا يصدم مذهبهم فيه اصلا من اصول الدين ليس من ضروره تصديق
الانبياء[75] منازعتهم فيه،

Kısm-ı sani: mezhebleri usul-i dinden asıla muarazet eylemeyen umurdır. Pes anlarla anda niza-i enbiya tasdiki zaruretden değildir. Yani tasdik, anları tekzib ve aks iktiza eylemez.

كقولهم: ان خسوف القمر عباره عن انمحاء ضوء القمر بتوسط الارض بينه
و بين الشمس من حيث انه يقتبس نوره من الشمس،

Mesela, husuf-ı kamer küre-i arz, cerem-i şems ile cerem-i kamer meyanına tavassut itmekle kamerin nurı zevalinden ibaretdir. Zira, kamer nurı güneşden ahz ider. Arz ise küredir ve sema cümle canibden zemini ihata itmişdir. Kaçan kamer zıll-ı arzda vaki olsa nur-ı şems andan münkat olur didikleri gibi … Zira umur- mezkure kavaid-i berahin-i hendesiye ve hesabiyeye delalet ider. Bir kimse ke ana matla olub, tahkikine kadir ola, sebebinden ve vaktinden ve kadrinden ve müddet-i bekasından haber vire, ana deyilse ke bu şer'e muhaliftir, ol yakin üzre istidlal itdiği umurda şübhe eylemez. Bil ke şer'de istişkal ider ke yakine muhalif şer' niçe olur deyu tevcihe başlar …

و ضرر الشرع ممن ينصره لا نطريقه اكثر من ضرره ممن يطعن فيه بطرىقه.
و هو كما قيل: عدو عاقل خير من صديق جاهل.[76]

Imam Ghazali has said in his *Incoherence of the Philosophers*, which I quote and translate verbatim: Let it be known that the dispute between the philosophers and the sects has three parts: [discussion of the self-subsistence of created things…], The second part is one where their doctrine does not clash with the principles of religion and does not require a defense of the prophets. They may say, 'The lunar eclipse is the loss of moonlight because the earth is between the sun and the moon. The earth being a sphere, it is surrounded by the heavens on all sides. Therefore, when the moon is under the earth's shadow, sunlight no longer reaches

75 In Marmura's edition, الرسل صلوات الله عليهم replaces الانبياء. For the complete passage (with minor variations) and its English translation, see Al-Ghazali, *The Incoherence of the Philosophers*, ed. and trans. M.E. Marmura (Provo, UT: 2000) 5–8.

76 Müteferrika, *Cihannüma* 18. Also see Topkapı Revan MS 1624, 5b–6a.

> it'.... This matter is a matter of geometrical and arithmetical proof. When one studies the demonstrations and is certain about the proofs ... and someone tells them that this is against religion, they will begin to ask how religion can refuse to admit a certainty. The harm to religion caused by improper defense is worse than that caused by a proper attack. As the saying goes 'A reasonable enemy is better than an ignorant friend'.

With this passage, Müteferrika was arguing for a separation between science and religion, echoing the physicians' call to separate the science of bodies from the science of religions. This passage was also a fairly selective interpretation of Al-Ghazali, who commanded great intellectual authority among Muslim scholiasts in a way that is similar to what St. Augustine is to the Christian world. While Ghazali's ultimate goal was to reach a theological synthesis that did not allow for a freethinking contingent, Müteferrika used it to argue for a social separation where naturalists could lead a separate existence from the theologians.

More pertinently, this specific deployment of al-Ghazali seemed to contradict Mehmed Emin El-Üsküdari's (d.1738) commentary on the same work, dated 1726.[77] The two men knew each other well. They were both employed by the court in the 1720s. Üsküdari was among Istanbul's most prominent scholiasts and had written one of the very few books (in this case, a 5-sheet memorandum) on theoretical astronomy at the time. The issue in question was barleycorn, a unit of length that Islamic astronomy had inherited from Babylonian astronomy.[78] Other astronomers around the court included Darendeli Mehmed, who had devised a perpetual solar-lunar with an 8-year cycle[79] and Esad of Ioannina.[80] Müteferrika and Üsküdari agreed that astronomical practice and religious creed were compatible, both men called for a deeper engagement with Islamic astronomy and both employed a rhetoric of proof and certainty. Müteferrika's disagreement with Üsküdari was on two issues: the first involved the role of experience in proof, and the second, the rejection of Ptolemaic astronomy. Üsküdari thought that experiential knowledge was limited and did

---

77 Gökdağ K., *Mehmed Emin el-Üsküdari ve Telhisu Tehafüti'l-Hükema Adlı Eseri*, Unpublished MA Thesis (Marmara Üniversitesi, 2008).

78 Ragep S.P., *Jaghmini's Mulakhkhas: An Islamic Introduction to Mathematical Astronomy* (New York: 2016) 288.

79 Navoni J.-B., "Rouz-namé ou Calendrier perpétuel des Turcs", in Hammer-Purgstall J. (ed.), *Fundgruben des Orients* 4 (1814) 38–67.

80 Aydüz S., "Lale Devri'nde Yapılan İlmi Faaliyetler", *Divan: İlmi Araştırmalar* 3, 1 (1997) 143–170.

not constitute proof; secondly Üsküdari operated entirely within the confines of geocentric astronomy, discussing classic Ghazalian issues such as celestial intellects.[81]

However, the differences between Üsküdari and Müteferrika were nothing when measured against a third view that rejected philosophy altogether. Specifically, he had Al-Suyuti and his contemporary follower, Nazmizade Murtaza of Baghdad, in mind.[82] Suyuti was fairly unique in Islamic history, because, unlike the famous Muslim astronomers that decorate the history of Ptolemaic astronomy, he had argued that the earth was flat. Although Müteferrika did not mention any names, his disdain was unequivocal. He said 'küre-i arz mistah zan idüb basit olmak fikrin idenler udhuke-i ukala ve serdefter-i agbiyadır' [those who defend that the earth is flat are ridiculously unreasonable and if one were to make a ledger of fools, these people would be the frontispiece].[83]

Al-Suyuti was a famous Mamluk scholar, and is recognized as the most prolific author in the history of Islam. Included in his massive oeuvre are history, prophetic lore, jurisprudence, and also invectives against Hellenistic sciences, including logic and astronomy. He was uniquely authoritative among scholars of the religious sciences in the sixteenth and seventeenth centuries. In the eighteenth century, Nazmizade, whom Müteferrika hardly liked, had translated Suyuti's Sunni Astronomy.

We know that the printing press that Müteferrika ran was a site of contention between the various elites, because Nazmizade's history of Baghdad—one that left out the entirety of Baghdad's philosophical splendor—was published at the printing press with a brief introduction about the virtues of history—one that had more praise for Katip Çelebi than for Nazmizade. *Gülşen-i Hulefa*, which continues to be a reference source for the history of Ottoman Baghdad was reduced to being an 'agreeable work by a virtuous Baghdadi scholar known as Nazmizade Murtaza'.[84] The governor of Baghdad, Ömer Pasha patronized this immaculately Sunni scholar, mainly because Baghdad was both a physical and an ideological frontier between Iran and the Ottoman Empire—and had been so for much of the seventeenth and the eighteenth centuries. Nazmizade's encomium to Ottoman rule was the mirror image of a late seventeenth-century

81 Özcan A., "Nazmizade Murtaza Efendi", *TDV İA* 32, 461–462.

82 Nazmizade's translation of Suyuti's *Sunni Astronomy* is *Terceme-i Heyet-i İslamiyye:* Süleymaniye MS Fatih 3390, Bağdatlı Vehbi 850 and Hacı Mahmud 2064.

83 Müteferrika, *Cihannüma* 19.

84 Nazmizade Murtaza, *Gülşen-i Hulefa*, ed. İbrahim Müteferrika (İstanbul: Sultanic Press, 1730) 3a.

Safavid geographical compendium, *Muhtasar-i Mufid*, which laid Shiite claims over Iraq.[85]

Conclusion

Where does this place Müteferrika on the religious map of the Ottoman Empire? While Müteferrika abhorred Nazmizade Murtaza's and Suyuti's foolish defenses of Sunni faith, he also conspicuously abstained from engaging with the Persian astronomers and mathematicians who would have had canonical status at the *medrese*—Tusi was absent, but so were people such Amili or Mir Damad. In other words, he still had his allegiance to Ottoman Sunnism, but his was an enlightened Sunnism that made room for the new sciences. As the opening quotation to this paper suggests, Müteferrika was open to Copernican or Cartesian ideas, but more importantly, to innovations by contemporary Ottoman scholars.

Ibrahim's invitation to scholiasts helps us explore the vastly complex landscape of science and religion in the Ottoman Empire. The many, hitherto unexplored layers include conversion, the Mamluk legacy, and finally, the career of Ottoman science over the seventeenth and the eighteenth centuries. It is clear that Müteferrika did not simply implant European radicalism on Ottoman soil, but rather gave voice to a view of natural knowledge that was native to the Ottoman Empire.

Selective Bibliography

Aydüz S., "Lale Devri'nde Yapılan İlmi Faaliyetler", *Divan: İlmi Araştırmalar* 3, 1 (1997) 143–170.

Bennett J., "The Mechanics' Philosophy and the Mechanical Philosophy", *History of Science* 24 (1986) 1–28.

Berkes N., *The Development of Secularism in Turkey* (Montreal: 1964).

Bouchard J., "Les relations épistolaires de Nicolas Mavrocordatos avec Jean Le Clerc et William Wake", *O Eranistes* 11 (1974) 62–92.

---

85 Karataş M., "Nazmi-zade Murteza'nın Gülşen-i Hulefa'sının Tenkitli Transkripsiyonu", Ph.D. Dissertation (Atatürk University, Erzurum, 2001). On *Muhtasar-i Mufid*, see Matthee R., "The Safavid-Ottoman Frontier: Iraq-i Arab as Seen by the Safavids", *International Journal of Turkish Studies* 9, 1–2 (2003) 157–173.

Demetracopoulos J.A., "Purchotius Graecus I: Vikentios Damodos's *Concise Ethics*", *Verbum – Analecta Neolatina* XII, 1 (2010) 41–67.

Dobre M. – Tammy N. (eds.), *Cartesian Empiricisms* (New York: 2013).

Erginbaş V., "Enlightenment in Ottoman Context: İbrahim Müteferrika and His Intellectual Landscape", in Roper G. (ed.), *Historical Aspects of Publishing in Languages of the Middle East* (Leiden: 2013) 53–100.

Frazee C.A., *Catholics and Sultans: The Church and the Ottoman Empire, 1453–1923* (Cambridge: 1983).

Göçek F.M., *East Encounters West: France and the Ottoman Empire in the Eighteenth Century* (Oxford: 1987).

Gutas D., "The Study of Arabic Philosophy in the Twentieth Century: An Essay on the Historiography of Arabic Philosophy", *British Journal of Middle Eastern Studies* 29, 1 (2002) 5–25.

Hagen, G., Ein osmanischer Geograph bei der Arbeit: Entstehung und Gedankenwelt von Katib Celebis Gihannüma (Berlin: Klaus Schwarz, 2003).

İbrahim Müteferrika, *Füyuzat-ı Mıknatısiye* (İstanbul: Sultanic Press, 1732).

Israel J.I., *Enlightenment Contested: Philosophy, Modernity, and the Emancipation of Man, 1670–1752* (Oxford: 2006).

Kalaycıoğulları İ., *Katip Çelebi'nin Cihannüma Adlı Eserine İbrahim Müteferrika'nın Yaptığı Ekler Doğrultusunda Çağdaş Bilimlerin Türkiye'ye Girişi*, Unpublished MA Thesis (Ankara University: 2003).

Karahasanoğlu S., *Politics and Governance in the Ottoman Empire: the Rebellion of 1730* (Cambridge, MA: 2009).

Katip Çelebi, *Cihannüma*, ed. and introduced by İbrahim Müteferrika (Istanbul, Sultanic Press: 1732).

Küçük B.H., "Science Studies and Early Modern Ottoman Science", *International Journal of Middle East Studies* 47, 3 (2015) 584–587.

Küçük B.H., "New Medicine and the *Hikmet-i Tabi'iyye* Problematic in Eighteenth-Century Istanbul", in Langermann T. – Morrison R. (eds.), *Texts in Transit in the Medieval Mediterranean* (College Park: 2016) 222–242.

Kurz M., *Ways to Heaven, Gates to Hell: Fazlizade Ali's Struggle with the Diversity of Ottoman Islam* (Berlin: 2011).

Livingstone D.N. – Withers C.W.J. (eds.), *Geography and Revolution* (Chicago: 2005).

Mayaud P.-N., *La condamnation des livres coperniciens et sa revocation* (Rome: 1997).

Menchinger E.L., "Free Will, Predestination and the Fate of the Ottoman Empire", *Journal of the History of Ideas* 77, 3 (2016) 445–466.

Mulsow M., "Socinianism, Islam and the Radical Uses of Arabic Scholarship", *Al-Qantara* 31, 2 (2010) 549–586.

Özervarli M.S., "Arbitrating between Ghazali and the Philosophers: The *Tahafut* Commentaries in the Ottoman Intellectual Context", in Tamer (ed.), *Islam and Rationality* vol. 1, 375–398.
Sabev, O., *Ibrahim Müteferrika ya da Ilk Osmanlı Matbaa Serüveni (1726–1746)* (Istanbul: 2006).
Schmaltz, T.M., "A Tale of Two Condemnations: Two Cartesian Condemnations in seventeenth-Century France", in Del Prete A. (ed.) *Descartes e i suoi Avversari: Incontri cartesiani II*. (Florence: 2004) 203–221.
Tamer G. (ed.), *Islam and Rationality: The Impact of al-Ghazali, Papers Collected on His 900th Anniversary*, 1 vol. (Leiden: 2015).
Tülay Artan, "18. Yüzyıl Başlarında Yönetici Elitin Saltanatın Meşruiyet Arayışına Katılımı", *Toplum ve Bilim* 83 (1999–2000) 292–322.
Varlık N., *Plague and Empire in the Early Modern Mediterranean World: The Ottoman Experience, 1347–1600* (Cambridge: 2015).
Warntz W., "Newton, the Newtonians, and the Geographia Generalis Varenii", *Annals of the Association of American Geographers* 79 (1989) 165–191.
Westman, R. S., *The Copernican Question: Prognostication, Skepticism, and Celestial Order* (Berkeley: UC Press, 2011).

CHAPTER 11

'Now Brought before You in English Habit': An Early Modern Translation of Galileo into English

Iolanda Plescia

> Long-dead translators are usually allowed to rest in peace.
> If their work was bad, it perishes through replacement or disuse; if good, it is preserved not for the sake of the translator but for that of the author. It is therefore customary and proper to say nothing of old translators.
>
> STILLMAN DRAKE, *A Kind Word for Salusbury*[1]

In the first half of the seventeenth century—a time when the divide between the so-called 'two cultures' did not exist as we know it—an Italian scientist remembered for his revolutionary theories embarked on a radical linguistic strategy that would eventually earn him a place as a literary author in his own right: an author, that is, who contributed to the development and enrichment of his country's language, making a lasting impact on its literary system. When Galileo Galilei chose to write science no longer exclusively in Latin, but in his native vernacular, Italian—and the Florentine variety at that, with its authoritative tradition as a literary language—he was hoping to reach a different, wider audience than the erudite university crowd. This broader audience did not necessarily comprehend the *lingua franca* of science at the time, but it was one that had 'eyes' and 'brains', the only organs that were truly required to look into the book of Nature and draw one's own conclusions.[2] Such an openly embraced communicative purpose would enable Galileo to newly articulate his 'discourse' in a truly pragmatic sense, that is, taking into account its intended

1 Drake S., "Galileo Gleanings II: A Kind Word for Salusbury", *Isis* 49, 1 (1958) 26–33 (p. 27).

2 See Galileo's famous letter to Paolo Gualdo (16 June 1612), in *Opere di Galileo Galilei, Edizione Nazionale*, ed. A. Favaro (Florence: 1901) vol. 11, 326–328 (p. 327). On Galileo's use of the vernacular and his relationship to the Italian literature and language, see the recent Battistini A., *Galileo: Profili di storia letteraria* (Bologna: 2011), chapters one and four in particular; and Bolzoni L., "Giochi di prospettiva sui testi: Galileo lettore di poesia", *Galilaeana. Journal of Galilean Studies* IV (2007) 157–175. For a different view, on what she sees as 'Galileo's occasional use of the vernacular', see Eisenstein E., *The Printing Press as an Agent of Change* (Cambridge: 1979) 529.

 | DOI 10.1163/9789004349261_013

audience as well as contexts of enunciation and reception, rather than simply enumerating mathematical laws and philosophical demonstrations. Galileo's decision to approach the most complex of scientific questions in a dialogic format, shaping them in a long-standing literary and philosophical form, attests to his commitment to a communicative practice which, coupled with the use of the vernacular, was to mark the beginning of a fundamental change in the way the educated public would view the relationship between language and things in Europe in the space of a few decades.[3]

In the second half of the same century, a much more obscure man—a tenuous memory of whom survives only thanks to the work of a handful of specialists, including rare book dealers—embarked on a similar project, believing that Englishmen had a right to be able to read the seminal texts of the new philosophy in their own language. This man was, as far as we know, able to make his living in London as a professional translator, and was the author of a remarkable feat, an English-language collection of epoch-making mathematical and scientific texts of extraordinary proportions. This article proposes to add to the picture of scientific translation in early modern England within the context of Anglo-Italian relations, by focusing in particular on one of the points of entry of Galilean science into England, represented by the work of the 'mysterious' Thomas Salusbury, the author of two volumes of *Mathematical Collections and Translations* published in the 1660s in London. The *Collections*, alongside texts by Kepler, Castelli, Foscarini, and others, contain the first published version in English—or indeed in any vernacular language—of Galileo Galilei's *Dialogo dei Massimi Sistemi* (1632).[4]

After a short consideration of Salusbury's life and work, I will move on to some of the paratextual materials of his monumental collection: these have plenty to reveal both about the translator's stance and the perceived status of the English language at the time, in its relationship with Latin and the more prestigious continental vernaculars. Finally, I offer some textual clues to the translator's attitudes toward language and his own role, concentrating not on the English translation of the *Dialogo*, which has received more critical attention, but on Salusbury's rendition of Galilei's *Letter to the Grand Duchess Cristina of Tuscany* (1615). This culturally-charged text proves to be a fertile ground for subtle interventions on the part of the translator, who clearly

3 For a discussion about the development of Dutch as a vernacular language for science, see the article by Charles van de Heuvel in this volume.

4 Drake S., "Galileo in English Literature of the Seventeenth Century", in *Essays on Galileo and the History and Philosophy of Science*, Vol. 1 (Toronto, Buffalo and London: [1967] 1999) 236–252 (p. 247).

aspires to the crucial role of cultural mediator, a skilled intermediary able to negotiate the arrival and comprehension of the texts in England.

'Were [...] I Any Thing but the Translatour': Thomas Salusbury's Life and Work

Ever since the beginning of the period that language historians call early modern English—often symbolically made to coincide with the introduction of the printing press to England by William Caxton (1475)—the language had been undergoing a process of lexical enrichment that made increasing use of external resources, namely borrowings and calques from foreign languages, as a way of adding to the expressive potential of the vernacular, which began at this time to be seen as a medium that might aspire to circulate concepts from all branches of knowledge. The pivotal role played in this respect by the great wave of translation from the classics in the English Renaissance is often emphasized. At the time, however, such a shift in the field of lexical innovation was not universally hailed as an improvement when it entailed importing 'hard words', which, purists pointed out, were obscure in meaning and application, and had no immediately recognizable connection with the vernacular.[5] The problem was far from resolved at the dawn of the age of experimental science in England. Although Latin was still mostly favoured as the vehicular language of choice for philosophical and scientific writing, the number of scientific translations into Latin used as a *lingua franca* slowly began to decrease.[6] Thomas Sprat's famous remarks on the linguistic confusion of his age

5 The debate is known as the 'inkhorn controversy', since long, complex words were often called 'inkhorn terms' in the 16th century, on account of the amount of ink they took to write: such words were considered by a number of influential learned men to be pedantic, but their main objection was to their obscurity. Among those who took a severe stand against the practice of importing opaque foreign words were the classical scholar Sir John Cheke, Roger Ascham, the author of *The Scholemaster*, who had served as the young Elizabeth I's tutor, and Thomas Wilson, author of the *Arte of Rhetorique*. A contrasting party of innovators who defended borrowing included George Pettie, the translator of *Civile Conversation* by Stefano Guazzo, and the eminent Latinist sir Thomas Elyot. For an account of the controversy, see Baugh A.C. – Cable T., *A History of the English language* (5th ed., London: 2002) 203–224.

6 Cf. Kelly L.G., *The True Interpreter: A History of Translation Theory and Practice in the West* (Oxford: 1979) 76. For a classic, compact treatment of lexical development and debates about language from the English Renaissance until the late modern period, see in particular Blake N.F., *A History of the English Language* (Basingstoke: 1996), chapters 7, 8, 9 (173–271), and Baugh A.C. – Cable T., *A History of the English language* (5th ed., London: 2002), chapters 8

testify to the degree to which anxiety about the national language was felt by intellectuals in the early years of the Restoration and at the birth of the Royal Society, and yet no particular language policy was put into place, with efforts to constitute a linguistic academy failing after a few attempts.[7] Sprat's call on 'some sober and judicious Men' to impose some sort of order on the 'mass' of words that—he believes—entered the language as a result of the increased need for communication during the Civil War certainly does not seem to be aimed at translators, who are rather suspected of having contributed to the problem in the first place.[8]

While attitudes towards language in England began to shift towards a lower threshold of tolerance for variation and a new interest in linguistic 'order'—a tendency that would turn into full-blown prescriptivism a few decades later[9]—at the same time the idea that the new philosophy should speak the native language of the people practising it was gaining ground, and a conviction was spreading throughout Europe that 'modern languages were appropriate to conveying a new philosophy'.[10] If on the continent Galileo Galilei's decision

---

and 9 (187–278). As for more specific contributions on the development of specialized varieties of English and scientific writing and discourse, see: Gotti M., *Specialized Discourse: Linguistic Features and Changing Conventions* (Bern: 2003) and *Investigating Specialized Discourse* (Bern: 2008); Taavitsainen I. – Pahta P. (eds.), *Medical and Scientific Writing in Late Medieval English* (Cambridge: 2004); Banks D., *The Development of Scientific Writing* (London: 2008); Skouen T – Stark R.J. (eds.), *Rhetoric and the Early Royal Society* (Leiden: 2015), chapters 1, 2, 3, 4, 8.

7 Skouen – Stark, *Rhetoric and the Early Royal Society* 246–247.

8 'It [English] receiv'd many fantastical Terms [...] and many outlandish Phrases, which several Writers, and Translators, in that great Hurry, brought in, and made free as they pleas'd; and withal it was inlarg'd by many sound and necessary Forms and Idioms [...]. And now, when Men's Minds are somewhat settled [...] if some sober and judicious Men would take the whole Mass of our Language into their Hands, [...] and would set a Mark on the ill Words, correct those which are to be retain'd, admit and establish the good, and make some Emendations in the Accent and Grammar; I dare pronouce, that our Speech would quickly arrive at as much Plenty, as it is capable to receive; and at the greatest Smoothness, which its Derivation from the rough *German* will allow it'. Sprat Thomas, *History of the Royal Society* (London, Printed by T.R. for J. Martin: 1667; facsimile reprint of 1772 ed., New York: 1999) 42.

9 On the early modern age's fundamental acceptance of variation, see Jonathan Hope's illuminating pages in *Shakespeare and Language: Reason, Eloquence, and Artifice in the Renaissance* (London: 2010) 1–39, 98–137.

10 Pantin I., "The role of translations in European scientific exchanges in the sixteenth and seventeenth centuries", in P. Burke P. – Po-chia Hsia R. (eds.), *Cultural Translation in Early Modern Europe* (Cambridge: 2007) 163–179 (166).

to employ his mother tongue aimed at carving out a new and broader readership for his most controversial of works, the work of his translator Thomas Salusbury seems to point to a similarly growing readership in early Restoration London: the address to the *Reader* with which the *Mathematical Collections* are prefaced openly states that the work was undertaken at the request of 'several [...] noble and learned Friends', 'gentlemen' who would support the costs of publication by subscribing to the publishing project.[11] Salusbury mentions financial difficulties and his impossibility to sustain the project by means of 'a private Purse, especially of mine', and proceeds to announce his decision to invite 'those Persons who had appeared desirous of the Book, to be contributary to their own Contentment, by subscribing towards the charge of this Publication'. The translator also refers to his printer, William Leybourn, as an associate who had been attracted by the scope of the book (having himself an interest in mathematics), but especially by the foreseeable profit ('my overtures of profit having interessed his diligence').[12] Though he does allude to some strain in the relationship—due perhaps to a number of printing errors which, as Stillman Drake has shown, were later corrected in a leaflet—everything in the appeal to the reader points to a business alliance from which Salusbury hoped to make an income.[13]

Thomas Salusbury is perhaps only known to a small circle of historians of science and language historians specifically interested in the translation of scientific texts, like Maurizio Gotti in Italy, who has done important work on the translation of Galileo's *Dialogo*.[14] Understandably, however, most of the

---

11 For the practice of translation around the Royal Society see also the chapters by Felicty Henderson, Jan van de Kamp, and Meghan C. Doherty in this volume.

12 Salusbury Thomas, *Mathematical Collections and Translations*, letter to the *Reader* (London, William Leybourn: 1661), fol. *2r. An electronic copy of the 1661 edition of the first tome of the *Mathematical Collections and Translations* is available on Early English Books Online (http://eebo.chadwyck.com), as well as the website of the Archimedes Project Digital Research Library (http://archimedes.mpiwg-berlin.mpg.de/cgi-bin/toc/toc.cgi?step=thumb&dir=salus_mathe_040_en_1667, last accessed 18.02.2017).

13 Drake, "A Kind Word for Salusbury" 26–27.

14 Along with the studies previously mentioned (in n. 6), see: Gotti M., "Lexical Choices in an Early Galilean Translation", in Coleman J. – Kay C.J. (eds.), *Lexicology, Semantics and Lexicography: Selected papers from the Fourth G.L. Brook Symposium* (Amsterdam – Philadelphia: 2000) 87–101; idem, "La traduzione di Thomas Salusbury della terminologia specialistica di Galileo", in Bernard M. – Rota I. – Bianchi M. (eds.), *Vivir es ver volver: Studi in onore di Gabriele Morelli* (Bergamo: 2009) 277–286. I would like to gratefully acknowledge that my own work on Salusbury was initially funded by a project coordinated by Maria Del Sapio Garbero (Roma Tre University) within a Socrates Acume2 program ('Interfacing Sciences, Literatures and the Humanities'). Cf. Plescia I., "'Strangers to Our

attention paid to Salusbury has come from scholars working in the first field, the history of science, most notably the already mentioned Stillman Drake, eminent Galileo scholar, and more recently Nick Wilding. The known facts of Salusbury's life are scarce and have mainly been pieced together by Drake in his introduction to the 1967 facsimile edition of the *Mathematical Collections and Translations*.[15] Salusbury seems to have been a royalist, born in Wales between 1620 and 1630, who left England during the Civil War, spent several years in France and Italy, and returned to London to earn his living as a writer and translator.[16] Only twelve letters signed by Salusbury have been discovered (in 1959, by the rare books dealer Jacob Zeitlin), but from them it seems possible to identify him with a Salusbury mentioned in Samuel Pepys' diary, married to a Susanna Birkenhead, and apparently close to the King.[17] This corroborates the story of his having had to flee to the Continent for several years during the Civil War as a consequence of his royalist sympathies, as well as his need to seek patronage and an occupation upon his return, perhaps with the hope of attracting the attention of some of the Royal Society Fellows.

Salusbury as Translator

While Latin was known and used by the members of the Royal Society, a good amount of translation work from continental vernaculars into English was done within and for the Society from its early years onwards, especially with regard to papers, letters and books received by its members.[18] Salusbury dedicated his work to John Denham, not yet a Fellow of the Society in 1661, but

Nation': Anglo-Italian relations and linguistic encounters in two early modern scientific translations", *Textus* 24, 3 (2011) 559–578.

15 Drake S. (ed.), *Mathematical collections and translations: in two tomes, London, 1661 and 1665 / by Thomas Salusbury; in facsimile with an analytical and bio-bibliographical introduction* (London – Los Angeles: 1967).

16 Nick Wilding has noted that 'there is no trace of him in the papers of the Royal Society'. Drawing on Drake, Wilding suggests he may have been 'socially excluded from certain sectors of the Republic of letters' (for reasons perhaps linked to his possible illegitimate status having been born out of wedlock), and considers his work to have been produced 'on the margins of the newly established Royal Society'. Wilding N., "The return of Thomas Salusbury's *Life of Galileo* (1664)", *British Journal for the History of Science* 41, 2 (2008) 241–265 (esp. 255 and 242).

17 Zeitlin J., "Thomas Salusbury discovered", *Isis* 50 (1959) 455–458.

18 For a recent and insightful review of the topic, see Felicity Henderson's "Faithful Interpreters? Translation theory and practice at the early Royal Society", *Notes and Records of the Royal Society* 67 (2013) 101–122.

a well-known poet who had, like Salusbury, suffered for his royalism, and had later been appointed Surveyor general by the King: 'To the Noble and most perfectly Accomplished *Sir John Denham*, Knight of the Noble Order of the Bath, and Surveyor General of his Majesties' Works'. Denham also had a strong interest in translation: he openly advocated for freedom of style in the case of literary translation (both in a poem written in praise of Richard Fanshawe's rendition of Guarini's *Pastor Fido*, and in his preface to his *Destruction of Troy*), and he was not the only intellectual to be admitted to the Royal Society who had reflected on issues of 'faithfulness' and linguistic practice: Abraham Cowley, Edward Sheburne, and of course John Dryden all had in common their status as 'prominent poets' as well as their being associated with the Society itself.[19] Cowley, Dryden and Denham were all Fellows—Denham himself was accepted into the Society in May 1663.

The first volume of Salubsury's *Mathematical Collections* was published in 1661. Its dedication has a clear goal, aside from the obvious one of choosing a credible patron for the book: to establish the importance of making science understandable in English, and underline the novelty of such an enterprise, while also advertising the names of the eminent scientists translated. If on the one hand Salusbury's letter to the Reader suggests the existence of a readership wide enough to be able to generate a profit through subscriptions, in his appeal to Denham the translator is also actively seeking the patronage of a man who had both literary inclinations and a connection with a more specialized readership:

> SIR,
> I humbly begge your Pardon for bringing this Book under your Protection. Were it a Work of my own, or I any thing but the Translatour, I should master my Thoughts to a meaner Dedication; But being a Collection of some of the greatest Masters in the World, and never made English till now, I conceived I might sooner procure their Welcome to a person so eminent for Noble Candor, as well as for all those Intellectual Excellencies wherewith Your Rich Soul is known to be furnished. I resolv'd to be as kind to this Book as I could, and [...] I at last concluded to prefix Your Name, whom His Majesty and all his Subjects, (who have a higher Sense and Judgment of Excellent Parts) know best able to defend my Imperfections. [...] therefore *Galileus, Kepler*, and those other Worthies in Learning are now brought before You in English Habit, having chang'd their Latine,

19 Henderson, "Faithful Interpreters?" 104–106.

> Italian and French, whereby they were almost Strangers to our Nation, unless to such as You, who so perfectly master the Originals.[20]

Whether or not Salusbury aspired to become a Royal Society Fellow himself is perhaps of less importance than his impassioned defence of the idea that the central tenets of the new science needed to be made available in English.

Scholars who have dealt with his work have considered Salusbury a 'faithful' translator, with the notable exception of Giorgio de Santillana, whose unfavourable opinion was immediately challenged by Drake. It is important to point out here that such a debate on what constitutes a 'faithful/unfaithful', 'good/bad', 'reliable' or 'obscure' translation is, understandably, heavily influenced by their perspective as historians of science—it could find no place, for example, in contemporary translation studies.[21] A more formal, linguistic study has been carried out on parts of Salusbury's Galilean translations by Maurizio Gotti, as mentioned above.[22] In his examination of a sample of the dialogue, Gotti also identifies 'faithfulness' as a major feature of Salusbury's lexical choices, 'not only in rendering the single lexemes, but in reproducing the whole of the original text': meaning that the translator takes care to use direct equivalents of Galileo's specialized terms and does not omit parts of the text, aware as he is of the literary quality of Galileo's writing.[23] In this case, Gotti's use of the notions of 'faithfulness' and 'equivalence' does not constitute a value judgement on the translation, but is to be read in the context of a diachronic linguistic analysis that aims to consider the ways in which specific problems of vocabulary were tackled by the translator. His conclusion is that the translation is characterized by 'clarity and precision' – the linguistic tenets of the Royal Society—and that Salusbury kept on his quest for exactness mainly by paraphrasing, coining new words through borrowing and calquing, and making the most of semantic innovation.[24]

---

20 Salusbury Thomas, *Mathematical Collections and Translations*, dedication page.

21 Drake's article "Galileo Gleanings II: A Kind Word for Salusbury" is in fact a response to Giorgio De Santillana's attack on Salusbury's work in his own revised edition of the 17th-century translation of the *Dialogue on the Great World Systems* (Chicago: 1953).

22 See notes 5 and 12.

23 Gotti, "Lexical Choices in an Early Galilean Translation" 91.

24 Gotti, "Lexical Choices in an Early Galilean Translation" 92–95. Although A.C. Crombie has claimed that the translation of the *Dialogo* was mainly based on the Latin edition, in *Styles of Scientific Thinking in the European Tradition*, 2 (London: 1994) 867, the general assumption, as well as my own, is that Salusbury did indeed translate from the original Italian, as stated on the title page of his *Mathematical Collections*; Salusbury specifically

My own perspective here, which draws on historical linguistics but is also concerned with painting a portrait of a specific translator, connects Salusbury's paratextual materials with his translations in order to speculate on the motivations and drive that led him to carry out such an impressive amount of work. Identifying Salusbury as a faithful translator concerned with his source text does not mean that he can be considered a wholly 'invisible' translator, to use Lawrence Venuti's important category somewhat broadly.[25] Salusbury has no ambition to produce a foreignizing or challenging translation, of course, nor does he wish to resist the call for transparency that Venuti identifies as a key feature in the history of Western (literary) translations: the cultural discourse on scientific expression of his age placed great emphasis on the importance of clear and simple prose. He is, however, particularly eager to claim a place in the history of translation and scientific transmission. If he respects the original texts in length and content, avoiding manipulation or summarization, he is also not shy about proclaiming the novelty and difficulty of his task, nor of sometimes signalling his presence within the text, for example by means of marginal glosses, which are used as modern textual notes and provide source quotations as well as clarifications. But the most innovative aspect of Salusbury's work is not so much his style in translating individual texts, as the role he fashions for himself as a selector and collector of knowledge: the translated texts are set in dialogue with one another, re-igniting a debate on the independence of the scientific method, across time and space.

The *Mathematical Collections and Translations*, 1661–1665

The *Mathematical Collections and Translations* were published between 1661 and 1665 and are divided in two volumes, each consisting of two separate parts [Fig. 11.1]. Tome I, part 1 contains Galileo's *System of the World* and his *Letter to the Grand-duchess Christina of Tuscany*, followed by extracts from Kepler, Diego de Zúñiga, the Spanish philosopher and Augustinian author of a reconciliation of the Copernican theory with the Scriptures, and Paolo Antonio Foscarini's letter to Father Fantonus (Sebastiano Fantoni), again an attempt at a compromise between Scriptures and heliocentrism: in fact, all of these materials are defined as 'reconcilings' of the sacred texts and the 'doctrine of the Earth's mobility', and thus may be regarded as important companion texts to

states in his letter to the Reader that he has corrected a number of mistakes in Bernegger's Latin version, and that his new translation is thus a necessary improvement.

25 Venuti L., *The Translator's Invisibility: A History of Translation* (London – New York: 2008).

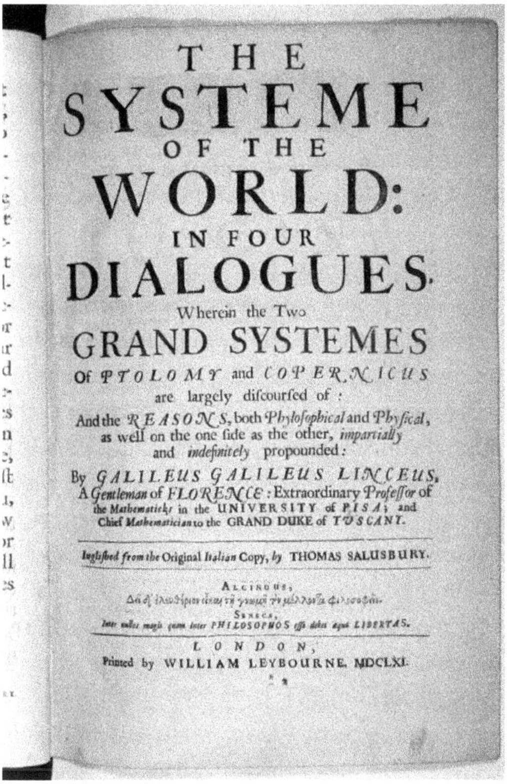
THE
SYSTEME
OF THE
WORLD:
IN FOUR
DIALOGUES.
Wherein the Two
GRAND SYSTEMES
Of PTOLOMY and COPERNICUS
are largely discoursed of:
And the REASONS, both Phylosophical and Physical,
as well on the one side as the other, impartially
and indefinitely propounded:
By GALILEUS GALILEUS LINCEUS,
A Gentleman of FLORENCE: Extraordinary Professor of
the Mathematicks in the UNIVERSITY of PISA; and
Chief Mathematician to the GRAND DUKE of TUSCANY.

Inglished from the Original Italian Copy, by THOMAS SALUSBURY.

ALCINOUS,
[illegible]
SENECA,
Inter nullos magis quam inter PHILOSOPHOS esse debet æqua LIBERTAS.

LONDON,
Printed by WILLIAM LEYBOURNE. MDCLXI.

FIGURE 11.1
Title page of *The Systeme of the World*, translated by Thomas Salusbury in Thomas Salusbury, *Mathematical Collections and Translations* (London, William Leybourn: 1661). Oxford, Bodleian Libraries.

Galileo's *Dialogo*. The second part of the first tome is almost entirely devoted to the writings of Benedetto Castelli (among which his *On the Menstruation of Running Waters*).

The selection of texts which make up the table of contents of the first tome clearly responds to an intent to showcase Galileo's work. The Italian scientist's writings are foregrounded by the mere fact of their being placed at the top of the list; in a 1667 edition of the book, Galileo's name is even more conspicuous, printed in large letters on the title page, which simply states: 'Mathematical Collections and Translations, in Two Parts. From the Original Copies of Galileus, and other Famous Modern Authors'. All of the texts in this first tome can be seen as creating a dialogic interchange: Galileo's *Letter to the Grand-duchess Christina* immediately follows his *System of the World*, and quite rightly so, as they were most probably disseminated outside of Italy together or in close succession. In fact, although the *Letter* had been written much earlier than the *Dialogo*, around 1615, it was not immediately printed and its existence was only mentioned in Galileo's correspondence; there is no mention of the *Letter* after the first years of its composition up until 1632. It re-emerged only after the *Dialogue* started to be circulated in Latin thanks to Matthias Bernegger's 1635 translation; and it was then, in 1636, also translated into Latin by Elia Diodati. This points to a possible circulation of the *Letter* as an accompanying text to the *Dialogue*, as indeed it was from a thematic point of view, since it tackles the theological questions raised by the Copernican theory that Galileo espoused in his work.

The texts that follow in the first tome are also ones that respond to the burning issues of the day: a selection from Johannes Kepler's Commentaries on Mars, or *Astronomia Nova* (1609), concentrates specifically on his discussion of the Scriptures; Diego de Zúñiga's *Commentary on Job* (1584) sought to achieve the same harmony between heliocentric theory and Scripture. A similar spirit animated Foscarini's letter to Father Fantonus—and Foscarini partly shared the fate of Galileo, for his 1615 booklet was banned by the Inquisition. The presence of Benedetto Castelli's work and letters can be explained in relation to Galileo, since Castelli had been his student and had participated in his teacher's experiments. By constructing the first tome as a collection of influential and learned voices dealing with the relation of theology to science, Salusbury offered his readers a remarkably large anthology of anti-Ptolemaic theory, as a way of backing up the authority of his central figure.

The first part of the second tome, of which very few copies have reached us, is devoted to technical treatises and also seems to be constructed as a companion to the main name in the table of contents. It contains Galileo's *Mathematical Discourses* and his *Mechanics* as well as his *Discourse of the Things that Move in or Upon the Water*, and once again accompanying texts are carefully selected to respond to or complement these works—for example, Descartes' *Mechanics*, Archimedes' *De incidentibus in fluido* with Niccolò Tartaglia's annotated commentary, as well as Tartaglia's own studies on water.[26] The contents of the second part of the second tome have only recently been described by Nick Wilding: the book had long been considered lost, until the only surviving and imperfect copy briefly resurfaced at the auction of the library of the Earls of Macclesfield at Shiburn Castle (2004–07), before being sold to a private collector. The book not only included Evangelista Torricelli's *Doctrine of Projects*, but also what Salusbury declared to be his own *Experiments of the Comparative Gravity of Bodies in the Aire and Water*, and, more excitingly, the first 'substantial' biography of Galileo Galilei published in any language, *Galilaeus Galilaeus*

26 The Latin title of Archimedes' *On floating bodies* is given by Salusbury in the table of contents.

His Life: In Five Books.[27] Indeed, the second tome seems to be entirely devised to set the stage for the chief attraction of the entire collection, as the very detailed table of contents attests. Readers are promised a thorough account of Galileo's life and education, his 'judgment in several Learnings', his 'opinions and doctrine', his 'manner of living', his 'morall Virtues', and by way of conclusion, 'certain Reflections upon his whole Life'. It is a great pity that the only known copy of this work is unavailable to scholars, since it is one in which it would have been possible to listen to Salusbury's own voice as he abandons his role as a translator to become a fully-fledged author.

Despite the remarkable scope of the collection, it is impossible to know whether Salusbury's work made a significant contribution to the dissemination of Galileo's ideas, and his *Dialogo* in particular, in England. The fact that such a limited number of copies of the collection have survived—it is generally thought that the bulk of published copies, especially of the second tome, perished in the Great Fire of London—would seem to testify to the contrary.[28] Drake argued in his survey of references to Galileo in seventeenth-century English literature that 'the English public was more responsive to the discoveries and opinions of Galileo [...] than is generally recognized', and that 'Galileo was more favourably received [...] by Englishmen than by men of any other nation outside of Italy', but he also acknowledges that there are not many references to the translation to be found in records mentioning Galileo in the years following its publication.[29] As for earlier work, Salusbury, who hoped to create interest by advertising the book as the first of its kind, may not have known of a previous, unpublished translation of the *Dialogue*, which Drake attributed to Joseph Webbe, a theorist of language instruction as well as a physician who had done his training in Padua.[30] This assumption is generally accepted, and the ECHO (European Cultural Heritage Online) website credits its online copy of the surviving British Library MS Harley 6320 to Webbe's name.[31] A Latin version of the *Dialogo* had already been produced by Matthias Bernegger and was printed at Leiden in 1635, and indeed Drake sees this as a possible explanation for the

27 Wilding, "The return of Thomas Salusbury's *Life of Galileo*" 241. It is in this article that Wilding describes the discovery of the only surviving copy of the biography, the contents of the volume, and the complex historical context that occasioned its production. I have here followed Salusbury's description of his work as consisting of two 'tomes' divided in two 'parts'.

28 Wilding, "The return of Thomas Salusbury's *Life of Galileo*" 245–250.

29 Drake, "Galileo in English Literature of the Seventeenth Century" 236–237; 250–251.

30 Drake, "Galileo in English Literature of the Seventeenth Century" 241.

31 See http://echo.mpiwg-berlin.mpg.de/ECHOdocuView?url=/permanent/library/YPQQ3K6M/index.meta, last accessed 18.2.17.

fact that the previous English translation had not seen the light of day in the 1630s, having become unnecessary. However, there is no certain evidence that Webbe's work had been intended for publication; it does not bear his mark or signature and appears to have been produced on commission for the private use of William Cavendish, Earl of Newcastle.[32] Also, interestingly, the Latin version of the *Dialogo* was re-published in London by Thomas Dicas in 1663, almost immediately *after* Thomas Salubsury's vernacular translation. The dates confirm, at the very least, an interest in Galileo in 1660s London, and the existence of a novel demand for his books. By the time the Royal Society and a new culture of experimentation had emerged, Galileo deserved, almost demanded, an English translation—but what must be stressed is that the linguistic culture had also changed, as debates emerged on the suitability of the English vernacular for scientific expression. It is significant that a mere thirty years after Bernegger's Latin version, a new English translation could be produced independently, with the support of an audience of subscribers, and that the translator entrusted this endeavour with the task of leaving his mark on the world.[33]

If the extent of the collection's circulation in the seventeenth century is uncertain, Salusbury's work nevertheless constitutes an interesting case in historical translation studies, in so far as it is a powerfully intertextual undertaking, through which a proposal for a European canon of modern philosophical and scientific writing is constructed. The translator's decision to close the collection with his own *Experiments* and *Life of Galileo* is equally telling of his ambitions, as he transcends his chosen role as mediator in the end and takes up his pen as an author whose *auctoritas* is mainly erected upon the illustrious company he has chosen to keep.[34]

---

32 The Webbe connection has been made mostly thanks to a reference in a letter from Hobbes to Cavendish, in which Hobbes expressed the hope that the ongoing translation would be finished soon (24 January 1634). On the letter and the probable private nature of the commission made by Cavendish, who 'had no concern for the dissemination of new knowledge', see Raylor T., "William Cavendish, Galileo, Hobbes and the Mechanical Philosophy", in Edwards P. – Graham E. (eds.), *Authority, Authorship and Aristocratic Identity in Seventeenth-century England* (Leiden – Boston: 2016) 183–186; see also Malcom N., *Aspects of Hobbes* (Oxford: 2002) 89.

33 Salusbury has been described as 'a key figure in the English exploitation of the publicity value of Galileo's being on the Index', in Eisenstein, *The Printing Press as an Agent of Change* 677.

34 Whether or not Salusbury was the sole translator of the entire body of work, or the actual author of the *Life of Galileo*—a fact which has been called into question (see, for example, Drake, "Galileo in English Literature of the Seventeenth Century" 249, as well as Wilding's extremely detailed history of the volume)—does not, in this sense, take away from his

The inadequate amount of information on the translation's circulation does remain a stumbling block, for it also makes it hard to say whether the collection had a direct influence on the development of scientific English.[35] However, looking at the endeavour from the perspective of the early modern shift in linguistic attitudes in England, the collection adds an important testimony to the growing belief in the possibilities of English as a language of science. Although Salusbury does not openly assert any principles for a theory of translation, his paratextual materials point to his sharp metalinguistic awareness as a translator. His proclamations on the novelty of his enterprise are certainly to be read within the norms of advertisement strategy, but the translator is clearly also taking great pride in introducing the English public to a figure of such intellectual standing as Galileo. Using the customary clothing metaphor, so popular among early modern intellectuals reflecting on translation, Salusbury promises that though the dress has been changed ('now brought before You in English Habit, having chang'd their Latine, Italian and French'), the semantic content of the texts has been reproduced for the benefit of his own people:[36]

> Mathematical learning [...] hath bin so sparingly imparted to our Countrymen in their *native English* [...] that in Compliance with [...] the Inclinations of such as are Mathematically disposed, more especially *those, who either want Time or Patience to look into the vulgar or unstudied Languages*, I did adventure upon this Work of Collecting and Translating [...].[37]

In the same letter to the Reader, Salusbury further states that he has been 'careful in contriving a *pleasant* and *beautiful* Impression' (my emphasis), in some way echoing the classical 'docere' and 'delectare' tenets;[38] he emphasizes that in undertaking his task, he has tried to respond to the current demand for English versions of texts dealing with mathematics, astronomy and natural philosophy. Indeed, Salusbury expects that his translations will have

---

will to fashion his authorial voice as mainly that of a translator with some ambition to add to the scientific debate.

35 Gotti, *Specialized Discourse: Linguistic Features and Changing Conventions* 205–206.

36 See, for example, Morini M., *Tudor Translation in Theory and Practice* (Farnham – Burlington: 2006) 36–37.

37 Salusbury T., "Letter to the Reader", in *Mathematical Collections and Translations*, fol. * 2 (my emphasis).

38 Cicero is the most famous proponent of such qualities as belonging to the good orator; see his *De Optimo Genere Oratorum*, in *On Invention. The Best Kind of Orator. Topics*, trans. H.M. Hubbell, Loeb Classical Library 386 (Cambridge, MA: 1949) 357.

something to add to his nation's culture from a specifically linguistic point of view: to the point that he openly assumes personal responsibility for having made a selection, based on 'those Authors and Treatices which *I judged* would most grace our Language' (my emphasis). Paradoxically, at the same time the translator also embraces the traditional rhetoric of humility that is expected of him: after all, what he is presenting are 'mere' translations—'Translations I own them to be, as not pretending to any thing more than the disposure and conversion of them'. But 'disposure' is here a key term, for it is indeed thanks to the idea of *dispositio*—again a significant allusion to classical rhetoric—that Salusbury is able to create an original product, whose macrostructure functions as a co-text that contextualizes and enables a correct interpretation of the main Galilean translation of the *Dialogue*. Seizing the opportunity to offer a summa of scientific thought to his contemporaries, the textual selector, collector and translator feels that the risks posed by the inherent 'disadvantages of Translations' are well worth running.

It is this ability that makes Thomas Salusbury a true cultural mediator between the British Isles and the ongoing theological debates on the new science in continental Europe. With nationalistic satisfaction Salusbury envisions his foreign guest as being finally set free: 'the English will be more hospitable, on the account of that Principle which induceth them to be civil to (I say not to dote on) Strangers'. Both the notion of Salusbury's 'faithfulness', and his casting of himself as a humble translator who has simply taken on a necessary task, need to be reconsidered.

The Letter to the Grand-duchess Christina

A detailed linguistic analysis of Salusbury's work is beyond the scope of this article, and the sheer proportions of the available textual material make even a preliminary survey of all the translation strategies employed a lengthy task, which must begin with a completely searchable electronic text. However, even a quick look at a more compact textual sample, such as Salusbury's translation of Galileo's *Letter to the Grand Duchess Cristina of Tuscany*,[39] has much to reveal about his stance as a translator. The letter, originally written in 1615, is itself modelled on the topic of an earlier letter that Galileo had addressed to Benedetto Castelli, and was not published by the scientist but circulated in scientific circles in manuscript form (it would only be published in Strasbourg in

39 The full title in Salusbury's table of contents is *Epistle to Her Serene Highnesse Christiana Lotheringa Grand Dutchesse of Tuscany*.

1636, both in Italian and in Latin translation). It is actually a reply to Cardinal Bellarmine's anti-Copernican position, expounded in a 1615 letter to Paolo Antonio Foscarini; Galileo's response was addressed to the Grand-duchess Christina as an attempt to explain more fully how the new science did not, in fact, clash with Scriptural passages, since the Bible's main aim is, as he says, certainly not to teach astronomy. It is this letter which contains the famous assertion thus translated by Salusbury: 'The intention of the Holy Ghost, is to teach us how we shall go to Heaven, and not how heaven goeth'.[40]

What scholars have referred to as Salubsury's fundamental 'faithfulness' is noticeable in this text, especially with respect to choices in vocabulary. Salusbury's translation of Galileo's main proposition is a good example of his attention in reproducing close lexical equivalents that echo the Latinate roots of the language he is dealing with:

> io ne' miei studi d'astronomia e di filosofia tengo, circa alla costituzione delle parti del mondo, che il Sole, senza mutar luogo, resti situato nel centro delle conversioni de gli orbi celesti, e che la Terra, convertibile in se stessa, se gli muova intorno.[41]

> I in my Studies of Astronomy and Philosophy hold, as to the Worlds Systeme, that the Sun, without changing place, is situate in the Centre of the Conversion of the Celestial Orbes; and that the Earth, convertible about its own Axis, moveth it Self about the Sun.[42]

The translator proves to be very capable of condensing text into a more effective expression in English when need be, reworking the literal sense of the words—so that 'la costituzione delle parti del mondo', that is, 'the constitution of the several parts of the world', becomes, more succinctly and efficiently, 'the Worlds Systeme'—a keyword since it is also the chosen translation for the title of the *Dialogo*. In light of this, Salusbury's decision to stick to close Latinate equivalents in vocabulary whenever possible points to a conscious effort to adhere to the source text.

---

40 For an exhaustive and updated review of the sources and publication history of the letter, see Ottavio Besomi's Introduction to his recent critical edition of the text: Besomi O. (ed.), *Lettera a Cristina di Lorena* (Rome – Padua: 2012). All the textual references to the original version that follow are to this edition.

41 Besomi (ed.), *Lettera a Cristina di Lorena* 37.

42 Salusbury, *Mathematical Collections* 428.

However, the strongly persuasive communicative function that prevails in this particular text-type evidently affords the translator some freedom. Although the text is, understandably, marked by elements of the epistolary style, in the Republic of letters it would have functioned as a proper treatise: Galileo quotes authorities, lays out his hypothesis, and demonstrates that it does not contradict a correct religious interpretation. While a line-by-line check confirms the general impression of Salusbury as an extremely careful translator, one who tends not to omit or summarize portions of text, and one who contrives to render unusual expressions and technical language as closely as possible in English, there are a few instances of subtle but significant alterations and adaptations that mostly have to do with the argumentative structure of the text.

One example is the way Salusbury deals with hedging expressions, that is, the linguistic devices that enable the writer to make statements that allow for tentativeness and the possibility of alternative explanations. The phenomenon, while used in a variety of communicative settings, is widely observed in scientific writing, as it enables authors to build an effective rhetorical strategy aimed at gaining acceptance by apparently mitigating their claims.[43] In some cases, Salusbury seems to play down instances of 'epistemic' modality—that is, those modal verbs and expressions that are concerned with possibility and the way the world is thought or perceived to be—in favour of a greater degree of assertiveness.[44] In one instance where the author inserts a hedging adverb, such as 'veramente' ('in truth', 'truly'), the translator uses an emphatic 'I protest':

> De' quali io veramente non farei maggiore stima di quel ch'io m'abbia fatto dell'altre contradizioni.[45]

> Of whom I protest that I would make no more account than I have done of those who heretofore have contradicted me.[46]

Galileo is referring to those among his opponents who, while not openly contradicting him, remained silent on the scientific debate, choosing to discredit

43 See Hyland K., "Talking to the Academy: Forms of Hedging in Science Research Articles", *Written Communication* 13, 2 (1996) 251–281, and Hyland K., *Hedging in Scientific Research Articles* (Amsterdam: 1998), for in depth introductions to hedging in contemporary scientific language.

44 For a classic introduction to linguistic modality, see Palmer F.R. *Mood and Modality* (Cambridge: 2001).

45 Besomi (ed.), *Lettera a Cristina di Lorena* 35.

46 Salusbury, *Mathematical Collections* 428.

him and damage his reputation in other ways. Though expressing roughly the same pragmatic value as *veramente*, the choice of 'I protest' allows the translator to re-iterate the first-person pronoun a second time, strengthening the force of the statement. In other instances, an impersonal form, such as 'pare' ('it appears that'), is rendered with a first person pronoun and verb, 'I conceive', which again strengthens the statement and also points to the authoritativeness of the writer.[47]

Such a tendency to increase emphasis by degrees is discernible in various passages of the translated text, though the final effect is often subtle and takes some close reading to be spotted. In the following sentence, in which Galileo states that any person through careful observation would be persuaded of the new discoveries, the translator has added an adverb:

> quelle novità, delle quali il senso stesso, quando avessero volute con attenzion riguardarle, gli avrebbe potuti render sicuri.[48]

> those Novelties; of which their very sense, had they but pleased to have intently beheld them, would have rendered them *thorowly* assured.[49]

Perhaps the most striking modal feature of the translated text is the way in which the various instances of the Italian verbs expressing obligation and necessity are articulated in English. The variety of forms used by Galileo might call for different renditions of the central modals *should, must*, and *ought to*, but in almost all of these instances Salusbury selects *ought to*, with some cases of the impersonal Italian form 'si deve' translated into the first person plural pronoun, as in 'we ought to'. Although 'we' can also convey an impersonal colouring, and is a natural sounding choice in English, Salusbury's insistent use of the first person plural pronoun inevitably carries with it an inclusive effect, calling upon the scientific community as a whole for action in implementing the methods of the new science:

> nelle dispute de' problemi naturali non si dovrebbe cominciare dall'autorità di luoghi delle Scritture, ma dalle sensate esperienze e dalle dimostrazioni necessarie.[50]

---

47 Salusbury, *Mathematical Collections* 433.

48 Besomi (ed.), *Lettera a Cristina di Lorena* 34.

49 Salusbury, *Mathematical Collections* 427. (my emphasis).

50 Besomi (ed.), *Lettera a Cristina di Lorena* 49.

> we ought not to begin at the authority of places of Scripture; but at Sensible Experiments and necessary Demonstrations.[51]

In a number of situations, it may be said that Salusbury is actually adapting the text rather than merely translating it, and doing so with a clear cultural goal in mind. Certain deliberate changes in verbal aspect are particularly interesting, such as the one which presents the scientific hypotheses that Galileo is putting forward as irrevocably proven facts: 'mentre si va scoprendo …' ('it is now being discovered …')[52] becomes 'it is now proved'.[53] In this particular case the lapse of time that has passed between the first composition of the letter (1615), its appearance in print (1636), and the 1660s translation may have a role to play—Salusbury may wish to stress that the passing of time and the development of experimental science have further clarified Galileo's positions; he is able to make a finite statement because he is writing in a place and time in which an established culture of experimentation has proven Galileo right. Instead of translating literally, considering the Italian text as situated in its own historical time and perhaps clarifying in a gloss as he does at other points, Salusbury chooses here to overstep his boundaries as translator in order to update the status of the debate.

In other instances, the translator's treatment of specific keywords is particularly telling—especially because, as we have seen, he generally tends to closely reproduce the roots of the Italian text in matters of vocabulary. For example, he translates 'animi', meaning something close to 'dispositions, inclinations', as 'humours', a word well-suited to an English scientific audience, which would have been familiar with the residual paradigm of humoral theory which had dominated early modern medical thought.[54] He uses a textual gloss to explain his use of the word 'curtain' in translating a biblical expression referring to the Heavens, so that his audience may recognize the more familiar English version of the Bible.[55] In another passage, in which Galileo refers to the possibility that his theories will be considered heretical by the authorities, the term 'dichiarazione' ('declaration'), is translated as 'censure': a much stronger word, which we find used in seventeenth-century English in relation to ecclesiastical law, and which reflects the unfolding of events surrounding Galileo's work

51 Salusbury, *Mathematical Collections* 433.
52 Besomi (ed.), *Lettera a Cristina di Lorena* 41.
53 Salusbury, *Mathematical Collections* 430.
54 Salusbury, *Mathematical Collections* 428.
55 Salusbury, *Mathematical Collections* 445.

more forcefully.[56] In general, Salusbury seems not to mince words when dealing with religious controversy, clearly adapting the text to his Protestant audience, who would not have looked upon the Catholic Church with any favour. 'Un simulato zelo di religione', literally a 'feigned' religious zeal, becomes outright 'Hipocritical'; the word 'damnable', which translates 'dannanda', is later swapped for 'Heretical'.[57]

I want to suggest—though I cannot make a definitive claim at this stage—that Salusbury is deliberately employing linguistic modality and certain domains of vocabulary to express a stronger, more emphatic mode of communication that is in keeping with the translator's conviction that the scientific culture of his contemporary England will more readily accept Galileo's arguments. The solutions the translator proposes are contextually and pragmatically adequate to the new language, but subtle shifts in meaning do occur. I am currently undertaking a larger project of systematic analysis of the text against the rest of the Galilean translations in the collection, as well as building a reference corpus made up of other coeval scientific translations to assess whether a generally consistent translation style or strategy can be seen at play. What this article has aimed to show is that the obscure Thomas Salusbury was far from a mere purveyor of scientific texts for his contemporaries, but a translator with a confident stance and clear awareness of his role.

Since the 1950s, when Stillman Drake, in defending Salusbury against Giorgio De Santillana's accusations of unfaithfulness to the original text, expressed the view that 'long-dead translators' are best left alone, the development of translation studies as a discipline in its own right, along with the cultural turn within the discipline itself, has authorized scholars to take an interest in such neglected figures. We now take stock of their work *not* only for the sake of the authors, but also for the translators' own sakes and the sake of a broader history of culture that can no longer ignore their contribution. The translator's is a complex, hybrid identity that he may at times even wish to cast off, but which he can never fully renounce—as, for instance, when Salusbury finds himself apologizing for having to include a passage in the *Letter* that shows perhaps too much deference to the Pope for his taste. As he does in other places, he chooses to add a gloss to the margin of the text and paradoxically draws even more attention to himself with this self-effacing comment: 'If this passage seem harsh, the Reader must remember that I do but Translate'.[58] The work of Thomas Salusbury, shrouded in mystery though his life may be, is not only

---

56 Salusbury, *Mathematical Collections* 429.

57 Salusbury, *Mathematical Collections* 430, 429, 432.

58 Salusbury, *Mathematical Collections* 455.

a testament to a rising confidence in the possibilities of the English language, but also a vindication of the role of the translator as one that carries its own special form of authority.

Selective Bibliography

Banks D., *The Development of Scientific Writing* (London: 2008).

Battistini A., *Galileo. Profili di storia letteraria* (Bologna: 2011).

Baugh A.C. – Cable T., *A History of the English language* (5th ed., London: 2002).

Besomi O. (ed.), *Lettera a Cristina di Lorena* (Rome – Padua: 2012).

Blake N.F., *A History of the English Language* (Basingstoke: 1996).

Bolzoni L., "Giochi di prospettiva sui testi: Galileo lettore di poesia", *Galilaeana. Journal of Galilean Studies* IV (2007) 157–175.

Cicero, *De Optimo Genere Oratorum*, in *On Invention. The Best Kind of Orator. Topics*, trans. H.M. Hubbell, Loeb Classical Library 386 (Cambridge, MA: 1949).

Crombie A.C., *Styles of Scientific Thinking in the European Tradition*, 3 vols. (London: 1994), vol. 2.

De Santillana G. (ed.), *Dialogue on the Great World Systems, in the Salusbury Translation* (Chicago: 1953).

Drake S., "Galileo Gleanings II: A Kind Word for Salusbury", *Isis* 49, 1 (1958) 26–33.

Drake S. (ed.), *Mathematical collections and translations: in two tomes, London, 1661 and 1665 / by Thomas Salusbury; in facsimile with an analytical and bio-bibliographical introduction* (London – Los Angeles: 1967).

Drake S., "Galileo in English Literature of the Seventeenth Century", in Drake S., *Essays on Galileo and the History and Philosophy of Science*, ed. N.M. Swerdlow – T.H. Levere, 3 vols. (Toronto, Buffalo and London: [1967] 1999), vol. 1, 236–252.

Eisenstein E., *The Printing Press as an Agent of Change* (Cambridge: 1979).

Galilei Galileo, *Opere di Galileo Galilei, Edizione Nazionale*, ed. A. Favaro, 20 vols. (Florence: 1901), vol. 11.

Gotti M., "Lexical Choices in an Early Galilean Translation", in Coleman J. – Kay C.J. (eds.), *Lexicology, Semantics and Lexicography: Selected papers from the Fourth G.L. Brook Symposium* (Amsterdam – Philadelphia: 2000) 87–101.

Gotti M., *Specialized Discourse: Linguistic Features and Changing Conventions* (Bern: 2003).

Gotti M., *Investigating Specialized Discourse* (Bern: 2008).

Gotti M., "La traduzione di Thomas Salusbury della terminologia specialistica di Galileo", in Bernard M. – Rota I. – Bianchi M. (eds.), *Vivir es ver volver: Studi in onore di Gabriele Morelli* (Bergamo: 2009) 277–286.

Henderson F., "Faithful Interpreters? Translation theory and practice at the early Royal Society", *Notes and Records of the Royal Society* 67 (2013) 101–122.
Hope J., *Shakespeare and Language. Reason, Eloquence, and Artifice in the Renaissance* (London: 2010).
Hyland K., "Talking to the Academy: Forms of Hedging in Science Research Articles", *Written Communication* 13, 2 (1996) 251–281.
Hyland K., *Hedging in Scientific Research Articles* (Amsterdam: 1998).
Kelly, L. G., *The True Interpreter: A History of Translation Theory and Practice in the West* (Oxford: 1979).
Malcom N., *Aspects of Hobbes* (Oxford: 2002).
Morini M., *Tudor Translation in Theory and Practice* (Farnham – Burlington: 2006).
Palmer F.R., *Mood and Modality* (Cambridge: 2001).
Pantin I., "The role of translations in European scientific exchanges in the sixteenth and seventeenth centuries", in Burke P. – Po-chia Hsia R. (eds.), *Cultural Translation in Early Modern Europe* (Cambridge: 2007) 163–179.
Plescia I., "'Strangers to Our Nation': Anglo-Italian relations and linguistic encounters in two early modern scientific translations", *Textus* 24, 3 (2011) 559–578.
Raylor T., "William Cavendish, Galileo, Hobbes and the Mechanical Philosophy", in Edwards P. – Graham E. (eds.), *Authority, Authorship and Aristocratic Identity in Seventeenth-century England* (Leiden – Boston: 2016).
Salusbury T., *Mathematical Collections and Translations*, 2 vols. (London: 1661–1665).
Skouen T – Stark R.J. (eds.), *Rhetoric and the Early Royal Society* (Leiden: 2015).
Sprat T., *History of the Royal Society* (London, Printed by T.R. for J. Martin: 1667; facsimile reprint of 1772 ed., New York: 1999).
Taavitsainen I. – Pahta P. (eds.), *Medical and Scientific Writing in Late Medieval English* (Cambridge: 2004).
Venuti L., *The Translator's Invisibility: A History of Translation* (London – New York: 2008).
Wilding N., "The return of Thomas Salusbury's *Life of Galileo* (1664)", *British Journal for the History of Science* 41, 2 (2008) 241–265.
Zeitlin J., "Thomas Salusbury discovered", *Isis* 50 (1959) 455–458.

CHAPTER 12

Language as 'Universal Truchman': Translating the Republic of Letters in the Seventeenth Century*

Fabien Simon

Let us start with a frontispiece [Fig. 12.1]. It shows a cosmopolitan encounter. In a small room, the allegories of the four known parts of the world are gathered. Europe, dressed as an English gentleman, is offering a parchment, with numbers written on it, to Asia and Africa, on his right side, and to America, standing in front of him and greeting or thanking him. This parchment symbolises a book published in 1657 entitled *Universal Character by which All the Nations in the World may understand one anothers Conceptions*. Its author is Cave Beck (c. 1623–1706), Master of Ipswich Grammar School, vicar of St Margaret's and correspondent of the Royal Society in Suffolk.[1] A poem clarifies the aim of the 'common voice' adopted during this diplomatic meeting:

Speech is the Index of the mind: Loe, here
Th' Index of Speech; the dumb Interpreter;
The Iliads in a Nut-shell; Tongues in Brief;
Babel revers'sd; The traveller's Relief;
Ferry of Nations Commerce [...][2]

'Dumb Interpreter', 'Tongues in Brief': these expressions allow all languages to be reduced to one, and this renewed unity is supposed to favour the propagation of sciences. Cave Beck explains that his universal character 'would

* We wish to thank the three anonymous referees, and Sietske Fransen, for their comments on this paper.

1 Beck Cave, *The Universal Character by which All the Nations in the World may understand one anothers Conceptions, Reading out of one Common Writing their own Mother Tongues* (London, printed by Tho. Maxey for William Weekley: 1657). On Cave Beck and his book: Salmon V., "Cave Beck: a Seventeenth Century Ipswich Schoolmaster and his 'Universal Character'", in Salmon V., *The Study of Language in 17th Century England* (Amsterdam – Philadelphia: 1988) 177–190; and remarks in Lewis R., *Language, Mind and Nature: Artificial Languages in England from Bacon to Locke* (Cambridge: 2007), especially 82–84 where the frontispiece is reproduced.

2 Beck, *Universal Character*, "Dedicatory poem" (signed '*Jos. Waite* MA').

 | DOI 10.1163/9789004349261_014

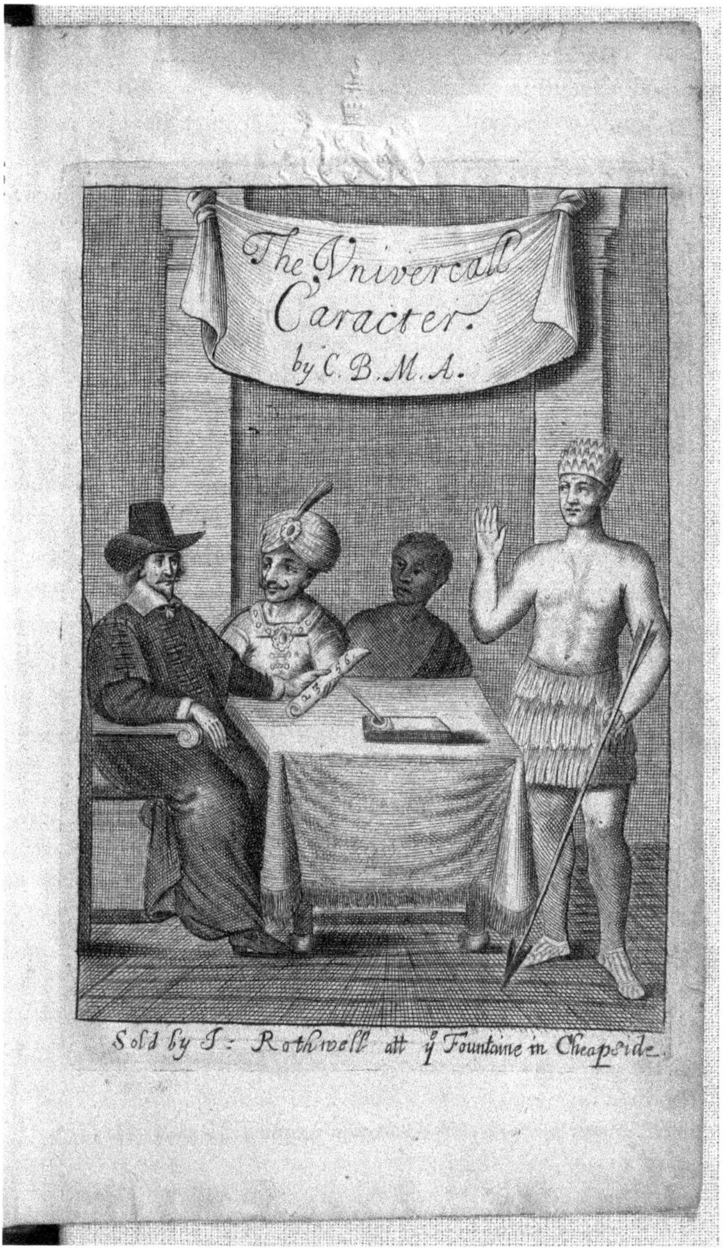

FIGURE 12.1 *Cave Beck, Frontispiece*: The Universal Character by which All the Nations in the World may understand one anothers Conceptions (*London, printed by Thomas Maxey for William Weekley: 1657*). *Engraving. Bibliothèque Nationale de France, RES 8-NFR-214.*

much advantage mankind in their civil commerce, and be a singular means of propagating all sorts of *Learning* and true Religion in the world'.[3] In his poem, the anonymous author emphasizes this dimension by referring to the touchstone of early seventeenth century English science, Francis Bacon: 'Great *Bacon*'s Soul, my friend, divides thee,/He found the Plat, and Thou the Husbandrie'.[4] Thanks to his language, Cave Beck supposedly prolonged or even surpassed Bacon's work.[5]

The 'Universal Character' is one of the many projects for a universal language flourishing in sixteenth- and seventeenth-century Europe. They took different shapes, from the quest for Adam's language, sometimes seen as being gestures, to the compiling of all the languages of the universe.[6] But here we are going to concentrate on the plans which explicitly claim to have invented a method to easily translate one language to another, be it a code to be read in several languages, like the *Polygraphia nova* by the Jesuit Athanasius Kircher; or the creation of a new 'philosophical' *a priori* language, to use the terminology of the linguists.[7] The most famous one is the 'Real Character' developed in the 1668 *Essay Towards a Real Character and a Philosophical Language* by the bishop of Chester and secretary of the Royal Society, John Wilkins. According to him, the first step is to organise all the knowledge of the world in a huge inventory, the book representing mainly tables and arborescences, based on 40 major *Genera*, divided into 251 *Differences*, themselves subdivided into 2030 *Species*.[8]

The European circulations of those plans delineated a province of the Republic of Letters in which the main focus was precisely universal language. It is possible to study the actors of those intellectual and cultural practices and their complex exchanges; the part played, for example, by go-betweens, such as Marin Mersenne or Nicolas-Claude Fabri de Pereisc; and the social

---

3 Beck, *Universal Character*, "To the Reader".

4 Beck, *Universal Character,* "Dedicatory poem" (signed '*Jos. Waite* MA').

5 On Bacon and language, see especially Lewis, *Language, Mind and Nature* 6–22.

6 I am referring here, among other plans, to: Bonifacio Giovanni, *L'arte de'cenni* (Vincenza, Francesco Grossi: 1616) and Duret Claude, *Thresor de l'histoire des langues de cest univers* (Cologny, M. Berjon pour la Société caldorienne: 1613).

7 Kircher Athanasius, *Polygraphia nova et universalis ex combinatoria arte detecta* (Rome, Varesius: 1663). On the link between translation and universal languages, some hints have already been given, for example, by Maillard J.-F., "Un avatar de la traduction: l'idéal d'une langue universelle à la Renaissance", in Contamine G. (ed.), *Traduction et traducteurs au Moyen Age* (Paris: 1989) 334–347 (on the French 16th-century plans mostly).

8 Wilkins John, *An Essay Towards a Real Character and a Philosophical Language* (London, S. Gellibrand and J. Martin: 1668).

organization of—and interaction between—the two main places of knowledge, Rome and London and their language planners' milieus.[9]

As in all the scientific networks of that era, the question of translation and of which language to use (notably in correspondence) is an issue. The act of translating, especially concerning sciences, is never neutral. It can be seen as a 'communication device' of which every wheel should be considered in its materiality: the actors (how did they learn the language they are able to translate?), and the tools (dictionaries, etc.).[10] But it is even more explicit, and the issue even more central, in the context of a network whose key concern was the creation of a utopian universal language. What was at stake was nothing less than choosing, not the language defining correct usage, but the language of sciences and truth, that of the Republic of Letters itself. It is indeed the created languages that were considered *in themselves* as universal 'translators': Robert Boyle spoke, for example, of the *Common Writing* of Francis Lodwick (1647) as a '*Universal Truchman or General Interpreter*'.[11] Universal languages are either the ideal translator and/or the negation of translation, rendering it obsolete. The goal of the language planners can be seen as a quest for a perfect language that would translate, to put it like John Wilkins, not words but things

9 For a study of those projects on a European scale to see the connections between those social milieus and their effects (and for the complete bibliography), see: Simon F., *Sortir de Babel. Une "République des Langues" en quête d'une langue universelle à la Renaissance et à l'Âge classique*, Ph.D. dissertation (Rennes 2 University: 2011); and, for example, Simon F., "Une "République des Langues". La quête de la langue universelle au XVIe–XVIIe siècle: une langue de la vérité à l'usage des Républicains des Lettres ?", in Dion N. – Masse S. – Plourde A.-A. (eds.), *Le Cosmopolitisme. Influences, voyages et échanges dans la République des Lettres (XVe–XVIIIe siècle)* (Paris: 2014) 257–288. For a general study in the long view of the plans: Eco U., *La Recherche de la langue parfaite dans la culture européenne*, trans. J.-P. Manganaro (Paris: 1994). The English language planners' milieu has been the most thoroughly studied, from the works of Dorothy Stimson to those of Rhodri Lewis, for example.

10 See the reflections of Patrice Bret and Jeanne Peiffer in Beaurepaire P.-Y. (eds.), *La communication en Europe de l'âge classique au siècle des Lumières* (Paris: 2014) chap. 3; Bret P. – Moerman E., chap. 8 "Sciences et arts", in Chevrel Y. – Cointre A. – Tran-Gervat Y.-M. (eds.) *Histoire des traductions en langue française, XVIIe et XVIIIe siècles, 1610–1815* (Lagrasse: 2014) 595–723.

11 Boyle Robert, *The Correspondence of Robert Boyle*, ed. M. Hunter – A. Clericuzio – L.M. Principe, 6 vols., (London: 2001) vol. 1, 52. On Lodwick, see Lodwick Francis, *On Language, Theology, and Utopia*, ed. F. Henderson – W. Poole (Oxford: 2011).

themselves, thus avoiding the crucial dilemma of translation—rude fidelity to the source-text and the betrayal of it in the target-language.[12]

Translation is considered in itself as a way of overcoming Babel's curse or, at least, an acceptance of its reality being considered as the existence of the diversity of languages, of which Babel can be a synonymous too. Translating in as many languages as possible is also related with the episode of the tongues of fire of the Pentecost, a major reference of authors who write about the diversity of languages.[13] One of the etymologies of the word 'translate'—at least for the French word 'traduire'—comes from the Latin 'trans-ducere' that leads to the idea of transmission, of a path—the translator being a 'passeur', a cultural broker of the text and its ideas.[14] In the light of this etymology, translation should be perceived as an effort to make knowledge available, to transmit knowledge, underlining the need to study precisely how this knowledge circulates and is appropriated in a different context, every translation implying 'negotiation'. The options are, according to Peter Burke: 'domesticating' a word *in* the language of translation, 'foreignizing' the word *from* the language translated, or 'classicizing' a word, making it part of the language of the past, which was still 'alive' in early modern Europe: Latin.[15] In each case, an idea of displacement, of transformation is at stake. The Italian definition of 'traslatare', given in the *Vocabolario degli Accademici della Crusca*, published in 1612, is built upon this idea of cultural transposition and geographical shifting: 'Trasportare di luogo a luogo. Lat *transferre*. Per ridurre le scritture, e i componimenti d'una lingua

12 For recent remarks on translation in early modern Europe, see for example: Barker S. – Hosington B. (eds.), *Renaissance Cultural Crossroads: Translation, Print and Culture in Britain, 1473–1640* (Leiden – Boston: 2013); Demetriou T. – Tomlinson R. (eds.), *The Culture of Translation in Early Modern England and France, 1500–1660* (New York: 2015); Le Blanc C. – Simonutti L. (eds.), *Le Masque de l'écriture: philosophie et traduction de la Renaissance aux Lumières* (Geneva: 2015). On translation in and around the Royal Society, see also the chapters by Meghan C. Dohert, Felicity Henderson, Jan van de Kamp, and Iolanda Plescia in this volume.

13 Dauphine J. – Jacquemier M. (eds), *Babel à la Renaissance. Actes du XIe colloque international de la société française d'étude du XVIe siècle*, new edition with a preface by Claude-Gilbert Dubois (Paris: 2007). And Céard J., "De Babel à la pentecôte: la transformation du mythe de la confusion des langues au XVIe siècle", *Bibliothèque d'Humanisme et de Renaissance* 42 (1980) 577–594. See also Simon F., "De Conrad Gessner (1555) à Claude Duret (1613), faire entendre la diversité, collectionner les langues", in Alazard F. – Geonget S. – Gerbier L. – Mellet P.-A. (eds.), *Dissensus. Pratiques et représentations de la diversité des opinions, 1500–1650*, Coll. Le Savoir de Mantice 26 (Paris: 2016) 205–224.

14 Vialon M. (ed.), *La Traduction à la Renaissance et à l'âge classique* (Saint-Etienne: 2001).

15 Burke P., "Translations into Latin in early modern Europe", in Burke P. – Po-chia Hsia R. (eds.), *Cultural Translation in Early Modern Europe* (Cambridge: 2007) 65–80; 80.

in un'altra, che oggi si dice anche' [to transport from one place to another. Lat. *Transferre*. To reduce scripture, and the components of a language to another, as it is said today too].[16] This last definition is all the more interesting: John Wilkins explicitly refers to this *Vocabolario* as a model for his enterprise of studying languages and building a new one:

> *The completing of such a design, being rather the work of a College and an Age, then of any single Person* [...] It has been sayd concerning that famous Italian Academy styled *de la Crusca*, consisting of many choice Men of great Learning, that they bestowed forty years in finishing their *Vocabulary*. And' tis well enough known, that those great Wits of the French Academy, did begin their Dictionary in the year 1639.[17]

Among the collective enterprises, in the Baconian sense of the term, constituting models for his own, Wilkins does not pick up scientific institutions, but linguistic ones. The French and the Florentine Academies are the models to follow. The second was born in the 1580s, around figures such as Lionardo Salviati (1540–1589), *l'Infarinato*, with the project of normalizing the vulgar tongue. Their emblem, appearing on the frontispiece of a book by Salviati, as early as 1584 (*Difesa dell'Orlando furioso dell'Ariosto, Stacciata prima*) is reproduced in the first edition of their *Vocabolario*. It is a *frullone* [a mechanical bolter], with the motto, '*il più bel fior ne coglie*' [it collects the most beautiful flower/flour] (*Rerum vulgarium Fragmenta*, LXXIII, 36) borrowed from a verse of Petrarca's *Canzoniere*. This brand new agricultural machine (the *frullone*) used to separate the wheat from the chaff is a metaphor of their linguistic task, purifying the Italian vernacular, especially 'florentinizing' it, thanks to the repositories shaped by Boccaccio, Dante and Petrarca.[18] With this reference, the bishop of Chester underlines the aim of his 'Real Character', that is first and foremost building a new language, a practical, even if 'philosophical' *i.e.* 'scientific', one. Besides the tables, a part of the book includes an *Alphabetical*

---

16 *Vocabolario degli Accademici della Crusca* (Venice, Giovanni Alberti: 1612); quoted in Vialon M., "Corpus de citations sur la traduction à la Renaissance et à l'Age classique", in Vialon M. (ed.), *La Traduction à la Renaissance et à l'âge classique* (Saint-Etienne: 2001) 239.

17 Wilkins, *Essay*, "Epistle dedicatory", fol. a3r–v.

18 On the *Accademia*: Sabatini Francesco (ed.), *Una lingua e il suo vocabolario* (Florence: 2014); on the emblem, 18.

Dictionary, to which William Lloyd actively collaborated.[19] Wilkins's 'client' and a member of the not-so-efficient 'Committee for improving the English Language', Thomas Sprat, also used the same references in his *ad hoc History of the Royal Society*, 'Sect. XIX. Modern Academies for Language'.[20] Like the Royal Society's project vis-à-vis English, and Wilkins' project concerning the language of science, Sprat put forward the fact that the *Académie française* managed to impose French as the new 'Greek', thanks to the correspondence of the language with the ideal of *plain style*, lying in the rejection of the 'Artifice of the Words'.[21] But instead of a specific language, a vernacular one, Wilkins's enterprise is aimed at finding a universal one, 'as things are better then words, as real knowledge is beyond elegancy of speech, as the general good of mankind, is beyond that of any particular Countrey or Nation'.[22] Wilkins's own *frullone* tends to purify language, but not only English, and the goal is to elaborate a transparent language, one in which science could be transmitted fluently and the study of nature, the Royal Society's goal, would be conveyed in explicit terms.

The counterpart is that, instead of the idea of transmission, the language planners considered their creations, as we are going to demonstrate, as translations *keeping knowledge away*, or at least making it obscure for the many and restricted to the few. A dimension which is already present in the definition of the 'interpreter [*Interpres*]' in Thomas Thomas' (1553–1588) *Dictionary*: '[Interpres] An interpretour, expounder, or declarer: a translater: one that is used to expound a strange language: a stickler betweene two at variance: a mediatour, a meane, a trouchman, a soothsayer, a divinour'.[23] This twofold definition is interesting in both respects: first describing the translator as someone who explains a 'strange language'; but, then, defining him as an augur too,

19 Dolezal F., *Forgotten but Important Lexicographers, John Wilkins and William Lloyd: A Modern Approach to Lexicography before Johnson* (Tübingen: 1985). Also Pauchard J., "Le dictionnaire alphabétique de William Lloyd et John Wilkins: une avancée majeure de la lexicographie au XVIIe siècle", in Groupe de linguistique appliquée des télécommunications (ed.), *Aspects méthodologiques pour l'élaboration de lexiques unilingues et multilingues: actes de GLAT-Bertinoro 2006* (Brest: 2006) 151–161.

20 Lewis, *Language, Mind and Nature* 147–148.

21 Sprat Thomas, *History of the Royal Society of London for the improving of natural knowledge* (London, printed for J. Martyn and J. Allestry: 1667; reprint London: 1959) 39–40.

22 Wilkins, *Essay*, fol. a3r–v.

23 Thomas Thomas, *Dictionarium summa Fide ac diligentia accerratissime emendatum*, (Cambridge, Thomas: 1587), *sub voce*; quoted in Vialon M., "Corpus de citations sur la traduction à la Renaissance et à l'Age classique" 232.

with the power of mastering a kind of secret language, or a language disclosing secrecies.

When we have a closer look at the mechanism of the social network constituting the 'Republic of Languages', we can see those universal languages as literary and social technologies meant to be appropriated by a small community of 'happy few', as we are going to see in the last point of our demonstration. The first one is going to focus on the Latin context of this circulation of a particular knowledge; and the second one will be centred on the tension between the insistence on universal translation on the one hand, and on cryptography as another form of genealogy of the language plans on the other hand.

Why Translate? Latin as the Lingua Franca of the Republic of Letters: Latin and its Social Space in Early Modern Europe

In seventeenth-century Europe, Latin remained, as in earlier centuries, a 'langue référentielle' (a reference language) and a 'European sign'.[24] It resisted the 'challenge to the Latin monopoly on *scientia*', and if important scientific works were starting to be published directly in the vernaculars, Latin was still the *lingua eruditorum vernacula*, the *lingua franca* of the learned.[25] The 'latinization', or 'counter-intuitive' translation from the vernaculars into Latin, was largely developed in the 1600s and illustrated the fact that international circulations of knowledge were mostly Latin.[26] Within the balance of 'scientific' languages, Latin was no longer hegemonic, but not yet supplanted by French for example. The new polyglot means of propagating scientific information, the periodical, could certainly be edited in English, Italian or French—with the 1665 Parisian *Journal des sçavans*—but also still in Latin, for its Uppsala or Leipzig (*Acta eruditorum*) versions, among others.[27] Within this frame, if this Babelization of erudite Europe certainly made the topic of translation all the

24 First quotation: Grévin B., *Le Parchemin des cieux. Essai sur le Moyen Âge du langage* (Paris: 2012) 18; second quotation: Waquet F., *Le Latin ou l'empire d'un signe, XVIe–XXe siècle* (Paris: 1998).

25 Blair A., "La Persistance du latin comme langue de science à la Renaissance", in Chartier R. and Corsi P. (eds.), *Sciences et langues en Europe* (Luxembourg: 2000), 19–39; 23. The expression '*lingua eruditorum vernacula*' is employed by a Swedish member of the Republic of Letters, quoted in Burke P., *Languages and Communities in Early Modern Europe* (Cambridge: 2004) 53.

26 Burke, "Translations into Latin".

27 Bret – Moerman, "Sciences et arts" 597.

more important, Latin remained a linguistic focus of the seventeenth century Republic of Letters.

That is why the language planners, even if they were precisely looking for a language that might replace Latin, were concerned with all the issues related to its place, especially as *lingua eruditorum*. Within a 'Latin frame', they had to make a stand. The Antique language was indeed their communication language, being used, for example, in 60% of the letters of Athanasius Kircher, and the metalanguage of many of the language plans.[28] In England, where English was, on the contrary, the metalanguage of almost every language plan (Wilkins's, Beck's, Lodwick's etc.), foreigners were complaining about this situation. Henry Oldenburg's correspondence gives a good view point on this situation: it includes many letters insisting on the necessity of translating works into Latin to make English books accessible to the community of the learned. It was the case with the aforementioned Thomas Sprat's *History of the Royal Society*. Franciscus de Le Boë Sylvius, a physician and medicine professor in Leiden, wrote—in Latin—to Oldenburg in December 1667, that it should also be made available 'to the whole world of literature [*literatorum orbi*] in the Latin tongue, in which, moreover, I myself [...] could more easily follow and understand its proceedings'. A few months later, the astronomer Hevelius echoed this statement, writing that 'the *History* would have been far more acceptable to foreigners if written in Latin'.[29] Latin stood for a transnational idiom surpassing linguistic frontiers. And, simultaneously, the mastery of Latin could be seen as a symbolic 'entrance fee' (in Bourdieu's meaning) into the Republic of Letters.[30]

Paradoxically, even the translation in Latin of Wilkins's *Real Character* was considered to facilitate its circulation. 'Omnium litteratorum censurae et animadversionibus hac ratione exponendus' [In order that it may in this way be submitted to the criticism and comment of all scholars], as Henry Oldenburg put it in a Latin letter to the Italian scholar Marcello Malpighi.[31] Even a universal language needs this conversion to be received widely. In the case of the

28 On Kircher's correspondence: Fletcher J., "Athanasius Kircher and his correspondence", in Fletcher J. (ed.), *Athanasius Kircher und seine Beziehungen zum gelehrten Europa seiner Zeit* (Wiesbaden: 1988) 139–195.

29 Oldenburg Henry, *The Correspondence of Henry Oldenburg*, ed. by A.R. Hall – M.B. Hall, 13 vols. (Madison 1965–1986) vol. 4, letter no. 731, 69 and no. 878, 444.

30 The notion of "entrance fee" is related to Bourdieu's "field" theory; see, for example: Bourdieu P., "Quelques propriétés des champs", in *Questions de sociologie* (Paris: 1980) 113–120; 115–116.

31 Oldenburg, *Correspondence*, letter no. 1051, vol. 5, 280.

Real Character, whose project of translation and its twists and turns have been studied in details by Rhodri Lewis, one does not know whether it was ever finished.[32] More generally, the universal language plans apparently needed to be translated to be understandable: Beck's was directly published in French too, whereas Pierre Besnier's 1674 *Réunion des langues* was being translated into English by Henry Roose in 1675.[33] And as far as the French language planner Antoine de Vienne Plancy is concerned, he knew about many of the English plans through an encyclopaedic compilation mentioning them, written in Latin by Johann Christoph Beckmann (*Historia orbis terrarum*). De Vienne Plancy's language was progressively unveiled in the periodical *l'Extraordinaire du Mercure*, a supplement to the *Mercure galant* published every six months, over the course of five years and on more than 177 pages, as a sort of 'feuilleton' of the universal language. Beckmann's book allowed him to assert his position vis-à-vis the English plans and to state that his own surpassed them.[34]

Translation was indeed a key feature of the circulation of knowledge within the Republic of Languages and was a recurring topic in the correspondence of its members. That's why the Minime Marin Mersenne suggested the creation of an international 'academy of translators' to give access to knowledge on a wider scale, using what he called the 'common language of Christian Europe', Latin.[35] Some plans even tried to build upon Latin, incarnating the ideal translation language, to create their own method or even their own language.

A New Latin as the Universal Language or Universal Language as the New Latin?

Some language planners could be considered as following Descartes' advice in a famous letter to Mersenne, on the 20th of November 1629, where he deemed

32 Lewis R., "The Publication of John Wilkins's *Essay* (1668): Some Contextual Considerations", *Notes and Records of The Royal Society* 56–2 (2002) 133–146; see also Slaughter M.M., *Universal Languages and Scientific Taxonomy in the Seventeenth Century* (Cambridge 1982) 176.

33 Beck Cave, *Le Charactere universel, par lequel toutes nations peuvent comprendre les conceptions l'une de l'autre, etc* (London, A. Maxey for Guillaume Weekley: 1657); Besnier Pierre, *A Philosophical Essay for the Reunion of the Languages or the Art of Knowing all by the Mastery of one* (Oxford, printed for Hen. Hall for James Good: 1675).

34 Vienne Plancy A. de, *Extraordinaire du Mercure galant* 14 (January 1681) 334–349; 19 (July 1682) 274–330; 31 (July 1685) 112–182; 32 (January 1686) 110–146. Beckmann Johann Christoph, *Historia orbis terrarum geographica et civilis, de variis negociis nostri potissimum et superioris seculi, aliisve rebus selectioribus* (Frankfurt, J.W. Fincelii: 1680) 390–393.

35 Mersenne Marin, *Correspondance du P. Marin Mersenne, religieux minime*, ed. by C. De Waard, 18 vols. (Paris: 1932–1988) vol. XI, no. 942bis (16 November 1640) 420.

creating a universal language unrealistic. He considered that the simplest and, in fact, the only way to attain it eventually would be to adopt one idiom already in use, especially Latin.[36] Two projects, among others, were aimed at renovating Latin itself and/or using it directly as a universal language.

The first one used Latin as the 'gateway' to all languages and was elaborated in 1674 by the Jesuit Pierre Besnier (1648–1705), in his *Réunion des langues*, where he exposed the theory more than the feasibility of the project. He was looking for a language that would allow one to learn all languages at the same time, by using reasoning and comparison. A member of the Republic of Letters, Besnier was renowned for his extraordinary memory and knowledge of many languages. He acted, for example, as a collaborator for the translation of the *New Testament* (Paris, 1697 and 1703), following the *Vulgate*, with Fathers Dominique Bouhours and Michel Letellier. He settled in the Levant in 1688.[37] First considering Hebrew as the possible reference language, but disqualifying it because it was not enough widespread and known in Europe, Besnier finally chose Latin, precisely because:

> La pluspart des autres Langues sont resserrées dans les bornes d'un Païs, ou d'un Royaume particulier: la Latine n'a pas ce desavantage; c'est, à proprement parler, la Langue de l'Europe [...] elle est par tout universellement connuë des Savans & des Gens de qualité, qui sont pour l'ordinaire les seules personnes, qui ayent besoin du secours des Langues étrangéres.[38]

> Most of the other languages are compressed within the landmarks of one country, or one particular Kingdom: Latin does not have this drawback; it is, strictly speaking, the Language of Europe [...] it is everywhere universally known by the learned and the persons of quality, who are usually the only persons, who are in need of the recourse of foreign languages.[39]

Latin had all the necessary qualities, defined by Aristotle, 'universality, certainty and proportion', and at the same time it had the advantages both of a dead language ('immutability') and of a living one ('universality').[40] Besnier's

36 Mersenne, *Correspondance* vol. II, no. 143.

37 De Backer A. – Sommervogel C., *Bibliothèque de la Compagnie de Jésus*, 8 vols. (Mansfield Centre: 1998) vol. 1, 1410.

38 Besnier Pierre, *La Reunion des langues, ou l'art de les apprendre toutes par une seule* (Paris, Sébastien Mabre-Cramoisy: 1674) 9–10.

39 Translations mine, unless otherwise stated.

40 Besnier, *La Reunion des langues* 9–10.

choice must be inscribed in the longer context of the elaboration of a linguistic knowledge in which Latin played the role of a linguistic paragon. During the process of 'grammatisation' of the languages in early modern Europe, it stood for the reference language, the grammatical model, even if in Besnier's time French, at least in France, tended to be the grammatical paradigm.[41] Latin was first and foremost the language of European scholars and the learned, who (according to the Jesuit) were the people needing language knowledge.

That was the argument of Philippe Labbé—another Jesuit who entered the Society in 1623 and was a teacher in Caen, Bourges and Paris, also publishing several books—when he chose to rely on Latin to shape a new universal language in his 1663 *Grammatica linguae universalis missionum et commerciorum*.[42] It was the best language, according to him, '*parce quelle est connuë des Europeens*' [because it is known by the Europeans]. His idea was to simplify it, make it closer to the French with, for example, an 's' for the plural, no declensions and simplified conjugations. For a "dog", you would say '*Can*', '*canen*' for a female, '*canu*' for a small dog and '*canou*' for a big one.[43]

Therefore this choice of the Latin tongue had two dimensions. The first was political: it was no coincidence that two Jesuits elected it as a renewed universal language in the second half of the seventeenth century. This attitude should be analysed in the context of the quarrel between the monarchy and the Jansenists, which had a clear linguistic dimension. It is explicitly underlined by Labbé in his preface. He directly took a major part in the polemic against the Jansenist enemy through, for example, the publication in 1661, of a book entitled *The Etymologies of several French words, against the misuses of the sect of the Port Royal Hellenists*.[44] He dedicated his work to the Académie française, as the authority legitimising language in France. And he described his task as saving the French from the 'ruin' provoked by the members of the Port-Royal sect, designated as 'poor housekeepers' looking for 'stinking and

41 See in particular: Auroux S., *La Révolution technologique de la grammatisation: introduction à l'histoire des sciences du langage* (Liège: 1994); and Bury E. (ed.), *Tous Vos Gens à Latin: le latin, langue savante, langue mondaine (XIVe–XVIIe Siècles)* (Genève: 2005) 13.

42 Balteau J. et al. (eds.), *Dictionnaire de biographie française* (Paris: 1932–2011), *sub voce*.

43 Labbé Philippe, *Grammaire de la langue universelle des Missions et du Commerce tres simple, tres courte et tres facile à apprendre à toutes sortes de personnes* … (Paris, J. Roger: 1663) 2 and then 23–24. The edition is both in French and Latin.

44 Labbé Philippe, *Les Étymologies de plusieurs mots françois, contre les abus de la secte des hellénistes du Port-Royal, sixiesme partie des Racines de la langue grecque* (Paris, G. et S. Bénard: 1661). On this idea, see for example: Wilding N., *Writing the Book of Nature: Natural Philosophy and Communication in Early Modern Europe*, Ph.D. dissertation (European University Institute: 2000) 272–274.

muddy swamp water' instead of the 'clear fountain water' they had in their background.[45] In the book, the split introduced by Port-Royal between French and Latin to make it closer to the Greek is seen as a break-up with the papacy, Labbé defending on the contrary the 'commerce developed between [the] French and Rome for 1200 years'.[46]

As was shown by the quotations before, the other dimension of the renewed use of Latin was the advantages of its social space, including the scholars, the learned, the members of the Republic of Letters.

Many other language plans tried to find an alternative to Latin. For it could be described as an outdated idiom, the language of a fallen Empire, no longer efficient enough. Wilkins devoted an 'appendix' of his *Real Character* to 'a Comparison betwixt this Natural *Philosophical Grammar* and that of other *Instituted* Languages, particularly the *Latin*'.[47] His conclusion was irrevocable: the 'Philosophical language' is superior, with 'no unnecessary Rules and no Exceptions', so that Latin would be positively replaced by his creation.[48] He used the example of the many synonyms of *liber* (meaning book as well as free or bark) to make this point. The idea was anyway to fit into the social space formerly circumscribed by the antique language and to conquer the Latin territory. The purpose was to define the frontiers of the Republic of Letters more strongly, by creating a new language enabling one to identify its members. Here translating means discriminating, as we are going to demonstrate, starting with the universal languages presenting themselves as 'universal translators', but not accessible to everyone.

Universal Languages, Between Universal Truchman and Cryptographic Code

Mercurius, Translator

The language planners considered their creations *in themselves* to be translators, interpreters. The motto of Kircher's *Polygraphy* is *linguarum omnium*

45 Labbé, *Les Étymologies*, fol. aiiii (v)–avi : 'On peut dire avec verité que ce sont de tres mauvais menagers, qui font de grands frais, pour faire venir de loin ce, qu'ils ont en abondance dans leurs propres maisons ; & ayment mieux puiser dans des marais éloignez une eau puante & bourbeuse, que d'étancher leur soif dans les claires fontaines de leur voisinage'.

46 Labbé, *Les Étymologies*, "To the Reader" (n.p.): 'Que le nom d'Hellenistes leur convient fort bien, à cause de l'affectation & du zele indiscret, qu'ils témoignent en faveur de la Langue Grecque au preiudice de la Latine, qu'il semble vouloir supprimer partout, & empescher le commerce que nos François ont eu avec Rome depuis pres de 1200 ans'.

47 Wilkins, *Essay* 441–442.

48 Wilkins, *Essay* 450.

ad unam reductio [reducing all languages to one], as well as *Unius Linguae ad omnes traductio* [with one language, translate them all].[49] George Dalgarno entitled his first reflection on the universal language a '*Swift Scribe* and *Faithful Interpreter*'.[50] Finally, Cave Beck, in a letter to Oldenburg in 1668, portrayed his own project as a '*Pocket Mercury to Travaylors*'.[51] This expression refers to the god of trade and travels, who is also the messenger of Mount Olympus, Greek Hermes. This omnipresent and protean figure of the Renaissance is affiliated to communication: the one who invented the art of naming, the one who informs, translates, transmits.[52] That is why it spreads as a common noun in Europe, in periodical titles for example, from the English *Mercurius politicus* to the French *Mercure galant*, cherished by de Vienne Plancy. Consequently it is not surprising that Beck should have chosen this reference to qualify his 'Universal Character'. His language was supposed to facilitate, as we have seen, scientific exchanges and be a 'transparent' means of communication, giving access to the Book of Nature itself, as was Wilkins's character, of which some tables were designed by the botanist John Ray.[53] But Beck's language was also supposed to be a 'Mercury' for the traveller, allowing him to avoid all problems usually inherent to every linguistic exchange, 'all Equivocal words, Anomalous variations, and superfluous Synonomas (with which all Languages are encumbred, and rendred difficult to the learner)'.[54] The idea was in fact to avoid translation itself and the person in charge, to skip intermediaries: 'Save the charges of hiring Interpreters: Besides, avoyding the danger of being mis-understood, or betrayed by Truch-men, misrelating his Expressions to Foraign Ears'.[55] The *Universal Character* would be the Truchman one would always have in one's

49 On the Polygraphy and its functioning, see, among other publications: McCracken G.E., "Athanasius Kircher's Universal Polygraphy", *Isis* 39, 4 (1948) 215–228 and Eco, *Recherche* 227–231.

50 Dalgarno G., *George Dalgarno on Universal Language: The Art of Signs (1661), the Deaf and Dumb Man's Tutor (1680), and the Unpublished Papers*, ed. D. Cram – J. Maat (Oxford: 2001) "Broadsheet 1: The Universal Character" (1657) 85.

51 Oldenburg, *Correspondence* vol. 5, letter no. 943 14–17.

52 Balavoine C. et al. (eds.), *Mercure à la Renaissance, Actes des journées d'étude des 4–5 octobre 1984* (Paris: 1988) 5.

53 On the idea of a 'transparent' scientific language in early modern sciences, see our synthesis with bibliographical references: Simon F., "Quelle est la langue de la science: dire efficacement la 'vérité' scientifique?", in Pérez L. – Simon F. – Thébaud-Sorger M. (eds.), *L'Europe des sciences et des techniques (XVe–XVIIIe): Un dialogue des savoirs* (Rennes: 2016) 257–267.

54 Beck, *The Universal Character*, "To the Reader".

55 Beck, *The Universal Character*, "To the Reader".

pocket: it takes the shape of a dictionary of 4000 words with their equivalent in numbers. Realising that it might be impossible to learn all the words, the author announced the printing of a 500-sentence *syllabus*, inspired by Comenius' *Janua linguarum*.[56] The reference to Latin was used once again, but this time the new pedagogy was applied to the learning of a new *lingua franca*, replacing the one born in the Antiquity.

It is no coincidence that Mersenne's aforementioned suggestion of a Latin translators academy occurred, in the 1640s, in the context of an exchange between the French Minim and the Palatinian Calvinist exile, Theodore Haak (1605–1690), himself a translator (from English into German notably). It is an epistolary exchange, consisting of 24 letters, in which many universal language plans are mentioned.[57] It actively contributed to the circulation of this knowledge: Haak sent pieces of information about Thomas Harrison's *Arca Studiorum* or John Johnson and William Bedell's 'hieroglyphs', and, as a literary counter-gift, Mersenne told him about the projects of the French language planner and inventor, from Toulouse, Le Maire, but also about Kircher in Rome, whom he visited in 1644.[58] Mersenne, described by Adrien Baillet as 'le centre des gens de lettres' [a 'center' for literati], acted, within this network, as a bridge between England, France and Italy.[59] He saw himself as an 'entremetteur', a go-between.[60] Cosmopolitan and polyglot, he played, in the 'Bourse of literary values', the same kind of part as the 'foreign exchange broker, responsible for exporting from one territory to another texts whose literary value they determine by virtue of this very activity', being a 'legislator of the Republic of Letters', a 'great intermediary' with 'power of consecration', as Pascale Casanova describes the role of contemporary translators.[61] It was thanks to Mersenne, that the plans of a small 'projector' from the South

56 Beck, *The Universal Character*, "To the Reader".

57 On Haak: Barnett P., *Theodore Haak, F.R.S. (1605–1690). The First German Translator of Paradise Lost* (The Hague: 1962). On the correspondence: Mersenne, *Correspondance* for the years 1640 (vol. VIII–XI) and 1648–1649 (vol. XV) and in a special annex vol. XI, 397 and *sq*.; and comments especially in: Salmon V., "Language planning in seventeenth-century England: its context and aims", in Salmon V., *The Study of Language* 148–151.

58 Mersenne, *Correspondance*, for example: On Harrison vol. VIII, no. 858, 305; on Bedell vol. XV, no. 1632 250; on Kircher vol. VIII, no. 799 720; On Le Maire vol. XI, no. 942, 420–423, vol. XV, no. 1654, 354. For details, see Simon, *Sortir de Babel* 575–596.

59 As quoted in Waquet F. – Bots H., *La République des Lettres* (Paris: 1997) 119.

60 The term is used to qualify Haak (Mersenne, *Correspondance* vol. XI, no. 952, 434) but it is also the way Mersenne described his own task in the same letter: 'en fin je voudrois que tout ce qui se passe de gentil et de subtil par mes mains, vous le puissiez voir aussi'.

61 Casanova P., *The World Republic of Letters*, trans. M.B. DeBevoise (Cambridge: 2007) 21–22.

of France—who, for example, received letters patent on the 27th of August 1644 recognizing him as the inventor of many machines and instruments and, in particular, of a *Méthode universelle pour traduire les langues* [A universal method for the translation of languages]—gained access to centres of scientific knowledge such as the Royal Society.[62]

Within this task as a broker, the universal language itself could be a major tool. It is what emerges, for example, from another of Mersenne's letter, in 1635, sent to a peer, a broker himself, Nicolas-Claude Fabri de Pereisc. Mersenne exposes his *Harmonie universelle* (1636) as a panacea for international intellectual commerce: 'Je me suis imaginé une sorte d'escripture et un certain idiome universel, qui vous pourroit servir à cet effect, en dressant un alphabet qui contient tous les idiomes possibles' [I imagined a sort of writing and universal idiom, that could help you for that, by building an alphabet that holds all the possible idioms].[63] Communicating by letters with all parts of the world, it would be a great advantage for Pereisc if he had adopted Mersenne's musical language, a perfect tool for the language 'passeur', since he was always trying to obtain information, alphabets, books, from the Levant especially, being one of the prominent 'Orientalists' of 17th century Republic of Letters.[64]

Discussions about languages, translation, a utopian means to communicate universally, lie at the core of the correspondences. These issues were indeed 'feeding them', providing opportunities to communicate, contributing to the circulation of ideas and knowledge. But who to communicate with precisely?

Translating as Encoding?

The comparison with Mercurius leads us to a different but related path. Mercurius was indeed a '*janus bifrons*', whose other face should be considered too. The metaphor is used in another type of works. In 1641 John Wilkins published a book entitled *Mercury, or the Secret and swift messenger*, mentioned in his letter to Oldenburg by Beck. It is a history of cryptography and its different methods over centuries (Arabic numerals, musical notes, etc.). Chapter 13

62 *Nouvelle biographie générale: depuis les temps les plus reculés jusqu'à 1850–1860*, 23 vols. (2 tomes per volume) (Paris: 1852–1866) vol. 30 559 (a patent we haven't found yet). One of the first meetings of the Royal Society is the setting of a concert of "Almerie", a musical instrument created by Le Maire (as mentioned in a letter from Oldenburg to Boyle in 1664); see, for example, Cohen A., "Jean Le Maire and la musique almérique", *Acta Musicologica* XXXV (1963) 175–181.

63 Mersenne, *Correspondance* vol. V, no. 422, 136–137.

64 Miller P.N., *Pereisc's Europe: Learning and Virtue in the Seventeenth Century* (New Haven – London: 2000).

deals with a 'universall Character, that may be legible to all nations and languages [...]'. His conclusion was that the discovery of such a character would 'conduce to the spreading and promoting of all Arts and Sciences'.[65] One possible genealogy of universal languages is this issue of secrecy and cryptography.[66] And Hans Sloane's personal copy of Wilkins's *Essay* contained, inserted between its pages, a sheet of paper depicting the key to a complex cryptographic system.[67] While the *Mercury*, an encyclopaedia on cryptography, was concerned with universal languages, conversely, the *Essay*, a universal language itself, could be used as a handbook for cryptographers.

This book and the paper inside may actually have been Hooke's personal copy, Sloane having bought some works from his library at an auction in 1703. Among the 3380 volumes of the *Bibliotheca Hookiana*, we were able to find at least another work linking cryptography and universal language.[68] It was indeed in his *Diary* at the date of Friday, June 16th 1693, that Hooke wrote he had purchased *Kircheri combinatoria* at an auction.[69] Kircher's 1669 *Ars magna sciendi sive combinatoria* is connected, through the reference to Lull especially, to the 1663 *Polygraphia*.[70] This last project too is deeply rooted in cryptography. Kircher explained that the first idea of working on it was suggested to him by a request from Emperors Ferdinand III and Leopold I, themselves fascinated by cryptography. The aim was to look for a cipher which could be used

65 Wilkins John, *Mercury, or the Secret and Swift Messenger, shewing how a man may with privacy and speed communicate his thoughts to a friend at any distance* (London, J. Maynard and T. Wilkins: 1641) 105–106.

66 On this issue, see especially: Strasser G.F., *Lingua Universalis: Kryptologie und Theorie der Universalsprachen im 16. und 17. Jahrhundert* (Wiesbaden: 1988).

67 As mentioned in Jardine L., *Ingenious Pursuits: Building the Scientific Revolution* (New York: 1999) 320 (as well as for the Hooke's hypothesis just below).

68 Rostenberg L., *The Library of Robert Hooke. The Scientific Book Trade of Restoration England* (Santa Monica: 1989). See also: http://www.hookesbooks.com for Robert Hooke's Books database.

69 Hooke Robert, *The Diary of Robert Hooke M.A., M.D., F.R.S. 1672–1680*, ed. H.W. Robinson – W. Adams (London: 1935) 250.

70 Kircher Athanasius, *Ars magna sciendi, in XII libros digesta* (Amsterdam, J. Janssonius a Waesberge and the widow of E. Weyerstraet: 1669). On the link between the two books: Simon, *Sortir de Babel* 680–689. Among the many "kircheriana" Hooke was possessing, next to the *Ars magna sciendi* (auct_BH_53—we indicate here the ID number of the Robert Hooke's Books Database—), we can find also the 1678 catalog of Kircher's Roman "Musaeum" (*auct_BH_2538;* on the Museum, see below). The Bibliotheca Hookiana contained also a 1564 edition of Trithemius's *Polygraphia* (*auct_BH_1117*), a major reference for Kircher, and Wilkins's Essay (*auct_BH_1861*).

in the correspondences circulating among their vast multilingual Empire.[71] Kircher's plan was also connected with the figure of Mercurius, this time in the shape of Hermes Trismegistos, considered to be the author of the *Corpus Hermeticum*. The mysteries of this ancient Egyptian knowledge (in particular the hieroglyphs) were what the Jesuit was trying to solve, seeing himself as an Oedipus, able to decipher their secrecies.[72] That is why his museum of the *Collegium Romanum*, a Roman social hotspot within the Republic of Letters, was full of obelisks, offering materials to Kircher's work on languages.[73] The relation between hieroglyphs and secrecy explains why their use is sometimes prescribed in handbooks of cryptography. In 1586 Blaise de Vigenère, for example, wrote a *Traité des chiffres ou secrètes manières d'écrire*, a book Kircher ordered in a 1633 letter to its patron, Pereisc.[74] The author very precisely described how those kinds of writings are meant to keep a certain knowledge within a network of 'enlightened' people. Thanks to this sort of secret language, knowledge could thus be kept away from the 'multitude', and only addressed to dignified people, 'consçachans', a neologism built upon the Latin *cum-sapere*, to 'know with'.[75] Knowing this, the fact that the first manuscript version of the *Polygraphy*—aimed at a few addressees like Pope Alexander VII—was designed to be written with the help of pictograms, described by Kircher as a sort of hieroglyphs, is meaningful.[76] But those pictograms could also be drawn closer to Kircher's Jesuit *ethos*, and the use of ciphers in the correspondence within the Society, advised as early as 1547 in Juan de Polanco instructions.[77]

71 The idea is referred to in the letter to Caramuel (see below). See, for example: Wilding, *Writing the Book of Nature* chapter 3.

72 On Kircher and hieroglyphs: David M., *Le Débat sur les écritures et l'hiéroglyphe aux XVIIe et XVIIIe siècle et l'application de la notion de déchiffrement aux écritures mortes* (Paris: 1965); and recently: Stolzenberg D., *Egyptian Oedipus. Athanasius Kircher and the Secrets of Antiquity* (Chicago: 2013).

73 Findlen P., "Scientific Spectacle in Baroque Rome: Athanasius Kircher and the Roman College Museum", *Roma moderna e contemporanea* 3 (1995) 625–665.

74 Bibliothèque Nationale de France, Ms. Fr. 9538, fol. 234v; quoted in Maillard J.-F., "Aspects de l'encyclopédisme au XVIe siècle dans le Traicté des chiffres annoté par Blaise de Vigenère", *Bibliothèque d'Humanisme et Renaissance* XLIV, 2 (1982) 235–268; 262, n. 103.

75 Vigenère Blaise de, *Traité des chiffres ou secrètes manières d'écrire* (Paris, Abel L'Angelier: 1586), fol. 3v: 'L'escriture ausurplus est double ; la commune dont on use ordinairement ; & l'occulte secrete, qu'on desguise d'infinies sortes, chacun selon sa fantaisie, pour ne la rendre intelligible qu'entre soy & ses consçachans [...] Afin de les garentir & subtraire du prophanement de la multitude, & en laisser la cognoissance aux gens dignes'.

76 We consulted the version held at the Vatican Library, BAV, MS Chigi J. VI. 225, fols. 11r–37v.

77 Archivio della Pontificia Universita Gregoriana (now APUG) 563, 68 and 239rv (for coded letters in Kircher's paper) and for a bundle in the Jesuit archives: Archivum Romanum

In both interpretations, the *Polygraphy* conceived as a universal language was also, and at the same time, a code, designed to disseminate knowledge only among chosen people. Kircher, even after the book was printed, kept sending it especially to people belonging to his network, accompanied by letters: 'the printed book paradoxically circulated in a manuscript economy'.[78] Kircher himself underlined the double dimension of his language. In the letter accompanying the book sent to Jean II Casimir of Poland, he explained it allowed the reader to: '[per] intendere e scrivere in qualunque lingue, ancor, que non se n'abbia cognitione; insieme con una cifra per aprire il proprio segreto, a chi si vuole, con renderlo impenetrabile a qualunque altro' [understand whatever language and write in it, even if you don't have any knowledge of it; at the same time it is a cipher to unveil the secret within to whom you want, and render it impenetrable to someone else].[79]

Even without its explicit connection with hieroglyphs, the printed polygraphy was still used to share secret knowledge only with chosen, dignified, people. The practice of cryptography can be circumscribed to the frame of literacy. But it reintroduced, into the openness of the written, the secrecy of a convention uniting two people exchanging letters or, more generally, possessing the key to a specific language, that is the book decrypting its system. Cryptography is indeed an 'art of exclusion included in an act of communication'.[80]

Translating a Social Network, or Universal Language as a Language of Social Distinction

Whereas translation is usually an effort to give access to knowledge, the language planners considered translation into the languages they created as a way *to not give access* to knowledge or at least to keep it in particular hands.

Societatis Iesu, *Fondo gesuitico* (FG), *Miscellanea 3. de administratione et disciplina Soc. Je. Opusc. Apologetica de eadem Soc.*, 21 *Chifraria del secolo XVI e del principi del XVII;* see Wicki J. (S.J.), "Die Miscellanea Epistolarum des P. Athanasius Kircher S.I. in missionarischer Sicht", *Euntes Docete* XXI (1968) 221–254 and 282–309; 295. And "Die Chiffre in der Ordenskorrespondenz der Gesellschaft Iesu von Ignatius bis General Oliva (ca. 1554–1676). Anhand der Kodex Fondo Gesuitico 678/21 (ARSI)", *Archivum Historicum Societatis Iesu* XXXII (1963) 133–178.

78 Wilding N., "'If You Have a Secret, Either Keep It, or Reveal It': Cryptography and Universal Language", in Stolzenberg D. (ed.), *The Great Art of Knowing. The Baroque Encyclopedia of Athanasius Kircher* (Stanford: 2001) 93–105; 100.

79 APUG 555, fols. 179r–180v.

80 Fraenkel B., "Comment ne s'adresser qu'à quelques-uns? Remarques sur la cryptographie de tradition alphabétique", *Extrême-Orient, Extrême-Occident* 30 (2008) 175–185; 176.

The very limited effective reception of those languages, not in the discussions and exchanges they aroused—of which we have given but a brief overview before—but in the use of the languages themselves, confirms this interpretation. It may indicate a failure of those universal languages but also a conscious strategy on the part of their creators.

Concerning Kircher's *Polygraphy*, the only letter using the method was written by the Cistercian Juan de Caramuel y Lobkowitz (1606–1682). He starts with a message written in Kircher's pasigraphy ('XXVIII. 10. XVI. 23 Å Kircher [...]') and then translates it, with the help of the Jesuit's work, into Latin, Italian, Spanish, German, French and Czech.[81] In spite of the many linguistic problems of this 'automatic' translation, it is interesting to note that the letter is in fact inscribed in Kircher's strategy. Caramuel was indeed a member of his network for many years (since 1644) and was one of the addressees of the book. In the missive enclosed, the Jesuit explained his language was meant for the Princes and 'curiosioris doctrinae amantibus' [the lovers of curious doctrines/ theories], a group Caramuel belonged to, being also designated as 'Litterariae Reipublicae arbitro' [mediator of the Republic of Letters]. And he, in fact, activated his own network to have the letter translated in all the languages mentioned.[82] Universal languages are meant to be appropriated by men belonging to the Republic of Letters, as a space precisely imagined and devised to be 'universal'. The translation is a map of Caramuel's network, each text reflecting a person's position. It is a language designed for the cosmopolitans, considering themselves 'citizens of the world'. The social technology associated with the language, and its reticular diffusion, is as important as the material technology itself (the shaping of the character, the grammar, etc.). And as the *Accademia degli Investigatori* he helped to create in Naples corresponded with the Royal Society, Caramuel's connections went as far as England. Maybe it is the reason why Wilkins mentioned the *Grammatica audax* (1654) of the Cistercian in his *Essay*, as a reference for his work?[83]

---

81 APUG 563, fols. 186r–188v and APUG 564, fols. 181r-184v; the letter is also reproduced in Alexander VII's manuscript version of the *Polygraphia*: BAV, Chigi J. VI. 225, fols. 59r–63v. It is mentionned in Wilding, "If You Have a Secret ..." 293 and *Writing the Book of Nature* 264–266; also in Strasser, *Kryptologie* 164–165. On Caramuel, see for example: Pissavino P. (ed.), *Le meraviglie del probabile. Juan Caramuel (1606–1682)* (Vigevano: 1990); and Ceñal Lorente R. (S.J.), "Juan Caramuel. Su epistolario con Athanasio Kircher S.J.", *Revista de Filosofia* 12 (1953) 101–147.

82 BAV, Chigi J. VI. 225, fols. 59v–60r.

83 Caramuel y Lobkowitz J., *Grammatica audax* (Francfort, Schönwetter: 1654); mentioned in Wilkins, Essay 297–298. On the Accademia: Torrini M., "Monsignor Juan Caramuel e l'Accademia napoletana degli Investiganti", in Pissavino, *Le meraviglie del probabile* 29–33.

In Wilkins's case precisely, the reception was very limited too. It is well-known that one of the few occurrences of the Real Character outside the *Essay* can be found on a plate of Robert Hooke's 1676 *Description of Helioscopes*. As was the case for its elaboration, the reception of the project remained a 'corporate' one, 'mediated' by the networks of the Royal Society.[84] But still the context should not be overlooked. The booklet was indeed published with five other short texts under the general title of *Lectiones Cutlerianae*, and here is how Hooke justified its publication:

> I have formerly in the Preface of my Micrographia given the World an account of the founding a Physico-Mechanical Lecture in the year 1665, by Sir John Cutler for the promoting the History of Nature and of Art. In prosecution thereof, I have collected many Observations both of the one and the other kind, and from time to time (as obliged) I have acquainted the Royal Society at their Publick Meetings, both at *Gresham Colledge* and *Arundel House* therewith, by *Discourses* and *Lectures* thereupon.[85]

The print was directly related to the experiences, for which Hooke was responsible, as 'Curator of Experiments'. And as such, he occupied an intermediary position between the genuine natural philosopher and the mechanic,

---

84 Among the other forms of reception of Wilkins's 'Real Character', we can mention, even if it is not possible to focus on them also here: the use of the system of tables/arborescences, with indirect reference to Wilkins, in Grew Nehemiah, *Musaeum regalis societatis, or a Catalogue and description of the natural and artificial rarities belonging to the Royal society and preserved at Gresham colledge* (London, printed by W. Rawlins for the author: 1681); and the adjunction of the 'Our Father' in Wilkins's 'philosophical language' to the collection of 'Our Fathers' established by Johannes F. Fritz and B. Schultze (probably leaning on a previous one by John Chamberlayne (1715)): Fritz Johannes Friedrich – Schultze Benjamin, *Orientalisch- und occidentalischer Sprachmeister, welcher nicht allein hundert Alphabete nebst ihrer Aussprache…* (Leipzig, Christian Friedrich Gessner: 1748); see Simon, *Sortir de Babel* 222–223 and 734–735.

85 Hooke Robert, *Lectiones Cutlerianae, or a Collection of Lectures: Physical, Mechanical, Geographical & Astronomical, Made before the Royal Society on several Occasions at Gresham Colledge. To which are added divers Miscellaneous Discourses by Robert Hooke, S.R.S.* (London, Printed for John Martyn: 1679) "To the Reader". And inside, the third text (each having title page and different pagination) is: *A Description of Helioscopes* (London, J. Martyn: 1676). On Hooke's *lectiones*, see Hunter M., "Science, Technology and Patronage: Robert Hooke and the Cutlerian Lectureship", *Establishing the New Science: The Experience of the Early Royal Society* (Woodbridge: 1989) 279–338.

employee of the Society.[86] The printed book was sort of an 'on the spot' report, a collection of texts and/or of observations, that the reader could rearrange. As Hooke wrote, he wanted to 'rather inrich the Store-house of Art and Nature with choice and excellent Seed, freed from the Chaff and Dross that do otherwise bury and corrupt it'.[87] It acted as a repository of experiments, the raw material of experimental science. The public dimension was reinforced by the publication. And in this view, the plates play a major part. So that it is important that Wilkins's character should appear precisely on 'tabula III' [Fig. 12.2], with the other unfolding plates of the *Lectiones*.[88] It is displayed on the plate next to one of Hooke's 'Schemes and delineatious Descriptions' ('When things cannot be well explained by words only (which is frequent in Mathematical and Mechanical Discourses)'),[89] representing his 'Universal joint'. This engraving can be inserted in Hooke's more general use of images in the shaping of experimental science, as a proof, or *evidentia*. They play the role of a 'visual alphabet of forms'.[90] Ordered in 1665 by the Royal Society, the *Micrographia* is an illustration of its motto '*Nullius in Verba*', the iconography presenting the reader with the new invisible world now directly accessible to his eyes thanks to the 'Optical glasses'. The plates are the results of microscopic observations, and offer a new visualisation technology; they are transformed into visual

86 Bennett J.A., "Robert Hooke as Mechanic and Natural Philosopher", *Notes and Records of the Royal Society of London* 35, 1 (1980) 33–48. And Bennett J.A. – Cooper M. – Hunter M. – Jardine L., *London's Leonardo: The Life and Work of Robert Hooke* (Oxford – New York: 2003).

87 Hooke, *Lectiones* "To the Reader" (n.p.). And in the 'Postscript' (see below) to Hooke, *Helioscopes* 29, the author, once again, insists on the public dimension of the lectures 'in the open Hall at *Gresham* Colledge'.

88 It is to be noted that the number of plates is quite variable according to the copies of Hooke's book: for example, in the exemplary kept at the BNF, V-6573 (3), the plate using the real character is lacking; whereas in the copy in the ETH Bibliothek, Zurich (Rar 4339, online: http://www.e-rara.ch/doi/10.3931/e-rara-2171), it is apparently a facsimile added at the digitized book. We are reproducing here the plate as it appears in the Madrid's exemplary (accessible, but for the moment without the unfolded plate, on Google Books). This variation of the form of the book is maybe linked, next to the variable conditions of conservation, to the relative rush with which Hooke published the booklets, especially the one to answer Huygens. On Hooke's attention to printing and engraving: Iliffe R., "Material doubts: Hooke, artisan culture and the exchange of information in 1670's London", *British Journal for the History of Science* 28 (1995) 285–318; 307.

89 Hooke, *Lectiones Cutlerianae*, "To the Reader".

90 Hamou P., *La mutation du visible: essai sur la portée épistémologique des instruments d'optique au XVII*e *siècle, Volume 2: Microscopes et télescopes en Angleterre de Bacon à Hooke* (Villeneuve-d'Ascq: 2001) 119–154; 153.

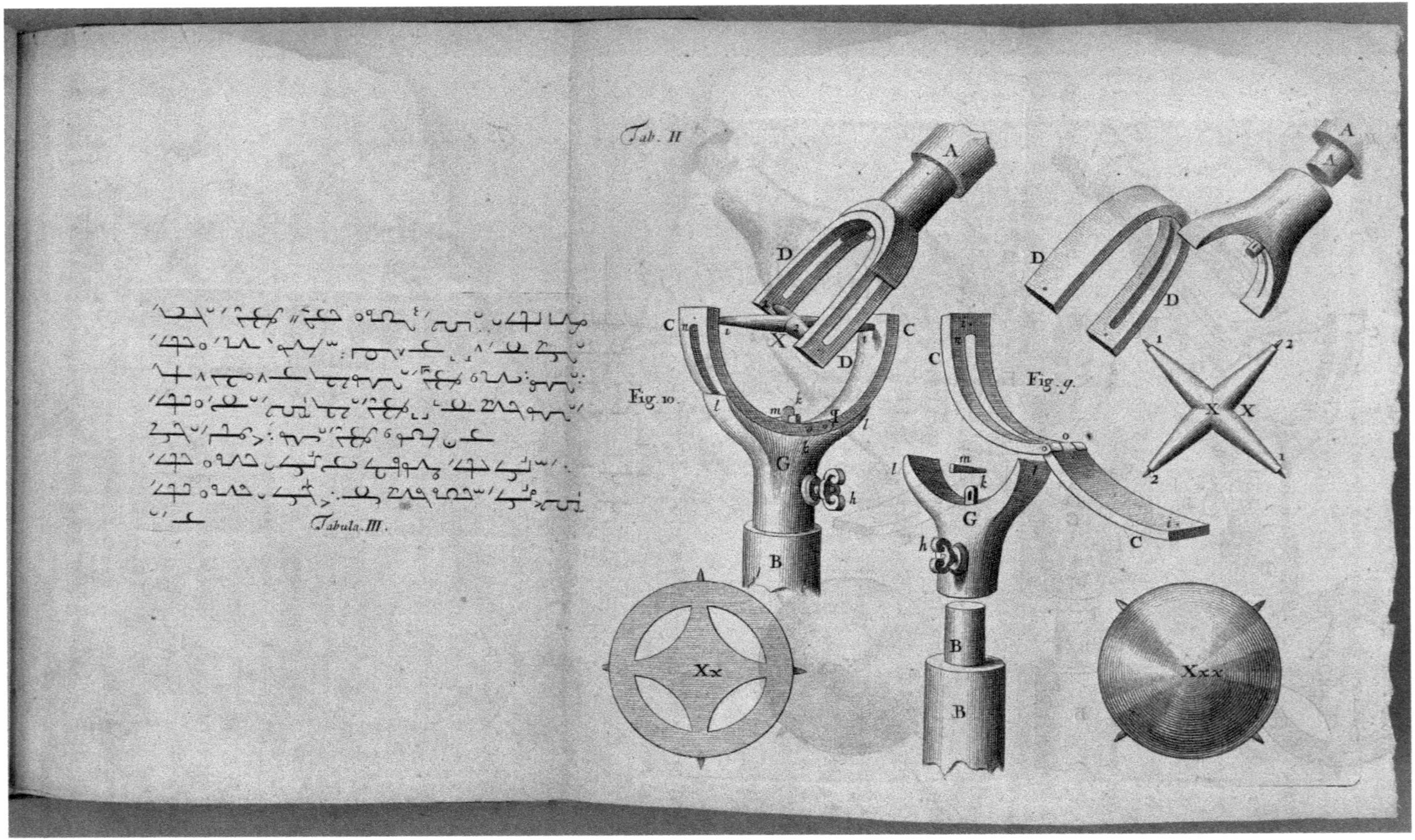

FIGURE 12.2 *Robert Hooke,* Lectiones Cutlerianae, or a Collection of Lectures (*London, Printed for John Martyn: 1679*): A Description of Helioscopes [...] (*London, J. Martyn: 1676*). *Engraving. Tab.* II (*'Universal Joint'*) *and* III (*'Real Character'*).

demonstrations, '*ad oculum* proofs'.[91] The images make a sort of vicarious experimentation possible. But besides the genuine alphabet of nature they provide, many of them are also filled with letters, allowing Hooke to caption them, to insist on their details, to connect text and image. The 'Real Character' should, so to speak, include both aspects in only one alphabet, acting also as an image, a transparent proof, making what it holds explicit. Wilkins's language should work as a map, on which every character is a 'you are here' sign in the path constituted by the tables. There are 40 characters referring to the 40 *Genera*, and variations on the right or the left side of the character indicate where to 'turn' at the 'crossroads' between Differences and Species in the arborescences. It is supposed to put its reader *in the middle* of science, to locate him on a cartography of universal knowledge.

In fact, it rather worked as a labyrinth. And though it was a plate, Hooke also considered it this way. So it must be underlined that Hooke, secondly, used the Real Character when mentioning one of his inventions concerning '*balance-spring regulator for a pocket-watch*', for which he is engaged in a scientific polemic with Christiaan Huygens (1629–1695).[92] Here is how he justified its use:

> I shall conclude this Tract with a short communication of the general ground of my Invention for *Pocket-Watches*, the number of particular ways being very great, which (that the true Lovers of Art, and they only may have the benefit of) I have set down in the *Universal* and *Real Character* of the late Reverend Prelate, my Honoured Friend Dr. *John Wilkins*, Lord Bishop of *Chester*, deceased. In which I could wish, that all things of this nature were communicated, it being a Character and Language for truly Philosophical, and so perfectly and thoroughly Methodical, that there seemeth to be nothing wanting to make it have the utmost perfection, and highest Idea of any Character or Language imaginable, as well for Philosophical as for common and constant use. And I have this further to desire of my Reader, who will be at the pains to decipher and understand

91 Aït-Touati F. – Gaukroger S., *Le Monde en images: Voir, représenter, savoir, de Descartes à Leibniz* (Paris: 2015) 68, 75–76, 84.

92 On the polemic, we are not going to develop here: Wright M., "Robert Hooke's longitude timekeeper", in Hunter M. – Schaffer S. (eds.), *Robert Hooke: New Studies* (Woolbridge: 1989) 63–118; Espinasse M., *Robert Hooke* (London: 1956) chap. 4 (evokes Wilkins on p. 78 and reproduces the plate); Johns A., *The Nature of the Book: Print and Knowledge in the Making* (Chicago – London: 1998) 521–531 (evokes Wilkins on p. 525 and reproduces the plate); Jardine, *Ingenious Pursuits* 'Time Trial' 144–157 and on Wilkins (with the plate) 321–322.

> this description, that he would only make use of it for his own information, and not communicate the explication thereof to any that hath not had the same curiosity with himself.[93]

We can perceive the tension between universality and cryptography. One of Hooke's purposes was to promote the already forgotten language created by Wilkins eight years before. But the use of the Real Character was also inscribed in the *habitus* of the Royal Society in the context of scientific debates, the purpose being to preserve an invention from the rivals.[94] In the same 'postscript', Hooke used one of the other privileged means, anagrams, another 'way of "place-keeping" for ideas that would be divulged later': his 'theory *of Elasticity or Springiness*' was coded as '*Ceiiinosssttuu*' and revealed only in 1678 as '*Ut tensio sic vis*' [the springiness equals (is proportional to) the strength].[95] This type of encoded message was indeed part of the scientific *ethos*, and already used, for example, between Galileo and Kepler. It was a way of securing priority over a discovery, and meanwhile being able to improve it. It was also a way to increase the impact of a discovery, dramatizing its unveiling, through the use of several layers of 'rhetorical ingeniosity'.[96] At the same time it is *part of* the discovery, in its core, as Fernand Hallyn showed: in the form of more regular cryptograms, the play with the letters evokes the playfulness of the 'Book of Nature' and its metaphorical alphabet.[97] The natural philosopher had to decrypt it, and the literary code was a sort of training for this necessary mathematical reading of Nature. The 'Real Character' should have functioned like that sort of code.

It was this 'revelation-veiling'[98] of the scientific discovery that Christiaan Huygens advocated as a rule for the Royal Society in a letter to Oldenburg and

---

93 Hooke, *Helioscopes* 30–31.

94 Iliffe R., "'In the Warehouse': Privacy, Property and Priority in the Early Royal Society", *History of Science* XXX (1992) 29–68; on this context especially: 41–52. Hooke's Postscript starts with the issue of publicity/publication, since he discovered the betrayal, says he, of Huygens and Oldenburg in a paper of the *Philosophical Transactions* 10, n°112 (24 mar. 1675) 272–273, leaning on one 'of the *French* Journal *de Scavans*' (25 fev. 1675) (Hooke, *Helioscopes* 26). Oldenburg then reacted in the *Transactions* (25 October 1676) considering Huygens acted 'publicly' whereas Hooke was 'secretive' (Iliffe, "'In the Warehouse"50).

95 Hooke, *Helioscopes* 31. For the quotation before: Jardine, *Ingenious Pursuits* 322.

96 Hallyn F., *Les Structures rhétoriques de la science. De Kepler à Maxwell* (Paris: 2004) 85–86 ; 90. See also: Dear P., "*Totius in verba*: Rhetoric and Authority in the Early Royal Society", *Isis* 76, 2 (1985) 144–161.

97 Hallyn, *Structures rhétoriques* 108.

98 Hallyn, *Structures rhétoriques* 96.

in an exchange with Robert Moray in 1669, where he reflected on the best language for sciences. He already associated scientific cryptography and Wilkins' character even if less directly.[99] Paradoxically, seven years later, Hooke obeyed Moray's advice to use the 'Real Character', but against Huygens. Anyway, within the context of the elaboration—especially by the Royal Society's members—of an 'experimental' science of which the 'modernity' and legitimacy were supposed to rely on its 'openness', universal language could also incarnate the still important place of 'secrecy'.[100]

Among the 'true Lovers of Art' for whom Hooke intended his cryptographic text, one at least tried to take up the challenge. In one of his letters to Richard Towneley, starting with observational issues concerning instruments, in November 1675, Astronomer John Flamsteed (1646–1719) expressed his puzzlement at the enigma of Hooke's plate:

> I perceive you understand the Universall Character and therefore should be very glad to understand the Secret, though I am apt to believe it will not performe much better then Monsieur Hugens his. Sir Jonas informd mee hee would write to you last weeke about the watches. I doubt not but hee has done it to your satisfaction.[101]

Flamsteed, sometimes called '*Astronomicus Regius*', closely related to the Royal Society, and directly involved in the crucial issue of defining a method to find longitudes at sea, was obviously interested in Hooke's and Huygens' discoveries. A few months before, in a letter mentioning the watch controversy, he wrote he had attended a meeting of the Society and evoked the code used by Huygens: 'The secret hee onely discovered to Mr Oldenburge in this sentence'.[102]

---

99 Huygens Christiaan, *Oeuvres complètes de Christiaan Huygens*, 23 vols. (The Hague: 1888–1950) vol. 6, 396–397 (letter no. 1721); Moray's answer 'defending' Wilkins, see vol. 6, 425 (no. 1730); and for the letter with a cipher to Oldenburg, see vol. 6, 354–355 (no. 1701).

100 So that reflections on the universal language plans can contribute to the historiographical debate on the openness of 'early modern sciences': Eamon W., "From the Secrets of Nature to Public Knowledge: The Origins of the Concept of Openness in Science", *Minerva* 23, 3 (1985) 321–47; Long P.O., *Openness, Secrecy, Authorship: Technical Arts and the Culture of Knowledge from Antiquity to the Renaissance* (Baltimore: 2001); Shapin S., *A Social History of Truth : Civility and Science in Seventeenth-Century England* (Chicago: 1994).

101 Flamsteed J., *The Correspondence of John Flamsteed: First Astronomer Royal*, ed. E.G. Forbes, 3 vols. (Bristol – Philadelphia: 1995) vol. 1 379–380 (letter no. 232): Flamsteed to Towneley (mentioned also in Johns, *Nature of the Book* 526 n. 169). On Flamsteed: Willmoth F., "Flamsteed, John (1646–1719)", in *Oxford Dictionary of National Biography* (Oxford: 2004).

102 Flamsteed, *The Correspondence* n° 207 329–330.

He was *de facto* one of Hooke's targets. But at the same time, the Real Character also functioned as an unbreakable code, that Flamsteed was unable to decipher, calling for the expertise of a friend. Eventually Towneley may have been unable to break code, at least there is no trace left of his 'translation'.[103]

Hooke's expression addressing the plate with the 'Real Character' only to the 'true Lovers of Art' recalls the words of Kircher keeping his own character for 'curiosioris doctrinae amantibus'—different, though connected, scientific contexts, same target. Their language is designed for the few.

As we have seen, besides the reception of the languages themselves, the reception of the universal language plans sets in motion a network of scholars, brokers and erudites, exchanging ideas on that issue. Their schemes being to be 'universal translators', ultimately the plans turned out to be more of a code designed for an elite. They were languages for the language planners, since, for example, Hooke and Caramuel, with his little-known *Apparatus philosophicus*, can be added to this circle.[104] Language planners were in competition, on a European scale, sometimes combining Jesuit networks with Royal Society networks. Wilkins disqualified Kircher's or Labbé's projects among others (including Beck's):

> To this I might adde something concerning the advantage of this Philosophical way, above those attempts towards a Universal Character which have been made by others. That of *Marks* or Letters by *Cicero*; that of *numbers* by an Ingenious Country-man of our own [in the margin: Mr. *Beck of Ipswich*], followed since by *Becherus*, and by *Athanasius Kircher*; together with that other attempt towards an Universal Language, by *Philip Labbé*. All which are in this one respect defective, because they are not Philosophical; upon which account they are much more difficult, and less distinct.[105]

His language was 'philosophical', that is to say fit for the natural philosopher of the Royal Society, which was not the case of the 'Jesuit' languages.

103 Although Lisa Jardine states her incapacity to translate the message (*Ingenious Pursuits* 322), a deciphering was proposed by Andrade E.N. da C., "The Real Character of Bishop Wilkins", *Annals of Science* 1, 1 (1936) 4–12; 12 (quoted also in Johns, *Nature of the Book* 525).

104 See: Sousedík S., "Universal Language in the Work of John Caramuel", *Acta Comeniana* XXXIII, 9 (1991) 149–158 and Velarde Lombraña J., "Proyectos de lengua universal ideados por españoles (1653–1954)", *Taula. Quaderns de pensament* 7, 8 (1987) 7–78 (15–25 on Caramuel). On Hooke and universal language plans, see among other publications: Hesse M.B., "Hooke's Philosophical Algebra", *Isis* 57, 1 (1966) 67–83.

105 Wilkins, *Essay* 452.

In the context of seventeenth century Europe, the monopoly of Latin was starting to be shaken, being challenged by the vernaculars. Meanwhile, Latin was also no longer the monopoly of the learned, with the diffusion of colleges and tools like the *Calepino* dictionary and its editorial success. Hence, the universal languages were created to offer a new tool meant to regain one feature of the members of the Republic of Letters: the mastery of languages of knowledge. Considered as codes and/or languages of distinction, that are hard to understand and require specific skills to be mastered, the universal languages reintroduced a kind of a monopoly. They helped to translate a social network, to build a 'learned community' (or learned *communities*)—the universal language tried to reinvest the social space deserted by Latin.

Selective Bibliography

Sources

Beck Cave, *The Universal Character by which All the Nations in the World may understand one anothers Conceptions, Reading out of one Common Writing their own Mother Tongues* (London, printed by Thomas Maxey for William Weekley: 1657).

Beck Cave, *Le Charactere universel, par lequel toutes nations peuvent comprendre les conceptions l'une de l'autre, etc* (London, A. Maxey for Guillaume Weekley: 1657).

Beckmann Johann Christoph, *Historia orbis terrarum geographica et civilis, de variis negociis nostri potissimum et superioris seculi, aliisve rebus selectioribus* (Frankfurt, J.W. Fincelii: 1680).

Besnier Pierre, *La Reunion des langues, ou l'art de les apprendre toutes par une seule* (Paris, Sébastien Mabre-Cramoisy: 1674).

Besnier Pierre, *A Philosophical Essay for the Reunion of the Languages or the Art of Knowing all by the Mastery of one*, trans. H. Roose (Oxford, printed for Hen. Hall for James Good: 1675).

Bonifacio Giovanni, *L'arte de'cenni* (Vincenza, Francesco Grossi: 1616).

Boyle Robert, *The Correspondence of Robert Boyle*, ed. M. Hunter – A. Clericuzio – L.M. Principe, 6 vols. (London: 2001).

Caramuel y Lobkowitz Juan, *Grammatica audax* (Frankfurt, Schönwetter: 1654).

Dalgarno George, *George Dalgarno on Universal Language: The Art of Signs (1661), the Deaf and Dumb Man's Tutor (1680), and the Unpublished Papers*, ed. D. Cram – J. Maat (Oxford: 2001).

Duret Claude, *Thresor de l'histoire des langues de cest univers* (Cologny, M. Berjon pour la Société caldorienne: 1613).

Flamsteed John, *The Correspondence of John Flamsteed: First Astronomer Royal*, ed. E.G. Forbes, 3 vols. (Bristol – Philadelphia: 1995).

Fritz Johannes Friedrich and Schultze Benjamin, *Orientalisch- und occidentalischer Sprachmeister, welcher nicht allein hundert Alphabete nebst ihrer Aussprache...* (Leipzig, Christian Friedrich Gessner: 1748).

Grew Nehemiah, *Musaeum regalis societatis, or a Catalogue and description of the natural and artificial rarities belonging to the Royal society and preserved at Gresham colledge* (London, printed by W. Rawlins for the author: 1681).

Hooke Robert, *Lectiones Cutlerianae, or a Collection of Lectures: Physical, Mechanical, Geographical & Astronomical, Made before the Royal Society on several Occasions at Gresham Colledge. To which are added divers Miscellaneous Discourses by Robert Hooke, S.R.S.* (London, Printed for John Martyn: 1679).

Hooke Robert, *A Description of Helioscopes* (London, J. Martyn: 1676).

Hooke Robert, *The Diary of Robert Hooke M.A., M.D., F.R.S. 1672–1680*, ed. H.W. Robinson – W. Adams (London: 1935).

Huygens Christiaan, *Oeuvres complètes de Christiaan Huygens*, 23 vols. (The Hague: 1888–1950).

Kircher Athanasius, *Ars magna sciendi, in XII libros digesta* (Amsterdam, J. Janssonis a Waesberge and the widow of E. Weyerstraet: 1669).

Kircher Athanasius, *Polygraphia nova et universalis ex combinatoria arte detecta* (Rome, Varesius: 1663).

Labbé Philippe, *Grammaire de la langue universelle des Missions et du Commerce tres simple, tres courte et tres facile à apprendre à toutes sortes de personnes* (Paris, J. Roger: 1663).

Labbé Philippe, *Les Étymologies de plusieurs mots françois, contre les abus de la secte des hellénistes du Port-Royal, sixiesme partie des Racines de la langue grecque* (Paris, G. et S. Bénard: 1661).

Lodwick Francis, *On Language, Theology, and Utopia*, ed. F. Henderson – W. Poole (Oxford: 2011).

Mersenne Marin, *Correspondance du P. Marin Mersenne, religieux minime*, ed. C. De Waard, 18 vols. (Paris: 1932–1988).

Oldenburg Henry, *The Correspondence of Henry Oldenburg*, ed. A.R. Hall – M.B. Hall, 13 vols. (Madison 1965–1986).

Sprat Thomas, *History of the Royal Society of London for the improving of natural knowledge* (London, printed for J. Martyn and J. Allestry: 1667; reprint London: 1959).

Thomas Thomas, *Dictionarium summa fide ac diligentia accuratissime emendatum*, (Cambridge, Thomas: 1587).

Vienne Plancy Antoine de, *Extraordinaire du Mercure galant* 14 (January 1681) 334–349; 19 (July 1682) 274–330; 31 (July 1685) 112–182; 32 (January 1686) 110–146.

Vigenère Blaise de, *Traité des chiffres ou secrètes manières d'écrire* (Paris, Abel L'Angelier: 1586).

Vocabolario degli Accademici della Crusca (Venice, Giovanni Alberti: 1612).

Wilkins John, *An Essay Towards a Real Character and a Philosophical Language* (London, S. Gellibrand and J. Martin: 1668).

Wilkins John, *Mercury, or the Secret and Swift Messenger, shewing how a man may with privacy and speed communicate his thoughts to a friend at any distance* (London, J. Maynard and T. Wilkins: 1641).

Scholarly Literature

Aït-Touati F. – Gaukroger S., *Le Monde en images: Voir, représenter, savoir, de Descartes à Leibniz* (Paris: 2015).

Andrade E.N. da C., "The Real Character of Bishop Wilkins", *Annals of Science* 1, 1 (1936) 4–12.

Auroux S., *La Révolution technologique de la grammatisation: introduction à l'histoire des sciences du langage* (Liège: 1994).

Balavoine C. et al. (eds.), *Mercure à la Renaissance, Actes des journées d'étude des 4–5 octobre 1984* (Paris: 1988).

Balteau J. et al. (eds.), *Dictionnaire de biographie française* (Paris: 1932–2011).

Barker S. – Hosington B. (eds.), *Renaissance Cultural Crossroads: Translation, Print and Culture in Britain, 1473–1640* (Leiden – Boston: 2013).

Barnett P., *Theodore Haak, F.R.S. (1605–1690): The First German Translator of* Paradise Lost (The Hague: 1962).

Beaurepaire P.-Y. (ed.), *La communication en Europe de l'âge classique au siècle des Lumières* (Paris: 2014).

Bennett J.A., "Robert Hooke as Mechanic and Natural Philosopher", *Notes and Records of the Royal Society of London* 35, 1 (1980) 33–48.

Bennett J.A. – Cooper M. – Hunter M. – Jardine L., *London's Leonardo: The Life and Work of Robert Hooke* (Oxford-New York: 2003).

Blair A., "La Persistance du latin comme langue de science à la Renaissance", in Chartier R. – Corsi P. (eds.), *Sciences et langues en Europe* (Luxembourg: 2000), 19–39.

Bourdieu P., "Quelques propriétés des champs", in *Questions de sociologie* (Paris: 1980) 113–120.

Bret P. – Moerman E., "Sciences et arts", in Chevrel Y. – Cointre A. – Tran-Gervat Y.-M. (eds.), *Histoire des traductions en langue française, XVIIe et XVIIIe siècles, 1610–1815* (Lagrasse: 2014) 595–723.

Burke P., "Translations into Latin in early modern Europe", in Burke P. – Po-Chia Hsia R. (eds.), *Cultural Translation in Early Modern Europe* (Cambridge: 2007) 65–80.

Burke P., *Languages and Communities in Early Modern Europe* (Cambridge: 2004).

Bury E. (ed.), *Tous Vos Gens à Latin: le latin, langue savante, langue mondaine (XIVe–XVIIe Siècles)* (Genève: 2005).

Casanova P., *The World Republic of Letters*, trans. M.B. DeBevoise (Cambridge: 2007).

Céard J., "De Babel à la pentecôte: la transformation du mythe de la confusion des langues au XVIe siècle", *Bibliothèque d'Humanisme et de Renaissance* 42 (1980), 577–594.

Ceñal Lorente R. (S.J.), "Juan Caramuel. Su epistolario con Athanasio Kircher S.J.", *Revista de Filosofia* 12 (1953) 101–147.

Dauphine J. – Jacquemier M. (eds), *Babel à la Renaissance. Actes du XIe colloque international de la société française d'étude du XVIe siècle*, new edition with a preface by Claude-Gilbert Dubois (Paris: 2007).

David M., *Le Débat sur les écritures et l'hiéroglyphe aux XVIIe et XVIIIe siècle et l'application de la notion de déchiffrement aux écritures mortes* (Paris: 1965).

Dear P., "*Totius in verba*: Rhetoric and Authority in the Early Royal Society", *Isis* 76, 2 (1985) 144–161.

De Backer A. – Sommervogel C., *Bibliothèque de la Compagnie de Jésus*, 8 vols. (Mansfield Centre: 1998).

Demetriou T. – Tomlinson R. (eds.), *The Culture of Translation in Early Modern England and France, 1500–1660* (New York: 2015).

Dolezal F., *Forgotten but Important Lexicographers, John Wilkins and William Lloyd: A Modern Approach to Lexicography before Johnson* (Tübingen: 1985).

Eamon W., "From the Secrets of Nature to Public Knowledge: The Origins of the Concept of Openness in Science", *Minerva* 23, 3 (1985) 321–347.

Eco U., *La Recherche de la langue parfaite dans la culture européenne*, trans. J.-P. Manganaro (Paris: 1994).

Espinasse M., *Robert Hooke* (London: 1956).

Findlen P., "Scientific Spectacle in Baroque Rome: Athanasius Kircher and the Roman College Museum", *Roma moderna e contemporanea* 3, 3 (1995) 625–665.

Fletcher J., "Athanasius Kircher and his correspondence", in Fletcher J. (ed.), *Athanasius Kircher und seine Beziehungen zum gelehrten Europa seiner Zeit* (Wiesbaden: 1988) 139–195.

Fraenkel B., "Comment ne s'adresser qu'à quelques-uns? Remarques sur la cryptographie de tradition alphabétique", *Extrême-Orient, Extrême-Occident* 30 (2008) 175–185.

Grévin B., *Le Parchemin des cieux. Essai sur le Moyen Âge du langage* (Paris: 2012).

Hallyn F., *Les Structures rhétoriques de la science. De Kepler à Maxwell* (Paris: 2004).

Hamou P., *La mutation du visible: essai sur la portée épistémologique des instruments d'optique au XVIIe siècle, Volume 2: Microscopes et télescopes en Angleterre de Bacon à Hooke* (Villeneuve – d'Ascq: 2001).

Hesse M.B., "Hooke's Philosophical Algebra", *Isis* 57, 1 (1966) 67–83.

Hunter M., "Science, Technology and Patronage: Robert Hooke and the Cutlerian Lectureship", *Establishing the New Science: The Experience of the Early Royal Society* (Woodbridge: 1989) 279–338.

Iliffe R., "*Material doubts: Hooke, artisan culture and the exchange of information in 1670's London*", *British Journal for the History of Science* 28 (1995) 285–318.

Iliffe R., "'In the Warehouse': Privacy, Property and Priority in the Early Royal Society", *History of Science* XXX (1992) 29–68.

Jardine L., *Ingenious Pursuits: Building the Scientific Revolution* (New York: 1999).

Johns A., *The Nature of the Book: Print and Knowledge in the Making* (Chicago – London: 1998).

Le Blanc C. – Simonutti L. (eds.), *Le Masque de l'écriture: philosophie et traduction de la Renaissance aux Lumières* (Geneva: 2015).

Lewis R., *Language, Mind and Nature: Artificial Languages in England from Bacon to Locke* (Cambridge: 2007).

Lewis R., "The Publication of John Wilkins's *Essay* (1668): Some Contextual Considerations", *Notes and Records of The Royal Society* 56, 2 (2002) 133–146.

Long P.O., *Openness, Secrecy, Authorship: Technical Arts and the Culture of Knowledge from Antiquity to the Renaissance* (Baltimore: 2001).

McCracken G.E., "Athanasius Kircher's Universal Polygraphy", *Isis* 39, 4 (1948) 215–228.

Maillard J.-F., "Un avatar de la traduction: l'idéal d'une langue universelle à la Renaissance", in Contamine G. (ed.), *Traduction et traducteurs au Moyen Age* (Paris: 1989) 334–347.

Maillard J.-F., "Aspects de l'encyclopédisme au XVIe siècle dans le Traicté des chiffres annoté par Blaise de Vigenère", *Bibliothèque d'Humanisme et Renaissance* XLIV, 2 (1982) 235–268.

Miller P.N., *Pereisc's Europe. Learning and Virtue in the Seventeenth Century* (New Haven – London: 2000).

Pauchard J., "Le dictionnaire alphabétique de William Lloyd et John Wilkins: une avancée majeure de la lexicographie au XVIIe siècle", in Groupe de linguistique appliquée des télécommunications (ed.), *Aspects méthodologiques pour l'élaboration de lexiques unilingues et multilingues: actes de GLAT-Bertinoro 2006* (Brest: 2006) 151–161.

Pissavino P. (ed.), *Le meraviglie del probabile. Juan Caramuel (1606–1682)* (Vigevano: 1990).

Rostenberg L., *The Library of Robert Hooke. The Scientific Book Trade of Restoration England* (Santa Monica: 1989).

Sabatini F. et. al. (eds.), *Una lingua e il suo vocabolario* (Florence: 2014).

Salmon V., "Cave Beck: a Seventeenth Century Ipswich Schoolmaster and his "Universal Character"", in Salmon V., *The Study of Language in 17th Century England* (Amsterdam-Philadelphia: 1988) 177–190.

Salmon V., "Language planning in seventeenth-century England: its context and aims", in Salmon V., *The Study of Language* 129–153.

Shapin S., *A Social History of Truth : Civility and Science in Seventeenth-Century England* (Chicago: 1994).

Simon F., "Quelle est la langue de la science: dire efficacement la 'vérité' scientifique?", in Pérez L. – Simon F. – Thébaud-Sorger M. (eds.), *L'Europe des sciences et des techniques (XVe–XVIIIe): Un dialogue des savoirs* (Rennes: 2016) 257–267.

Simon F., "De Conrad Gessner (1555) à Claude Duret (1613), faire entendre la diversité, collectionner les langues", in Alazard F. – Geonget S. – Gerbier L. – Mellet P.-A. (eds.), *Dissensus: Pratiques et représentations de la diversité des opinions, 1500–1650*, Coll. Le Savoir de Mantice 26 (Paris: 2016) 205–224.

Simon F., "Une 'République des Langues': La quête de la langue universelle au XVIe–XVIIe siècle: une langue de la vérité à l'usage des Républicains des Lettres?", in Dion N. – Masse S. – Plourde A.-A. (eds.), *Le Cosmopolitisme. Influences, voyages et échanges dans la République des Lettres (XVe–XVIIIe siècle)* (Paris: 2014) 257–288.

Simon F., *Sortir de Babel. Une "République des Langues" en quête d'une langue universelle à la Renaissance et à l'Âge classique*, Ph.D. Dissertation (Rennes 2 University: 2011).

Slaughter M.M., *Universal Languages and Scientific Taxonomy in the Seventeenth Century* (Cambridge: 1982).

Sousedík S., "Universal Language in the Work of John Caramuel", *Acta Comeniana* XXXIII. 9 (1991) 149–158.

Stolzenberg D., *Egyptian Oedipus. Athanasius Kircher and the Secrets of Antiquity* (Chicago: 2013).

Strasser G.F., *Lingua Universalis: Kryptologie und Theorie der Universalsprachen im 16. und 17. Jahrhundert* (Wiesbaden: 1988).

Torrini M., "Monsignor Juan Caramuel e l'Accademia napoletana degli Investiganti", in Pissavino, *Le meraviglie del probabile* 29–33.

Velarde Lombraña J., "Proyectos de lengua universal ideados por españoles (1653–1954)", *Taula. Quaderns de pensament* 7, 8 (1987) 7–78.

Vialon M. (ed.), *La Traduction à la Renaissance et à l'âge classique* (Saint-Etienne: 2001).

Waquet F., *Le Latin ou l'empire d'un signe, XVIe–XXe siècle* (Paris: 1998).

Waquet F. – Bots H., *La République des Lettres* (Paris: 1997).

Wicki J. (S.J.), "Die Miscellanea Epistolarum des P. Athanasius Kircher S.I. in missionarischer Sicht", *Euntes Docete* XXI (1968) 221–254 and 282–309.

Wicki J. (S.J.), "Die Chiffre in der Ordenskorrespondenz der Gesellschaft Iesu von Ignatius bis General Oliva (ca. 1554–1676). Anhand der Kodex Fondo Gesuitico 678/21 (ARSI)", *Archivum Historicum Societatis Iesu* XXXII (1963) 133–178.

Wilding N., "'If You Have a Secret, Either Keep It, or Reveal It': Cryptography and Universal Language", in Stolzenberg D. (ed.), *The Great Art of Knowing. The Baroque Encyclopedia of Athanasius Kircher* (Stanford: 2001) 93–105.

Wilding N., *Writing the Book of Nature: Natural Philosophy and Communication in Early Modern Europe*, Ph.D. dissertation (European University Institute: 2000).

Willmoth F., "Flamsteed, John (1646–1719)", in *Oxford Dictionary of National Biography* (Oxford: 2004).

Wright M., "Robert Hooke's longitude timekeeper", in Hunter M. – Schaffer S. (eds.), *Robert Hooke: New Studies* (Woolbridge: 1989) 63–118.

Index Nominum

www.ingramcontent.com/pod-product-compliance
Ingram Content Group UK Ltd.
Pitfield, Milton Keynes, MK11 3LW, UK
UKHW021421091026
15624UKWH00010B/347

9 789004 349254